Annual Reports in Organic Synthesis—1977

Annual Reports in Organic Synthesis

ANNUAL REPORTS IN ORGANIC SYNTHESIS—1970
John McMurry and R. Bryan Miller, Eds.

ANNUAL REPORTS IN ORGANIC SYNTHESIS—1971
John McMurry and R. Bryan Miller, Eds.

ANNUAL REPORTS IN ORGANIC SYNTHESIS—1972
John McMurry and R. Bryan Miller, Eds.

ANNUAL REPORTS IN ORGANIC SYNTHESIS—1973
R. Bryan Miller and Louis S. Hegedus, Eds.
John McMurry, Series Editor

ANNUAL REPORTS IN ORGANIC SYNTHESIS—1974
Louis S. Hegedus and Stephen R. Wilson, Eds.
R. Bryan Miller, Series Editor

ANNUAL REPORTS IN ORGANIC SYNTHESIS—1975
R. Bryan Miller and L. G. Wade, Jr., Eds.

ANNUAL REPORTS IN ORGANIC SYNTHESIS—1976
R. Bryan Miller and L. G. Wade, Jr., Eds.

ANNUAL REPORTS IN ORGANIC SYNTHESIS—1977
R. Bryan Miller and L. G. Wade, Jr., Eds.

Annual Reports in Organic Synthesis—1977

edited by

R. Bryan Miller
University of California, Davis, California

L. G. Wade, Jr.
Colorado State University, Fort Collins, Colorado

ACADEMIC PRESS   New York San Francisco London   1978
A Subsidiary of Harcourt Brace Jovanovich, Publishers

COPYRIGHT © 1978, BY ACADEMIC PRESS, INC.
ALL RIGHTS RESERVED.
NO PART OF THIS PUBLICATION MAY BE REPRODUCED OR
TRANSMITTED IN ANY FORM OR BY ANY MEANS, ELECTRONIC
OR MECHANICAL, INCLUDING PHOTOCOPY, RECORDING, OR ANY
INFORMATION STORAGE AND RETRIEVAL SYSTEM, WITHOUT
PERMISSION IN WRITING FROM THE PUBLISHER.

ACADEMIC PRESS, INC.
111 Fifth Avenue, New York, New York 10003

United Kingdom Edition published by
ACADEMIC PRESS, INC. (LONDON) LTD.
24/28 Oval Road, London NW1 7DX

LIBRARY OF CONGRESS CATALOG CARD NUMBER: 71-167779

ISBN 0-12-040808-2

PRINTED IN THE UNITED STATES OF AMERICA

CONTENTS

PREFACE xi
JOURNALS ABSTRACTED xiii

I. **CARBON–CARBON BOND FORMING REACTIONS** 1
   A. Carbon–Carbon Single Bonds (*see also:* I.E., I.F., I.G.) . 1
     1. Alkylation of Aldehydes, Ketones, and Their
        Derivatives (*see also:* I.A.7) 1
     2. Alkylations of Nitriles, Acids, and Acid Derivatives .. 5
     3. Alkylation of β-Dicarbonyl and β-Cyano-carbonyl
        Systems 13
     4. Alkylation of N, S, or Se Stabilized Carbanions 16
     5. Alkylation of Organometallic Reagents (*see also:*
        I.F.2, I.G.) 25
     6. Other Alkylation Procedures and Reviews 35
     7. Nucleophilic Addition of Electron Deficient Carbon
        (*see also:* I.G.) 36
        a. 1,2-Additions 36
           1) Aldol-Type Condensations 36
               a) Intermolecular 36
               b) Intramolecular 44
           2) Addition of N, S, or Se Stabilized Carbanions .. 46
           3) Grignard-Type Additions 52

CONTENTS

	b. Conjugate Additions	64
	1) Enolate-Type Additions	64
	2) Organometallic Reagents	66
	3) Other Conjugate Additions	72
	8. Other Carbon–Carbon Single Bond Forming Reactions	76
B.	Carbon–Carbon Double Bonds (*see also:* III.G)	84
	1. Wittig-Type Olefination Reactions	84
	2. Eliminations	93
	a. Alcohols and Derivatives	93
	b. Halides	99
	c. Other Eliminations	102
	3. Other Carbon–Carbon Double Bond Forming Reactions	104
	4. Allene Forming Reactions	110
C.	Carbon–Carbon Triple Bonds	112
D.	Cyclopropanations	114
	1. Carbene or Carbenoic Addition to a Multiple Bond	114
	2. Other Cyclopropanations	117
E.	Thermal Reactions (*see also:* VI.B)	123
	1. Cycloadditions	123
	2. Other Thermal Reactions	130
F.	Aromatic Substitutions Forming a New Carbon–Carbon Bond	135
	1. Friedel—Crafts-Type Reactions	135
	2. Coupling Reactions (*see also:* I.G)	138
	3. Other Aromatic Substitutions	148
G.	Synthesis *via* Organometallics	150
	1. Organoboranes	150
	2. Carbonylation Reactions	160
	3. Other Syntheses *via* Organometallics	165
	4. Reviews	190
II.	**OXIDATIONS**	192
A.	C–O Oxidations	192
	1. Alcohol → Ketone, Aldehyde	192
	2. Alcohol, Aldehyde → Acid, Acid Derivative ..... (1976, 194)
B.	C–H Oxidations	195
	1. C–H → C–O	195
	2. C–H → C–Hal	197
	3. Other C–H Oxidations	200
C.	C–N Oxidations	201
D.	Amine Oxidations	202
E.	Sulfur Oxidations	203

CONTENTS

	F.	Oxidative Additions to C–C Multiple Bonds	205
		1. Epoxidations	205
		2. Hydroxylation	208
		3. Other	209
	G.	Phenol → Quinone Oxidation	212
	H.	Oxidative Cleavages	214
	I.	Photosensitized Oxygenations	215
	J.	Dehydrogenation	216
	K.	Other Oxidations and Reviews	216
III.	**REDUCTIONS**		218
	A.	C=O Reductions	218
	B.	Nitrile Reductions	227
	C.	Reduction of Sulfur Compounds	228
	D.	N–O Reductions	231
	E.	C–C Multiple Bond Reductions	233
		1. C=C Reductions	233
		2. C≡C Reductions	239
		3. Reduction of Aromatic Rings	241
	F.	Hydrogenolysis of Hetero Bonds	243
		1. C–O → C–H	243
		2. C–Hal → C–H	245
		3. C–S → C–H	248
		4. C–N → C–H (see also: III.H)	249
	G.	Reductive Eliminations	250
	H.	Reductive Cleavages	251
	I.	Hydroboration (reduction only)	252
	J.	Other Reductions and Reviews	254
IV.	**SYNTHESIS OF HETEROCYCLES**		255
	A.	Aziridines	255
	B.	Furans, *etc.*	256
	C.	Indoles	259
	D.	Lactams	263
	E.	Lactones	270
	F.	Pyridines, Quinolines, *etc.*	281
	G.	Pyrroles, *etc.*	289
	H.	Other Heterocycles with One Heteroatom	294
	I.	Heterocycles with Two or More Heteroatoms	298
		1. Heterocycles with 2 N's	298
		a. 5-Membered	298
		b. 6-Membered (see also: VI.A.15)	304
		c. Other	310
		2. Heterocycles with 1 N and 1 O	310

      3. Heterocycles with 1 N and 1 S 313
      4. Heterocycles with 1 S and 1 O .... (1976, 300)
      5. Heterocycles with 3 N's 316
      6. Other Heterocycles 319
  J. General Reviews.................................. 320

V. PROTECTING GROUPS 321
  A. Hydroxyl 321
  B. Amine 324
  C. Sulfhydryl 329
  D. Carboxyl 329
  E. Ketone, Aldehyde................................. 333
  F. Phosphate 339
  G. Pi Bond...................................... (1976, 321)
  H. Miscellaneous Protecting Groups 340

VI. USEFUL SYNTHETIC PREPARATIONS 341
  A. Functional Group Preparations 341
      1. Acids, Acid Halides, *etc.* (*see also:* II.A.2) 341
      2. Alcohols, Phenols (*see also:* III.A) 343
      3. Alkyl, Aryl Halides (*see also:* II.B.2) 346
      4. Amides (*see also:* IV.D, VI.A.17) 353
      5. Amines (*see also:* III.B, III.D) 359
      6. Amino Acids and Derivatives 365
      7. Carbenes (*see also:* I.D)........................ 369
      8. Enamines 371
      9. Epoxides (*see also:* II.F.1)...................... 373
     10. Esters (*see also:* IV.E) 374
     11. Ethers 378
     12. Ketones and Aldehydes (*see also:* II.A.1, III.F.1, III.F.4) 381
     13. Nitriles (*see also:* II.D, II.J) 389
     14. Nitro (*see also:* II.D) 395
     15. Nucleotides, *etc.* (*see also:* IV.1.1a,b) 396
     16. Olefins, Acetylenes (*see also:* I.B, I.C, II.J, III.G) .. 399
     17. Peptides (*see also:* V.B, V.D) 402
     18. Vinyl Halides, Vinyl Ethers, Vinyl Esters 406
     19. Sulfur Compounds (*see also:* II.E, III.C) 411
  B. Ring Enlargement and Contraction 420
      1. Enlargement 420
      2. Contraction 424
  C. Multi-Step Transformations 425
      1. Masked-Carbonyl Systems 425
      2. Other 430

CONTENTS

VII. OTHER COMPLETELY MISCELLANEOUS REACTIONS 433

VIII. MISCELLANEOUS REVIEWS 435

AUTHOR INDEX 445

PREFACE

One of the most difficult problems facing chemists today is that of "keeping up with the literature." For several reasons, the problem is particularly severe for the synthetic organic chemist. Bits of information of potential use to the synthetic organic chemist are scattered throughout common chemistry journals and can be found in any paper, not just those dealing strictly with synthesis. Thus a synthetic chemist must read a large number of journals. He must organize and index what he reads to make the information available for future reference. All synthetic chemists do this; but the task is becoming more difficult each year as the flow of information increases.

The problem however is shared to some extent by all. Most organic chemists are at some time faced with the problem of synthesizing a desired material and, for many, the problems are formidable. Nonspecialists faced with a synthetic problem are most likely not to have kept pace with the developments in synthetic chemistry that may well solve their problems and will not have the necessary information in their files.

Thus, we felt that an organized annual review of synthetically useful information would prove beneficial to nearly all organic chemists, both specialist and nonspecialist in synthesis. It should help relieve some of the information-storage burden of the specialist and should aid the nonspecialist who is seeking help with a specific problem to become rapidly aware of recent synthetic advances. Ideally also, such a review should be minimally priced to be within the means of potential users including graduate students, and it should appear as promptly as possible after the close of the abstracting period.

PREFACE

In producing *Annual Reports in Organic Synthesis—1977*, we have abstracted 47 primary chemistry journals, selecting useful synthetic advances. We have tried to present the information in an organized manner, emphasizing rapid visual retrieval. Only the common journals received by our libraries have been abstracted. Any journal received after March 1, 1978 will be covered in the next volume. We have also exercised selectively in choosing which papers to abstract. Our general guidelines have been to include all reactions and methods that are new, synthetically useful, and reasonably general. Each entry is comprised primarily of structures, accompanied by very few comments. The purpose of this is to aid the reader in scanning the book. The mind is capable of absorbing a whole picture in an instant, but is considerably slowed by having to read sentences. If the pictures presented catch the reader's interest, he should then seek details from the original paper.

For the second year we have included an author index to aid the user. However, this year the index includes only corresponding authors, senior authors, or first authors instead of a comprehensive author index as appeared in the volume last year. We still have no subject index. To include one would have greatly increased both the cost of the book and the delay time before publication. Instead, we have chosen to use an extensive table of contents. Chapters I–III are organized by reaction type and constitute a major part of the book. The organization of these sections is self-explanatory, and there should be no difficulty in locating a new method of oxidation or a new cyclopropanation procedure. Chapter IV deals with methods of synthesizing heterocyclic systems. Chapter V covers the use of new protecting groups and is also self-explanatory. Chapter VI is divided into three main parts and covers those synthetically useful transformations that do not fit easily into the first three chapters. The first part deals only with functional group synthesis. The second and third parts of Chapter VI are self-explanatory. The third part involves useful multistep sequences, the individual steps of which may be well known. Future volumes of this series will maintain the present table of contents as much as possible. If no entry is found for a particular section, the last volume in which one appears will be cited in the table of contents—see II.A.2.

Any undertaking of this type involves a series of compromises. We have chosen to emphasize reasonable cost, rapid publication, and rapid visual retrieval of information at the admitted expense of detail and beauty. This volume is the eighth in an annual series. We welcome suggestions for improvement of future volumes.

The arduous task of drawing the multitude of structures appearing in this review was carried out by Ms. Linda Benedict and Ms. Sandi Hanson. We thank them very much for their efforts. Also we thank Joel Slade and James McKearn for aid in proofreading portions of the manuscript.

R. Bryan Miller
L. G. Wade, Jr.

JOURNALS ABSTRACTED

Accounts of Chemical Research
Acta Chemica Scandinavica
Angewandte Chemie International Edition in English
Annales de Chimie
Australian Journal of Chemistry
Bulletin of the Chemical Society of Japan
Bulletin de Societes Chimiques Belges
Bulletin de la Societe Chimique de France
Canadian Journal of Chemistry
Chemical Communications
Chemical and Pharmaceutical Bulletin
Chemical Reviews
Chemical Society Reviews
Chemische Berichte
Chemistry and Industry
Chemistry Letters
Collection of Czechoslovakian Chemical Communications
Comptes Rendus Hebdomadaires de Seances de l'Academie des Sciences (C)
Doklady Chemistry
Experientia
Fortschritte der Chemischen Forschung
Gazzetta Chimica Italiana
Helvetica Chimica Acta
Indian Journal of Chemistry
Israel Journal of Chemistry
Journal of the American Chemical Society
Journal of the Chemical Society (Perkin I)

JOURNALS ABSTRACTED

Journal of the Chemical Society (Perkin II)
Journal of General Chemistry (USSR)
Journal of Heterocyclic Chemistry
Journal of the Indian Chemical Society
Journal of Organic Chemistry
Journal of Organic Chemistry (USSR)
Journal of Organometallic Chemistry
Journal fur Praktische Chemie
Liebig's Annalen der Chemie
Monatschefte fur Chemie
Pure and Applied Chemistry
Recueil des Travaux Chimiques des Pays-bas
Russian Chemical Reviews
Steroids
Synthesis
Synthetic Communications
Tetrahedron Letters
Zeitschrift fur Chemie
Zeitschrift fuer Naturforschung, Teil B

I.A.1-1 J. W. Huffman and P. G. Harris, Synth. Commun., 7, 137 (1977).

$$\mathrm{PhCOCHMe_2} \xrightarrow[\text{2) MeI}]{\text{1) Ph}_3\text{CK, DME}} \mathrm{PhCOCMe_3}$$

Facile preparation of potassium triphenylmethide (Ph_3CH + KH) and its use as a strong base for alkylations is described.

I.A.1-2 S. C. Goyal and S. M. Gupta, Ann. Chim., Ser. 15, 2, 57 (1977).

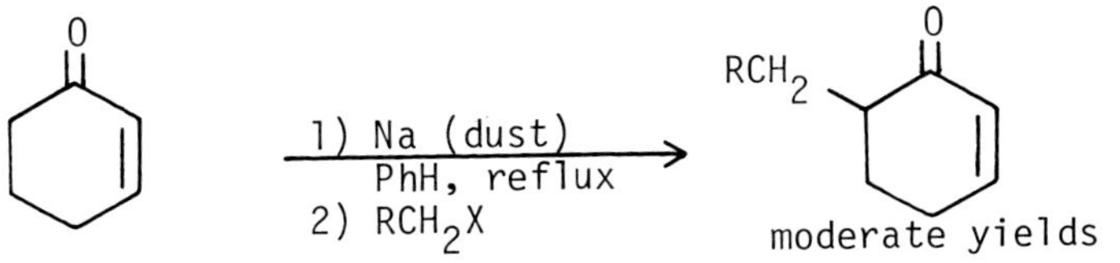
O
1) Na (dust)
PhH, reflux
2) RCH2X
RCH2
O
moderate yields

I.A.1-3 E. C. Taylor and J. L. LaMattina, Tetrahedron Lett., 2077 (1977).

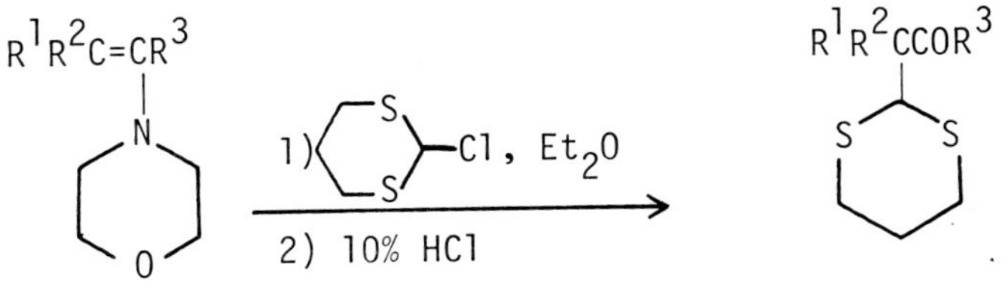
R1R2C=CR3
N
O
1)
S
S
Cl, Et2O
2) 10% HCl
R1R2CCOR3
S
S

R^1	R^2	R^3	% Yield
Me	Me	H	73
Me	H	Et	65
$-(CH_2)_4-$		H	40

ISBN 0-12-040808-2

I.A.1-4   J.-F. Le Borgne, J. Organometal. Chem., 122, 123 (1977); see also: ibid, 129 (1977).

$$R^1R^2CHCH=N- \xrightarrow[\substack{\text{2) } Br(CH_2)_nX \\ \text{3) } H_3O^+}]{\text{1) } Et_2NLi, HMPT} X(CH_2)_nCR^1R^2CHO$$

X = Cl, 50-75%
X = Br, 32-56%

$$R^1R^2CHCH=N- \xrightarrow[\substack{\text{2) } 1/2 \text{ eq. } Br(CH_2)_nBr \\ \text{3) } H_3O^+}]{\text{1) } Et_2NLi, HMPT} OHCCR^1R^2(CH_2)_nCR^1R^2CHO$$

$R^1=R^2=$Me, n=4, 70%

$R^1=$Et, $R^2=$H, n=6, 71%

I.A.1-5   R. M. Jacobson, R. A. Raths, and J. H. McDonald, III, J. Org. Chem., 42, 2545 (1977).

$$R^2CH_2CR^1=NR \xrightarrow[\substack{\text{2) } CH_2=C(OMe)CH_2Br \\ \text{3) } (CO_2H)_2, H_2O, THF}]{\text{1) LDA, THF}} R^1COCHR^2CH_2COMe$$

43-84%

Other compounds with active CH_2 groups were alkylated with the acetonyl alkylating agent.

I.A.1-6   M. E. Jung and R. B. Blum, <u>J. Org. Chem.</u>, 3791 (1977).

$$CH_2=CHOAc \xrightarrow[\text{2) RX, HMPA, heat}]{\text{1) MeOSnBu}_3} RCH_2CHO$$

R	% Yield
Me	61
Et	40
$PhCH_2$	68
$CH_2=CHCH_2$	41

I.A.1-7   T. H. Chan, I. Paterson, and J. Pinsonnault, <u>Tetrahedron Lett.</u>, 4183 (1977).

$$R^1C(OSiMe_3)=CR^2R^3 \xrightarrow[TiCl_4]{t\text{-BuCl}} R^1COC(\underline{t}\text{-Bu})R^2R^3$$

R^1	R^2	R^3	% Yield
Ph	H	H	43
H	$\text{-(CH}_2\text{)}_5$		40
$\text{-(CH}_2\text{)}_3$		H	59

I.A.1-8 S. Hashimoto, A. Itoh, Y. Kitagawa, H. Yamamoto, H. Nozaki, *J. Am. Chem. Soc.*, $\underline{99}$, 4192 (1977).

$$Me_2C=C(OSiMe_3)(CH_2)_2CHMe=CHCH_2OAc \xrightarrow[CH_2Cl_2,\ 0°C]{MeAl(O_2CCF_3)_2}$$

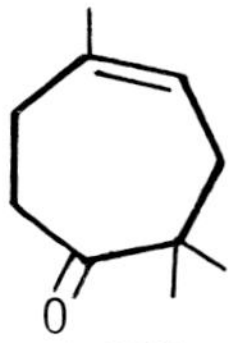

95%

$$Me_2C=C(OSiMe_3)(CH_2)_2CH(OAc)CH=CMe_2 \xrightarrow[CH_2Cl_2,\ 0°C]{MeAl(O_2CCF_3)_2}$$

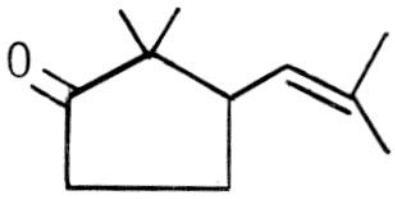

83%

Other systems also studied.

I.A.1-9 L. M. Jackman and B. C. Lange, *Tetrahedron*, $\underline{33}$, 2737 (1977).

Review: "Structure and Reactivity of Alkali Metal Enolates."

I.A.1-10 J. K. Rasmussen, *Synthesis*, $\underline{1977}$, 91.

Review: "O-Silylated Enolates — Versatile Intermediates for Organic Synthesis."

CARBON–CARBON BOND FORMING REACTIONS

I.A.2-1 Y. Masuyama, Y. Ueno, and M. Okawara, <u>Chem. Lett.</u>, <u>1977</u>, 835.

$$NCCH_2SePh \xrightarrow{RCH_2X,\ TBA} NCCH(SePh)CH_2R$$

$$81\text{-}96\%$$

R = Me, 1° alkyl, allyl or benzyl

Products can be easily converted to NCCH=CHR by treatment with NCS, NBS, or 35% H_2O_2.

I.A.2-2 N. Ono, R. Tamura, R. Tanikaga, and A. Kaji, <u>Synthesis</u>, <u>1977</u>, 690.

$$TsOCH_2CN\ +\ RX\ \xrightarrow[PhH]{DBU}\ TsOCHRCN$$

$$41\text{-}96\%$$

Use of DBU in benzene allows for selective monoalkylation.

I.A.2-3 M. Fedorynski, I. Gorzkowska, and M. Makosza, <u>Synthesis</u>, <u>1977</u>, 120.

$$CH_2=CHOAc\ +\ PhCHRCN\ \xrightarrow[PhCH_2NEt_3Cl]{aq.\ NaOH}\ PhCR(CN)CHMeOAc$$

R = Me, 64%

R = OMe, 56%

I.A.2-4 A. Debal, T. Cuvigny, and M. Larcheveque, Tetrahedron Lett., 3187 (1977).

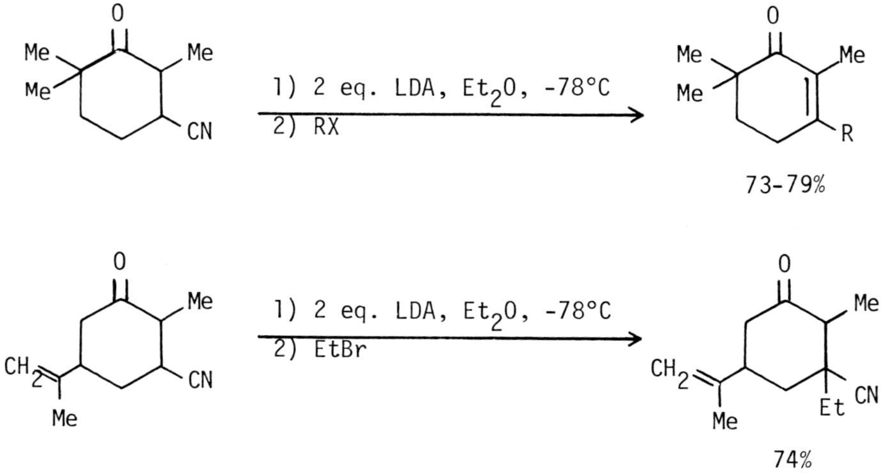

[cyclohexanone with gem-dimethyl, Me, and CN substituents]
1) 2 eq. LDA, Et$_2$O, -78°C
2) RX
⟶ [cyclohexenone with gem-dimethyl, Me, and R substituents]
73-79%

[cyclohexanone with Me, CN, and isopropenyl substituents]
1) 2 eq. LDA, Et$_2$O, -78°C
2) EtBr
⟶ [cyclohexanone with Me, Et, CN, and isopropenyl substituents]
74%

I.A.2-5 R. R. Schmidt and J. Talbiersky, Angew. Chem. Int. Ed. Engl., 16, 853 (1977).

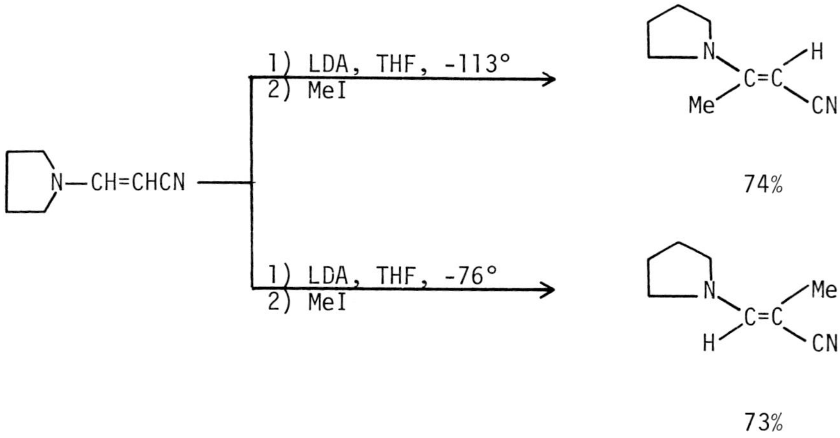

pyrrolidine-N-CH=CHCN

1) LDA, THF, -113°
2) MeI
⟶ [pyrrolidine-N-C(Me)=C(H)(CN)]
74%

1) LDA, THF, -76°
2) MeI
⟶ [pyrrolidine-N-C(H)=C(Me)(CN)]
73%

I.A.2-6   D. Savoia, C. Trombini, and A. Umani-Ronchi, <u>Tetrahedron Lett.</u>, 653 (1977); see also: H. Hart, B.-L. Chen, and C.-T. Peng, <u>ibid</u>, 3121 (1977).

$$R^1CH_2Y \xrightarrow[2)\ R^2X]{1)\ C_8K,\ THF} R^1R^2CHY$$

R^1	Y	R^2X	% Yield
H	CN	$\underline{n}$-$C_8H_{17}Br$	65
H	CN	$PhCH_2Cl$	42
Ph	CN	$\underline{i}$-PrBr	58
Ph	CO_2Et	MeI	71
Ph	CO_2Et	$\underline{i}$-PrBr	45

I.A.2-7   W. E. Parham and D. W. Boykin, <u>J. Org. Chem.</u>, **42**, 260 (1977).

$$Me_2C=CBrCO_2H \xrightarrow[2)\ EtI]{1)\ 2\ eq.\ n\text{-}BuLi \atop THF,\ -100°} Me_2C=C(Et)CO_2H$$

79%

I.A.2-8 R. P. Woodbury and M. W. Rathke, J. Org. Chem., **42**, 1688 (1977).

$$\text{MeCONMe}_2 \xrightarrow[\text{2) RX}]{\text{1) LDA, THF, -78°C}} \text{RCH}_2\text{CONMe}_2$$

RX	% Yield
MeI	62
i-PrBr	26
PhCH$_2$Br	99

Reactions with epoxides are also successful.

I.A.2-9 S. D. Carter and R. J. Stoodley, J. Chem. Soc., Chem. Commun., **1977**, 92.

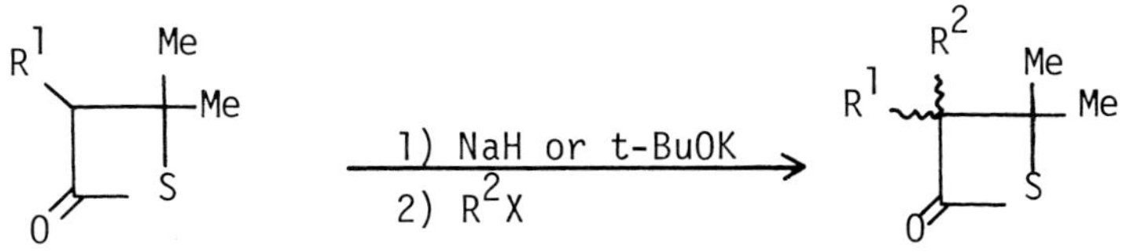

$$R^1 = \text{(oxazoline-fused β-lactam with CH}_2\text{Ph)}$$

Starting β-lactam (with R^1, Me, Me, S, O) $\xrightarrow[\text{2) R}^2\text{X}]{\text{1) NaH or t-BuOK}}$ product with R^2 substituent

R^2X = MeI, 74%

R^2X = BrCH$_2$CO$_2$t-Bu, ~50%

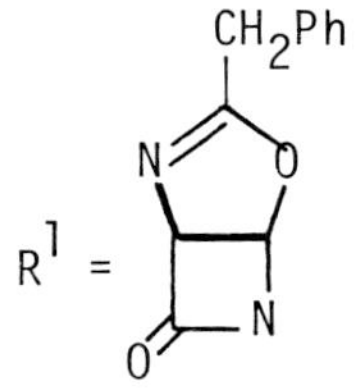

I.A.2-10 J. A. MacPhee and J.-E. Dubois, J. Chem. Soc., Perkin Trans. I, 1977, 694.

$$R^1R^2CHCO_2Et \xrightarrow[\text{2) }R^3X]{\text{1) LDA, HMPA}} R^1R^2R^3CCO_2Et$$

A study is reported on the effect of the nature of the alkylating agent and the order of introducing alkyl groups upon the yields of alkylation of α-mono- and α-disubstituted esters.

I.A.2-11 R. K. Boeckman, Jr., M. Ramaiah, and J. B. Medwid, Tetrahedron Lett., 4485 (1977).

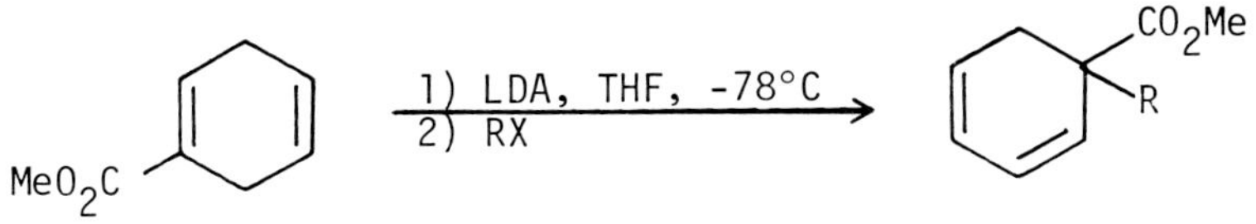

RX	% Yield
MeI	88
$Me_2C=CHCH_2Br$	52
n-C_8H_{17}I	96
$ClCH_2O_2Ct$-Bu	78

I.A.2-12   C. A. Bunnell and P. L. Fuchs, J. Am. Chem. Soc., **99**, 5184 (1977).

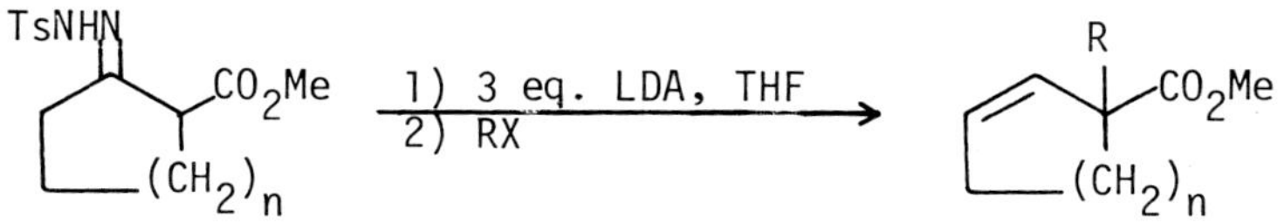

[Scheme: TsNH-substituted cyclic compound with CO_2Me and $(CH_2)_n$ ring, treated with 1) 3 eq. LDA, THF; 2) RX, giving cycloalkene with R, CO_2Me, $(CH_2)_n$]

n	RX	% Yield
1	MeI	64
1	$PhCH_2Br$	49
2	MeI	79
3	MeI	80

I.A.2-13   M. P. Zimmerman, Synth. Commun., **7**, 189 (1977).

$$MeOCH_2CH=CHCO_2Me \xrightarrow[\text{2) } RCH_2X]{\text{1) LDA, THF, HMPA, } -78°} MeOCH=CHCH(CH_2R)CO_2Me$$

42-81%

Products are very easily hydrolyzed to the aldehydes in contrast to the analogous methylthio compounds.

I.A.2-14 L. J. Ciochetto, D. E. Bergbreiter, and M. Newcomb, J. Org. Chem., 42, 2948 (1977).

$$R^1CH(OH)CO_2Et \xrightarrow[2)\ R^2X]{1)\ LDA,\ THF} R^1R^2C(OH)CO_2Et$$

R^1	R^2X	% Yield
Ph	n-BuBr	79
Ph	i-PrI	57
Ph	$BrCH_2CO_2Et$	69
Me	$n-C_{10}H_{21}I$	69

I.A.2-15 P. Bey and J. P. Vevert, Tetrahedron Lett., 1455 (1977).

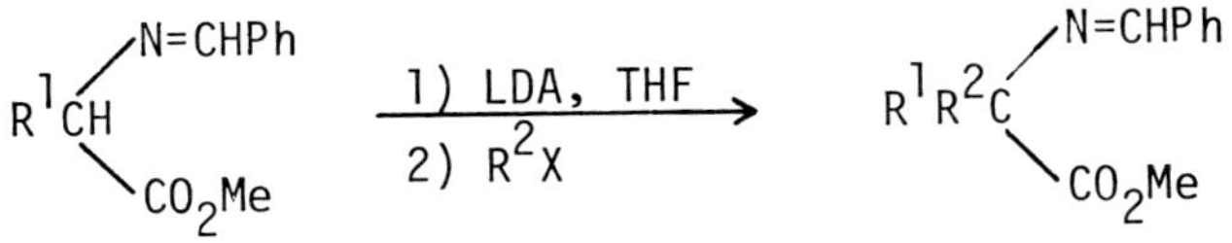

$$R^1CH\begin{smallmatrix}N=CHPh\\ \\CO_2Me\end{smallmatrix} \xrightarrow[2)\ R^2X]{1)\ LDA,\ THF} R^1R^2C\begin{smallmatrix}N=CHPh\\ \\CO_2Me\end{smallmatrix}$$

76-95%

I.A.2-16 J. J. Fitt and H. W. Gschwend, J. Org. Chem., **42**, 2639 (1977).

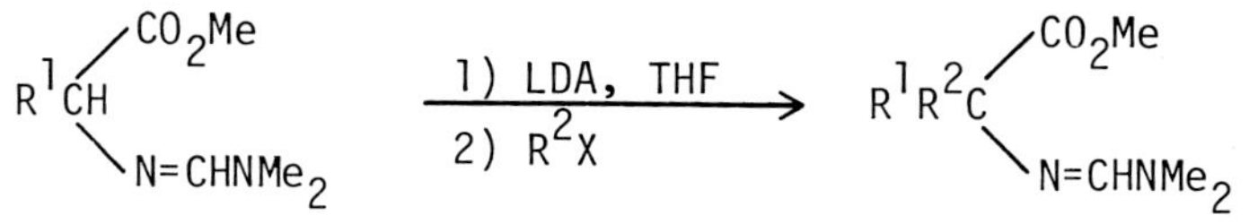

$$R^1CH(CO_2Me)(N=CHNMe_2) \xrightarrow[2) R^2X]{1) LDA, THF} R^1R^2C(CO_2Me)(N=CHNMe_2)$$

R^1	R^2X	% Yield
H	$PhCH=CHCH_2Br$	65
$MeSCH_2CH_2$	$p\text{-}ClC_6H_4CH_2Cl$	84
Ph	MeI	83
$PhCH_2$	$\underline{n}\text{-}PrI$	80

I.A.2-17 K. Hirai, Y. Iwano, and Y. Kishida, Tetrahedron Lett., 2677 (1977).

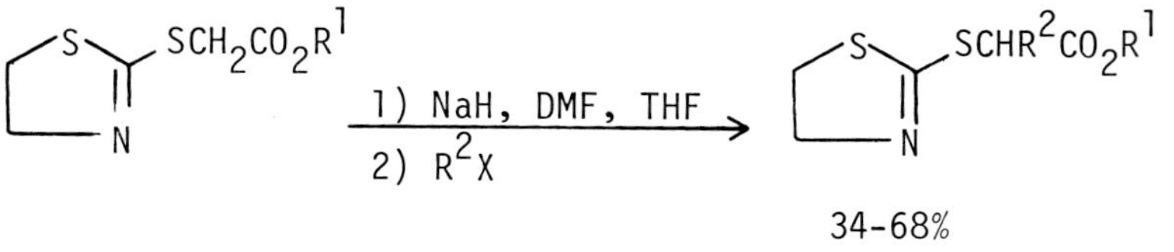

$$\underset{N}{\overset{S}{\diagdown}}C\text{-}SCH_2CO_2R^1 \xrightarrow[2) R^2X]{1) NaH, DMF, THF} \underset{N}{\overset{S}{\diagdown}}C\text{-}SCHR^2CO_2R^1$$

34-68%

The product can be converted into either $R^2CH_2CO_2R^1$ (Zn, AcOH) or $R^2CHICO_2R^1$ [MeI, DMF, $CaCO_3$, Hg(cat)] providing a C-2 elongation procedure.

CARBON—CARBON BOND FORMING REACTIONS

I.A.3-1 J. H. Clark and J. M. Miller, *J. Chem. Soc., Chem. Commun.*, **1977**, 64; see also: *ibid*, *J. Chem. Soc. Perkin Trans. I*, **1977**, 1743.

$$R^1COCH_2COR^2 \xrightarrow[Et_4NF, \ CHCl_3]{R^3I} R^1COCHR^3COR^2$$
$$91\text{-}95\%$$

Use of Et_4NF allows high yields of mono-C-alkylation products to be isolated.

I.A.3-2 P.-E. Sum and L. Weiler, *Can. J. Chem.*, **55**, 996 (1977).

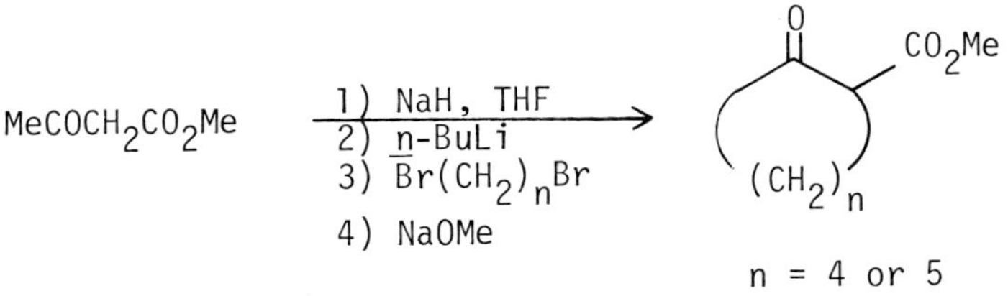

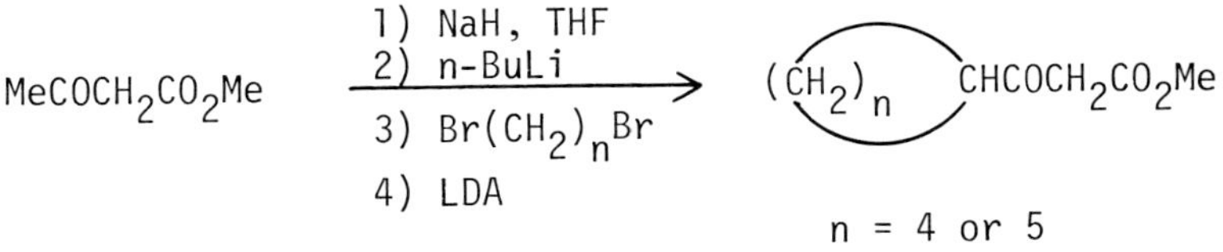

$$MeCOCH_2CO_2Me \xrightarrow[\substack{1) \ NaH, \ THF \\ 2) \ n\text{-}BuLi \\ 3) \ Br(CH_2)_nBr \\ 4) \ NaOMe}]{}$$

(cyclic product with CO_2Me, $(CH_2)_n$)

n = 4 or 5

$$MeCOCH_2CO_2Me \xrightarrow[\substack{1) \ NaH, \ THF \\ 2) \ n\text{-}BuLi \\ 3) \ Br(CH_2)_nBr \\ 4) \ LDA}]{} (CH_2)_n \ CHCOCH_2CO_2Me$$

n = 4 or 5

I.A.3-3 C. G. Kruse, N.L.J.M. Broekhof, A. Wijsman and A. van der Gen, Tetrahedron Lett., 885 (1977).

[1,3-dithiane]-Cl + NaCH(CO_2Et) → [1,3-dithiane]-CH($CO_2Et)_2$

80%

I.A.3-4 T. Tsujikawa and M. Hayashi, Chem. Pharm. Bull., 25, 3147 (1977).

5,5-dimethyl-1,3-cyclohexanedione + [cyclic enol ether/thioether]-($CH_2)_n$ $\xrightarrow{H^+, PhH, reflux}$ product

X = S, n = 1; 53%
X = O, n = 2; 40%

I.A.3-5 M. Hájek and J. Málek, Synthesis, 1977, 454.

XCH_2COR^1 + $R^2CH=CH_2$ $\xrightarrow{\text{Metal oxide}}_{\text{heat}}$ $R^2CH_2CH_2CHXCOR^1$

R^1	X	R^2	Metal Oxide	% Yield
OEt	CN	C_8H_{17}	CuO	84
OEt	CN	C_8H_{17}	NiO_2	62
OEt	CN	C_8H_{17}	PbO_2	72
OEt	MeCO	C_6H_{13}	AgO	57
OEt	CO_2Et	C_8H_{17}	MnO_2	87
OEt	CO_2Et	C_8H_{17}	Ag_2O	67
Me	MeCO	C_6H_{13}	PbO_2	49

I.A.3-6 W. G. Dauben and D. J. Hart, J. Am. Chem. Soc., 99, 7307 (1977).

$$\triangleright\!\!\!\!\triangleleft\!\!\begin{array}{c}CO_2Et\\+\\PPh_3\ BF_4^-\end{array} + R^1COCHR^2CHO \xrightarrow{\text{NaH}}_{\text{HMPT}}$$

(cyclopentene product with R^2, COR^1, and CO_2Et substituents)

R^1, R^2	% Yield
$-CHMe(CH_2)_3$	35
$-(CH_2)_5-$	44
$-(CH_2)_6-$	44

A full paper is presented.

I.A.3-7 N. Ono, R. Tamura, J.-I. Hayami, and A. Kaji, Chem. Lett., 1977, 189.

$$Na^+\ R^1\bar{C}YCO_2Et\ +\ R^2R^3C(Cl)NO_2 \xrightarrow[\text{2) 120°C}]{\text{1) HMPA, }h\nu,\ 20°C} R^1CY=CR^2R^3$$

R^1	Y	R^2	R^3	% Yield
Et	CO_2Et	Me	Me	63
n-Bu	COMe	Me	Me	60
n-Bu	CO_2Et	$-(CH_2)_5-$		42

I.A.4-1 R. Schlecker and D. Seebach, Helv. Chim. Acta, 60, 1459 (1977).

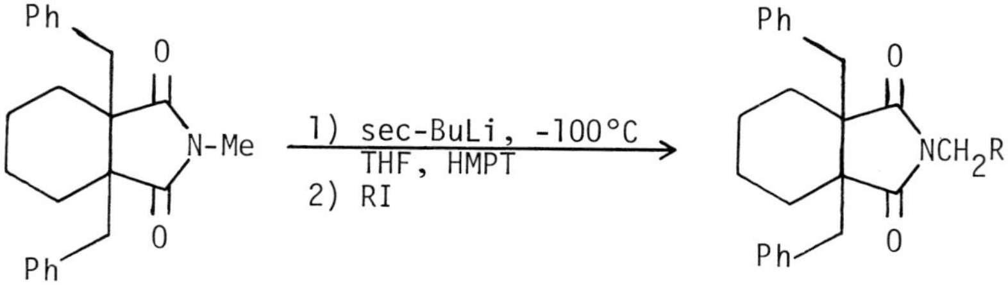

[Structure: bicyclic succinimide with Ph, Ph substituents and N-Me] → 1) sec-BuLi, -100°C, THF, HMPT; 2) RI → [Product structure with NCH$_2$R]

R = Me, 71%
R = C$_6$H$_{13}$, 59%

Sterically less hindered succinimides gave mainly dimers as products.

I.A.4-2 P. Beak, B. G. McKinnie, and D. B. Reitz, Tetrahedron Lett., 1839 (1977).

$$\text{ArCOYMe} \xrightarrow[\text{2) R}^2\text{X}]{\text{1) R}^1\text{Li, THF, -78°C}} \text{ArCOYCH}_2\text{R}^2$$

R^2X	Y	% Yield
MeI	MeN	77
MeI	S	86
CH$_2$=CHCH$_2$Br	S	94

I.A.4-3. D. Seebach, D. Enders, and B. Renger, Chem. Ber., **110**, 1852 (1977); see also: B. Renger, H.-O. Kalinowski, and D. Seebach, ibid, 1866 (1977); B. Renger and D. Seebach, ibid, 2334 (1977).

$$R^1N(NO)CH_2R^2 \xrightarrow[2)\ R^3X]{1)\ LDA,\ THF,\ -78°C} R^1N(NO)CHR^2R^3$$

R^1	R^2	R^3	% Yield
$\underline{n}$-C_6H_{13}	$\underline{n}$-C_5H_{11}	$PhCH_2$	> 95%
$\underline{t}$-Bu	$PhCH_2$	$PhCH_2$	70
$+CH_2+_3$		$\underline{n}$-Pr	52
$+CH_2+_4$		$\underline{n}$-C_5H_{11}	46

I.A.4-4 D. Seebach, R. Henning, F. Lehr, and J. Gonnermann, Tetrahedron Lett., 1161 (1977).

$$R^1CH_2NO_2 \xrightarrow[\substack{2)\ R^2X \\ 3)\ HOAc}]{1)\ 2\ eq.\ LDA,\ -78°C \\ THF,\ HMPT} R^1R^2CHNO_2$$

35-80%

$$Me_2CHNO_2 \xrightarrow{\text{same as above}} R^2CH_2CHMeNO_2$$

R = $PhCH_2$, 40%

$$MeO_2CCH_2CH_2NO_2 \xrightarrow{\text{same as above}} MeO_2CCHR^2CH_2NO_2$$

74-80%

I.A.4-5 B. L. Burt, D. J. Freeman, P. G. Gray, R. K. Norris, and D. Randles, Tetrahedron Lett., 3063 (1977).

$$Bu_4\overset{+}{N}CMe_2\overset{-}{NO_2} \text{ (aq.)} \xrightarrow[CH_2Cl_2 \text{ or PhH}]{ArCHXY} ArCHYCMe_2NO_2$$

Ar	X	Y	% Yield
$\underline{p}$-O$_2$NC$_6$H$_4$	Cl	H	73
$\underline{p}$-O$_2$NC$_6$H$_4$	Cl	Cl	62
Ph	Cl	H	— (80% yield of PhCHO)

I.A.4-6 U. Schöllkopf, K.-W. Henneke, K. Madawinata and R. Harms, Justus Liebigs Ann. Chem., 1977, 40.

$$R^1R^2CHNC \xrightarrow[2) R^3X]{1) \text{ n-BuLi, THF, -70°}} R^1R^2R^3CNC$$

R^1	R^2	R^3	% Yield
H	H	CH$_2$=CHCH$_2$	27
H	H	$\underline{n}$-Pr	39
Ph	H	PhCH$_2$	81
⸺(CH$_2$)$_2$⸺		PhCH$_2$	76

CARBON—CARBON BOND FORMING REACTIONS

I.A.4-7   U. Schöllkopf, D. Stafforst, and R. Jentsch, Justus Liebigs Ann. Chem., 1977, 1167.

$$PhCH=CHNC \xrightarrow[\substack{1)\ n\text{-BuLi, }-110°C \\ \text{Trapp mixture} \\ 2)\ MeI,\ -70°C}]{} PhCH=CMeNC$$

The alkenyl isocyanides can act as acyl anion synthons.

I.A.4-8   U. Schöllkopf, Angew. Chem. Int. Ed. Engl., 16, 339 (1977).

Review: "Recent Applications of α-Metalated Isocyanides in Organic Synthesis."

I.A.4-9   T. M. Dolak and T. A. Bryson, Tetrahedron Lett., 1961 (1977).

$$R^1R^2CHSPh \xrightarrow[\substack{1)\ Li,\ HMPA \\ THF,\ -78°C \\ 2)\ R^3X}]{} R^1R^2R^3CSPh \quad 74\text{-}82\%$$

I.A.4-10 K.-H. Geiss, D. Seebach, and B. Seuring, Chem. Ber., 110, 1833 (1977).

$$PhCH_2SH \xrightarrow[\begin{array}{c}1)\ 2\ eq.\ n\text{-}BuLi\\ THF,\ TMEDA\\ 2)\ R^1X\\ 3)\ R^2X\end{array}]{} PhCHR^1SR^2$$

$R^1 = Me,\ R^2 = Me,\ 76\%$

$R^1 = \underline{n}\text{-}Bu,\ R^2 = Me,\ 75\%$

$$CH_2=CHCH_2SH \xrightarrow[\begin{array}{c}1)\ 2\ eq.\ n\text{-}BuLi\\ THF,\ TMEDA\\ 2)\ R^1X\\ 3)\ R^2X\end{array}]{} R^1CH_2CH=CHSR^2 + CH_2=CHCHR^1SR^2$$

A B

$R^1 = \underline{n}\text{-}C_6H_{13},\ R^2 = Me,\ 80\%,\ A/B = 64/36$

$R^1 = i\text{-}Pr,\ R^2 = PhCH_2,\ 85\%,\ A/B = 59/41$

I.A.4-11 C. Huynh, V. Ratovelomanana, and S. Julia, Bull. Soc. Chim. Fr., 1977, 710.

$$R^1R^2C=CR^3CHR^4SR^5 \xrightarrow[\begin{array}{c}ClCH_2SPh\\ \overline{50\%\ NaOH\ (aq.)}\\ Aliquat\ 336\end{array}]{} PhSCH(SR^5)CR^1R^2CCR^3=CHR^4$$

R^1	R^2	R^3	R^4	R^5	% Yield
H	H	H	H	Et	46
Me	Me	H	H	Me	53
Me	Me	H	H	$PhCH_2$	76
Me	Me	H	Me	Ph	68

I.A.4-12 M. Wada, H. Nakamura, T. Taguchi, and H. Takei, Chem. Lett., 1977, 345.

$$\text{PhSCH=CHCH}_2\text{OMe} \xrightarrow[\text{2) RX}]{\text{1) LDA, THF, -78°C}} \text{PhSCHRCH=CHOMe}$$

RX	% Yield
MeI	72
n-BuI	80
PhCH$_2$CH$_2$Br	73
i-PrI	38
CH$_3$=CHCH$_2$Br	70

The 3-methoxy-1-phenylthio-1-propene serves as a useful β-formylvinyl anion equivalent.

I.A.4-13 K. Fuji, M. Ueda, and E. Fujita, J. Chem. Soc., Chem. Commun., 1977, 814.

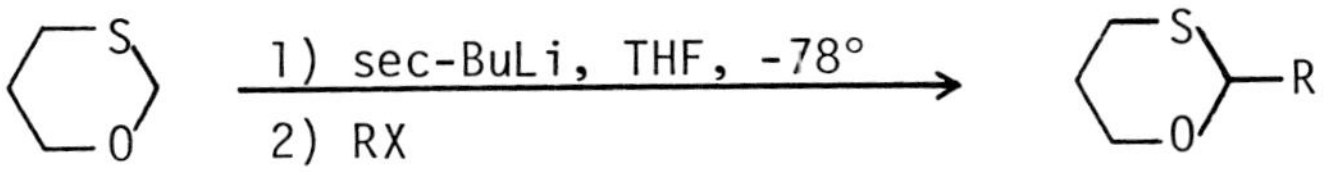

$$\underset{\text{O}}{\overset{\text{S}}{\bigcirc}} \xrightarrow[\text{2) RX}]{\text{1) sec-BuLi, THF, -78°}} \underset{\text{O}}{\overset{\text{S}}{\bigcirc}}\text{-R}$$

RX	% Yield
MeI	99
n-PrI	83
i-PrI	35
PhCH$_2$Br	26

I.A.4-14 R. D. Balanson, V. M. Kobal, and R. R. Schumaker, J. Org. Chem., **42**, 393 (1977).

Me—N(S)(S) $\xrightarrow{\text{1) n-BuLi, THF, -78°C}}_{\text{2) RX}}$ Me—N(S)(S)—R

RX	% Yield
$\underline{n}$-C_6H_{13}I	100
$\underline{i}$-PrI	85
$\underline{n}$-C_6H_{13}CHIMe	80

The products hydrolyze to the corresponding aldehydes more easily than the 1,3-dithiane analogues.

I.A.4-15 D. Seebach and M. Kolb, *Jusuts Liebigs Ann. Chem.*, **1977**, 811.

$R^2CH_2CHR^1$=(S)(S) $\xrightarrow{\text{1) n-BuLi or LDA}}_{\text{2) } R^3X}$ $R^3CH=CHR^1$-(S)(S) with R^3

60-99%

$R^2CH_2CH=CHCR^1$=(S)(S) $\xrightarrow{\text{1) LDA}}_{\text{2) } R^3X}$ $R^2CH=CHCH=CR^1$-(S)(S) with R^3

62-90%

General terms for umpolung reactivity are defined.

I.A.4-16 D. Seebach, R. Bürstinghaus, B.-T. Gröbel, and M. Kolb, Justus Liebigs Ann. Chem., 1977, 830.

$$R^1CH=C\underset{S}{\overset{S}{\diagdown}}\diagup \quad \xrightarrow[\text{2) MeI}]{\text{1) } R^2Li} \quad R^1R^2CH\underset{Me}{\diagdown}\underset{S}{\overset{S}{\diagup}}$$

I.A.4-17 R. Bürstinghaus and D. Seebach, Chem. Ber., 110, 841 (1977).

$$Me_3SnCH(SMe)_2 \quad \xrightarrow[\text{2) RI}]{\text{1) LDA, HMPTA, THF, }-40°C} \quad Me_3SnCR(SMe)_2$$

R = Me, 67%

R = n-Pr, 61%

I.A.4-18 B.-T. Gröbel and D. Seebach, Chem. Ber., 110, 867 (1977).

$$CH_2=C(Br)SiMe_3 \quad \xrightarrow[\text{2) } n\text{-}C_5H_{11}I]{\text{1) } t\text{-BuLi, THF, }-78°C} \quad CH_2=C(SiMe)_3C_5H_{11}$$

83%

$$CH_2=C(SPh)SnMe_3 \quad \xrightarrow[\text{2) MeI}]{\text{1) } n\text{-BuLi, THF, }-78°C} \quad CH_2=C(SPh)Me$$

78%

I.A.4-19   U. Klein and W. Sucrow, Chem. Ber., 110, 1611 (1977).

$$MeSO_2NMe \xrightarrow[2)]{1) \ LDA, \ THF} \xrightarrow{\overset{O}{\underset{R^2}{\diagup\!\!\!\diagdown} R^1}} R^1R^2C(OH)CH_2CH_2SO_2NMe_2$$

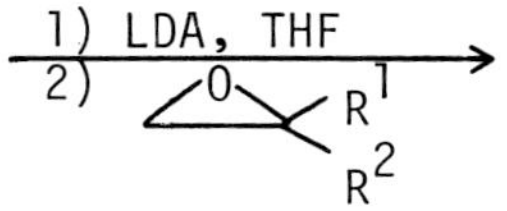

$$R^1, R^2 = -(CH_2)_5-, \ 83\%$$

$$R^1 = R^2 = Me, \ 71\%$$

I.A.4-20   D. Savoia, C. Trombini, and A. Umani-Ronchi, J. Chem. Soc., Perkin Trans. I, 1977, 123.

$$PhSO_2CH_2CH=CH_2 \xrightarrow[2) \ RX]{1) \ n\text{-}BuLi, \ -60°C \atop THF, \ TMEDA} PhSO_2CHRCH=CH_2$$

$$R = \underline{n}\text{-}Bu, \ 78\%$$

$$R = PhCH_2, \ 75\%$$

The product can be easily converted into RCH=CHMe.

I.A.4-21   B.-T. Gröbel and D. Seebach, Synthesis, 1977, 357.

Review: "Umpolung of the Reactivity of Carbonyl Compounds Through Sulfur-Containing Reagents."

CARBON−CARBON BOND FORMING REACTIONS

I.A.5-1 P. Coutrot, C. Laurenco, J. F. Normant, P. Perriot, P. Savignac and J. Villieras, <u>Synthesis</u>, <u>1977</u>, 615.

$$(EtO)_2P(O)CR^1Cl_2 \xrightarrow[\text{2) } R^2X]{\text{1) n-BuLi}} (EtO)_2P(O)CR^1R^2Cl$$

R^1	R^2X^*	% Yield
Cl	MeBr	81
Cl	$CH_2=CHCH_2Br$	80
Cl	n-BuBr	80
Me	EtI	67
n-Bu	$CH_2=CHCH_2Br$	66

*Alkyl bromides except MeBr require 1 equivalent of HMPT.

I.A.5-2 T. Kauffmann and R. Joussen, <u>Chem. Ber.</u>, <u>110</u>, 3930 (1977).

$$XCHR^1Cu + R^2I \longrightarrow XCHR^1R^2$$

X	R^1	R^2	% Yield
$Ph_2P(O)$	H	$PhSO_2CH_2$	75
$Ph_2P(O)$	n-Pr	$PhSO_2CH_2$	68
$Ph_2P(O)$	H	Ph	50
Ph_2P	H	Ph	33
$PhSO_2$	H	2-Quinolinyl	45

I.A.5-3 J. Villieras, P. Perriot, and J. F. Normant, Bull. Soc. Chim. Fr., 1977, 765.

$$R^2CCl_2H \xrightarrow[\text{Et}_2\text{O, THF, -90°C}]{\text{1) n-BuLi, TMEDA}} R^1CCl_2R^2$$
$$\text{2) } R^2X$$

R^1	R^2X	% Yield
H	$\underline{n}$-C_7H_{13}Br	86
H	$ClCH_2CH_2Br$	84
Me	$\underline{n}$-BuI	68

I.A.5-4 E. Piers and J.R. Grierson, J. Org. Chem., 42, 3755 (1977).

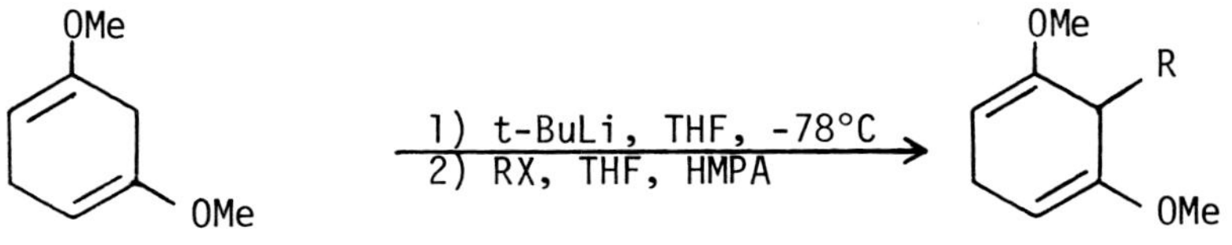

$$\xrightarrow[\text{2) RX, THF, HMPA}]{\text{1) t-BuLi, THF, -78°C}}$$

88-99%

CARBON−CARBON BOND FORMING REACTIONS

I.A.5-5 M. Pohmakotr and D. Seebach, Angew. Chem. Int. Ed. Engl., 16, 320 (1977).

$$PhCOCH_2CH_2CH=CH_2 \xrightarrow[\substack{2) \text{ sec-BuLi, TMEDA} \\ -78° \text{ to } 0°C \\ 3) \text{ RX}}]{1) \text{ KH, THF, } 0°C} PhCOCH_2CH=CHCH_2R$$

only (Z) isomer formed

R	% Yield
Me	60-72
n-Bu	59
Me$_2$C(OH)CH$_2$	32

I.A.5-6 B. M. Trost and L. H. Latimer, J. Org. Chem., 42, 3212 (1977).

MeO-[indanone] → 1) 2 eq. LDA, THF, -78°C 2) RI → MeO-[indanone with R at 3-position]

R = Et, 89%
R = PhCH$_2$OCH$_2$CH$_2$, 78%

I.A.5-7 P. Beak and B. G. McKinnie, J. Am. Chem. Soc., 99, 5213 (1977).

$$ArCO_2Me \xrightarrow[2) \text{ MeI}]{1) \text{ sec-BuLi, TMEDA, THF}} ArCO_2Et$$

84%

Ar = 2,4,6-triisopropylphenyl

I.A.5-8 T. Kauffmann, H. Fischer, and A. Woltermann, Angew. Chem. Int. Ed. Engl., 16, 53 (1977); see also: T. Kauffmann, R. Joussen and A. Woltermann, ibid, 709 (1977).

$$Ph_2\overset{O}{\underset{\|}{As}}Me \xrightarrow[\text{2) RX}]{\text{1) LDA, THF, -40°}} Ph_2\overset{O}{\underset{\|}{As}}CH_2R$$

RX = n-BuBr, 72%

RX = PhCH$_2$Br, 61%

Products can easily be converted into organic bromides.

I.A.5-9 T. Kauffmann, H. Ahlers, H.-J. Tilhard, and A. Woltermann, Angew. Chem. Int. Ed. Engl., 16, 710 (1977).

$$CH_2=CHM \xrightarrow{\text{RLi, THF}} R(CH_2)_2M$$

R	M	% Yield
Et	AsPh$_2$	37
n-Bu	AsPh$_2$	95
t-Bu	AsPh	57
sec-Bu	SePh	25

Products can easily be converted to alkyl halides thus providing an effective 2C-chain elongation of lithioalkanes.

I.A.5-10 K. Uchida, K. Utimoto, and H. Nozaki, Tetrahedron, 33, 2987 (1977).

$$R^1C\equiv CSiMe_3 \xrightarrow[\text{2) MeLi}]{\text{1) (cyclohexyl)}_2\text{BH}} \underset{H}{\overset{R^1}{>}}C=C\underset{Li}{\overset{SiMe_3}{<}} \xrightarrow{R^2X \text{ or CuI, } R^2X}$$

$$\underset{H}{\overset{R^1}{>}}C=C\underset{R^2}{\overset{SiMe_3}{<}}$$

R^1	R^2X	% Yield
$\underline{n}$-C_6H_{13}	Me	94
$\underline{n}$-C_6H_{15}	CH_2=CClCH$_2$Cl	71
THPO(CH$_2$)$_4$	EtI, CuI	85

I.A.5-11 R. K. Boeckman, Jr. and K. J. Bruza, Tetrahedron Lett., 4187 (1977); see also: O. Riobe, A. Lebouc, and J. Delaunay, Comptes rendus, Ser. C, 284, 281 (1977).

$$\underset{(CH_2)_n}{\overset{R^1}{R^2}\diagdown\!\!\diagup\!\!\diagdown O} \xrightarrow[\text{2) R}^3X]{\text{1) t-BuLi, 0°C}} \underset{(CH_2)_n}{\overset{R^1}{R^2}\diagdown\!\!\diagup\!\!\diagdown O\diagup\! R^3}$$

n	R^1	R^2	R^3X	% Yield
1	H	H	Me$_2$C=CHCH$_2$Br	67
1	H	H	$\underline{n}$-C_6H_{13}I	64
2	H	H	$\underline{n}$-C_6H_{13}I	54
2	MeO	Me	Me$_2$C=CHCH$_2$Br	57

I.A.5-12 L. Duhamel and J. M. Poirier, J. Am. Chem. Soc., 99, 8356 (1977).

$$R^2CBr=CR^1NR^3R^4 \xrightarrow[\text{2) } R^5X]{\text{1) n-BuLi, THF, -78°C}} R^2R^5C=CR^1NR^3R^4$$

R^1	R^2	R^3	R^4	R^5X	% Yield
H	t-Bu	Et	Et	MeI	45
H	Ph	Me	Me	EtI	40
Ph	$-(CH_2)_3-$		Me	n-BuI	55

I.A.5-13 K. Kitatani, T. Hiyama, and H. Nozaki, Bull. Chem. Soc. Japan, 50, 1600 (1977).

$$R^1 \underset{R^2}{\overset{X\quad X}{\triangle}} \xrightarrow[\text{2) } R^4I,\ -20°C]{\text{1) } R^3_2CuLi,\ Et_2O,\ -40° \text{ to } -78°C} R^1 \underset{R^2}{\overset{R^3\quad R^4}{\triangle}}$$

X	R^1	R^2	R^3	R^4	% Yield
Br	Ph	H	n-Bu	MeI	100
Br	Ph	H	sec-Bu	MeI	43
Br	Ph	H	t-Bu	MeI	20
Br	Ph	H	n-Bu	EtI	65
Br	$-(CH_2)_4-$		n-Bu	MeI	82
Cl	$-(CH_2)_3C(OCH_2CH_2O)-$		n-Bu	MeI	78

Use of Me_2CuLi or $(CH_2=CH)_2CuLi$ gives only the coupled product with no subsequent alkylation.

CARBON—CARBON BOND FORMING REACTIONS

I.A.5-14 H. Yamamoto, K. Kitatani, T. Hiyama, and H. Nozaki, J. Am. Chem. Soc., **99**, 5816 (1977).

$$\underset{R^1\ R^2}{\overset{R^4\ Br}{\triangle}}R^3 \quad \xrightarrow[2)\ R^5X]{1)\ n-Bu_2CuLi,\ THF} \quad \underset{R^1\ R^2}{\overset{R^4\ R^5}{\triangle}}R^3$$

R^1	R^2	R^3	R^4	R^5X	% Yield
H	Ph	H	H	MeI	91
H	Ph	H	H	$CH_2=CHCH_2Br$	97
H	$\underline{n}-C_6H_{13}$	H	H	MeO_2CCH_2Br	75
\-(CH$_2$)$_4$\-		H	$CH_2=CHCH_2$	MeI	96
\-(CH$_2$)$_4$\-		Me	H	$CH_2=CHCH_2Br$	65

I.A.5-15 K. Kitatani, H. Yamamoto, T. Hiyama, and H. Nozaki, Bull. Chem. Soc. Japan, **50**, 2158 (1977).

$$Ph\overset{Br\ Br}{\triangle} \quad \xrightarrow[2)\ MeI\ or\ Me_2SO_4]{1)\ n-BuLi\ (excess)\ THF,\ -95°C\ to\ 0°C} \quad Ph\overset{Me\ Bu}{\triangle}$$

40-58%

I.A.5-16   K. Maruyama and Y. Yamamoto, J. Am. Chem. Soc., 99, 8068 (1977).

$$R^1R^2C=CHCHR^3X \xrightarrow[Et_2O \text{ or } THF]{R^4CuBF_3} R^1R^2R^4CCH=CHR^3$$

R^1	R^2	R^3	X	R^4	% Yield
Ph	H	H	Br	Me	90
Ph	H	H	Br	n-Bu	94
Me	H	H	Cl	n-Bu	90
Me	Me	H	Br	Me	82
H	H	Me	Cl	n-Bu	90

I.A.5-17   Y. Gendreau, J. F. Normant and J. Villieras, J. Organomet. Chem., 142, 1 (1977).

$$CH_2=CHCHR^1SR^2 \xrightarrow[\text{THF, } 20°C]{\underline{n}-C_7H_{15}MgCl \atop CuBr \text{ (cat.)}} \underline{n}-C_7H_{15}CH_2CH=CHR^1$$

R^1 = H, R^2 = Ph, 70%

R^1 = Me, R^2 = Et, 55%

$$R^1R^2C=CHCH_2\overset{+}{S}Me_2 \; X^- \xrightarrow[CuBr \text{ (cat.), THF}]{R^3MgCl(Br), \; 0°C} \underset{A}{R^1R^2C=CHCH_2R^3} + \underset{B}{R^1R^2R^3CCH=CH_2}$$

R^1	R^2	R^3	% Yield	A:B
H	H	n-C$_7$H$_{15}$	70	-
Ph	H	EtMe$_2$C	76	95:5
n-Bu	Et	n-Bu	50	54:46
n-Bu	Et	CH$_2$=C=CH	58	100:0

I.A.5-18 T. Mukaiyama, M. Imaoka and T. Izawa, <u>Chem. Lett.</u>, <u>1977</u>, 1257.

$R^1CH=CHCHOHR^2$ + [pyridinium salt: 2,6-dimethyl-4-methyl-N-ethyl-fluoropyridinium BF_4^-] $\xrightarrow{\text{1) Et}_3\text{N} \quad \text{2) R}^3\text{MgBr}}$ $R^1R^3CHCH=CHR^2$

R^1	R^2	R^3	% Yield
Ph	H	n-Bu	84
Et	H	$PhCH_2CH_2$	94
H	Me	$PhCH_2CH_2$	76

When R^3 is Ph the product is $R^1CH=CHCHR^2R^3$.

I.A.5-19 H. Westmijze and P. Vermeer, <u>Synthesis</u>, <u>1977</u>, 784.

$\left[\begin{array}{c} R^2 \\ R^1 \end{array} C=C \begin{array}{c} H \\ CuX^1 \end{array} \right]$ MgHal $\xrightarrow[\text{THF}]{X^2CN}$ $\begin{array}{c} R^2 \\ R^1 \end{array} C=C \begin{array}{c} H \\ CN \end{array}$

R^1	R^2	X^1	X^2	% Yield
i-Pr	H	Br	Cl	90
t-Bu	H	t-Bu	Cl	91
Et	Ph	Br	Ts	95
n-Bu	Me	n-Bu	Cl	93

I.A.5-20 M. Commercon-Bourgain, J. F. Normant, and J. Villieras, Comptes rendus, Ser. C, <u>285</u>, 211 (1977).

$$R^1CH=CHCH_2\overset{+}{N}Et_3 \quad Br^- \quad \xrightarrow{\underset{THF-5\% \; CuBr}{R^2MgCl}} \quad R^1CH=CHCH_2R^2$$

R^1	R^2	% Yield
H	$\underline{n}$-C_7H_{15}	92
Me	Ph	85
Ph	Et	30

I.A.5-21 K. Itoh, T. Yogo, and Y. Ishii, Chem. Lett., <u>1977</u>, 103; see also: K. Itoh, M. Fukui, and Y. Kurachi, J. Chem. Soc., Chem. Commun., <u>1977</u>, 500.

$$CH_2 \!\!\underset{O}{\overset{}{\diamondsuit}}\!\!=\!O \quad \xrightarrow{\underset{NiCl_2, \; THF}{Me_3SiCH_2MgCl}} \quad Me_3SiCH_2\overset{\overset{CH_2}{\|}}{C}CH_2CO_2H \quad 95\%$$

I.A.5-22 K. Abe, T. Sato, N. Nakamura, and T. Sakan, Chem. Lett., <u>1977</u>, 817.

$$R = \underline{n}\text{-}C_6H_{13}, \; 66\%$$
$$R = Ph, \; 60\%$$

I.A.5-23 M. Hojo, R. Masuda, T. Saeki, K. Fujimori, and S. Tsutsumi, Tetrahedron Lett., 3883 (1977).

$$MeS(O)CH_2SMe \xrightarrow{\text{RMgX}}_{\text{THF}} RCH(SMe)_2$$

R	% Yield
Ph	66
PhCH$_2$	41
$\underline{n}$-C$_6$H$_{13}$	54
PhCH=CH	39

I.A.6-1 A. J. Birch and J. Slobbe, Aust. J. Chem., 30, 1045 (1977).

$$\underset{\text{NMe}_2}{\underset{\text{CO}_2\text{H}}{\text{C}_6\text{H}_4}} \xrightarrow[\substack{2)\ \text{RX}\\3)\ 4\ \underline{M}\ \text{HCl, 0°C}\\4)\ 1\ \underline{M}\ \text{HCl, THF, reflux}}]{1)\ \text{Li, NH}_3(\ell)} \quad \text{cyclohexenone with R} \quad 45\text{-}80\%$$

I.A.6-2 T. Nakai, T. Mimura, and A. Ari-izumi, Tetrahedron Lett., 2425 (1977).

$$R^1R^2C(OH)CR^3{=}CH_2 \xrightarrow[\substack{\text{KH or NaH}\\2)\ \text{distillation}}]{1)\ ClC(S)NMe_2} R^1R^2C{=}CR^3CH_2SC(O)NMe_2$$

37-76%

The product can be converted into $R^1R^2C{=}CR^3CHO$ by a two-step process.

I.A.7.a.1a-1 W. E. Parham and D. W. Boykin, J. Org. Chem., 42 260 (1977).

$$Me_2C=CBrCO_2H \xrightarrow[\text{2) } R^1COR^2]{\text{1) 2 eq. n-BuLi, THF, -100°}} Me_2C=C(CO_2H)C(OH)R^1R^2$$

$R^1 = R^2 = Ph, 64\%$

$R^1 = R^2 = -(CH_2)_5-, 30\%$

I.A.7.a.1a-2 A. P. Krapcho, D. S. Kashdan, and E.G.E. Jahngen, Jr., J. Org. Chem., 42, 1189 (1977).

$$R^1R^2CHCO_2H \xrightarrow[\substack{\text{2) } R^3COCl, -70°C \\ \text{3) 2 eq. HCl (aq.)} \\ \text{4) 150-200°C}}]{\text{1) 2 eq. LDA, THF, 50°C}} R^1R^2CHCOR^3$$

30-70%

R^1 and R^2 = alkyl

R^3 = alkyl or aryl

I.A.7.a.1a-3 P. Hullot, T. Cuvigny, M. Larcheveque, and H. Normant, Can. J. Chem., 55, 266 (1977).

$$R^1R^2CHCONMe_2 \xrightarrow[\text{2) } R^3COR^4]{\substack{\text{1) Et}_2\text{NH, Li} \\ \text{HMPT, PhH}}} R^3R^4C(OH)R^1R^2CCONMe_2$$

R^1	R^2	R^3	R^4	% Yield
H	H	Ph	H	68
H	Et	H	H	91
H	Et	Me	H	76
Me	Me	Me	Me	61
Me	Me	Me	$Me_2C=CH$	39
H	H	$-(CH_2)_5-$		86

I.A.7.a.1a-4   R. P. Woodbury and M. W. Rathke, J. Org. Chem., 42, 1688 (1977).

$$\text{MeCONMe}_2 \xrightarrow[\text{2) R}^1\text{COR}^2]{\text{1) LDA, THF, -78°C}} R^1R^2C(OH)CH_2CONMe_2$$

R^1	R^2	% Yield
Me	H	98
Me	Me	99
$-(CH_2)_5-$		97

I.A.7.a.1a-5   W. A. Kleschick, C. T. Buse, and C. H. Heathcock, J. Am. Chem. Soc., 99, 248 (1977).

$$\text{EtCO}\underline{t}\text{-Bu} \xrightarrow[\substack{\text{2) PhCHO, 5 sec.} \\ \text{3) NH}_4\text{Cl soln.}}]{\text{1) LDA, THF, -78°C}} \text{Ph-CH(OH)-CH(Me)-CO}\underline{t}\text{-Bu} \quad 78\%$$

no threo product detected

$$\underset{\text{Me}}{\text{H}}\text{C}=\text{C}\underset{\text{OSiMe}_3}{\underline{t}\text{-Bu}} \xrightarrow[\text{PhCHNMe}_3\text{F}]{\text{PhCHO, THF}} \text{Ph-CH(OSiMe}_2\text{)-CH(Me)-CO}\underline{t}\text{-Bu} \quad 52\%$$

no erythro product detected

Criteria for stereoselection in the aldol condensation are given.

I.A.7.a.1a-6 K. Tanaka, N. Yamagishi, R. Tanikaga, and A. Kaji, Chem. Lett., 1977, 471.

$$HSCHR^1CO_2Et \xrightarrow[\substack{2)\ R^3COR^4 \\ 3)\ ClCO_2Et}]{1)\ LDA,\ TMEDA,\ -78°C} R^3R^4C=CR^1CO_2Et$$

R^1	R^2	R^3	% Yield
H	Me	Et	64
H	Me	Ph	58
Me	Me	Me	57
H	$-(CH_2)_4-$		74

I.A.7.a.1a-7 G. M. Ksander, J. E. McMurry, and M. Johnson, J. Org. Chem., 42, 1180 (1977).

$$R^1COCHR^2COCO_2Et \xrightarrow[\substack{2)\ R^3CHO \\ 3)\ KHCO_3,\ CH_2Cl_2}]{1)\ LiH\ or\ NaH,\ THF} R^1COCR^2=CHR^3$$

R^1	R^2	R^3	% Yield
Et	Me	H	40
EtO	Pr	Me	71
$-(CH_2)_4-$		H	87
$-OCH_2CH_2-$		Me	68

I.A.7.a.1a-8 R. Couffignal and J.-L. Moreau, *J. Organomet. Chem.*, **127**, C65 (1977).

$$R^1R^2CHCOR^3 \xrightarrow[\text{2) } R^4CO_2CO_2Et]{\text{1) LDA}} R^4COCR^1R^2COR^3$$

R^1	R^2	R^3	R^4	% Yield
H	H	OEt	n-Pr	67
Me	Me	OEt	Et	76
H	H	Ph	Et	49
Me	H	Et	Ph	67

I.A.7.a.1a-9 R. Noyori, K. Yokoyama, J. Sakata, I. Kuwajima, E. Nakamura, and M. Shimizu, *J. Am. Chem. Soc.*, **99**, 1265 (1977).

$$R^1R^2C=C(OSiMe_3)R^3 \xrightarrow[\text{2) } H_2O]{\text{1) } R^4CHO,\ THF,\ n\text{-}Bu_4NF\ (5\text{-}10\ \text{mol \%})} R^3COCR^1R^2CH(OH)R^4$$

R^4 = Ar, 70-95%

R^4 = alkyl, 35-50%

I.A.7.a.1a-10 A. K. Beck, M. S. Hoekstra, and D. Seebach, Tetrahedron Lett., 1187 (1977).

$R^1COCH_2R^2$ + mesityllithium ⟶ $\xrightarrow{R^3COCl}$ $R^1COCHR^2COR^3$

$R^1C(OSiMe_3)=CHR^2$ $\xrightarrow{MeLi}$

R^1	R^2	R^3	% Yield
Et	Me	$\underline{i}$-Pr	55
$\mathrm{-(CH_2)_4-}$		Ph	83
$\mathrm{-(CH_2)_5-}$		$-CH_2CH_2NO_2$	73
$\mathrm{-(CH_2)_6-}$		$\underline{i}$-Pr	74

I.A.7.a.1a-11 T. Inoue, T. Uchimaru, and T. Mukaiyama, Chem. Lett., 1977, 153.

R^1CH_2COMe $\xrightarrow[\text{2,6-lutidine, Et}_2\text{O}]{\text{1) 9-BBN triflate, -78°C}}$ $MeCOCHR^1C(OH)R^2R^3$
2) R^2COR^3
3) H_2O_2, MeOH, pH 7 buffer

R^1	R^2	R^3	% Yield
$PhCH_2$	Ph	H	88
$PhCH_2$	Me	Me	63
Et	C_5H_{11}	H	67

I.A.7.a.1a-12 K. Maruoka, S. Hashimoto, Y. Kitagawa, H. Yamamoto, and H. Nozaki, J. Am. Chem. Soc., 99, 7705 (1977).

$$R^1COCBrR^2R^3 + R^4COR^5 \xrightarrow[\text{THF, } -20°C]{\text{Zn, Et}_2\text{AlCl} \atop \text{CuBr (cat.)}} R^1COCR^2R^3C(OH)R^4R^5$$

R^1	R^2	R^3	R^4	R^5	% Yield
Ph	H	H	Ph	H	95
EtO	H	H	$-(CH_2)_5-$		93
$-(CH_2)_4-$		H	i-Pr	H	93
$-(CH_2)_4-$		Me	Ph	H	100

I.A.7.a.1a-13 Y. Jasor, M. Gaudry, M. J. Luche, and A. Marquet, Tetrahedron, 33, 295 (1977).

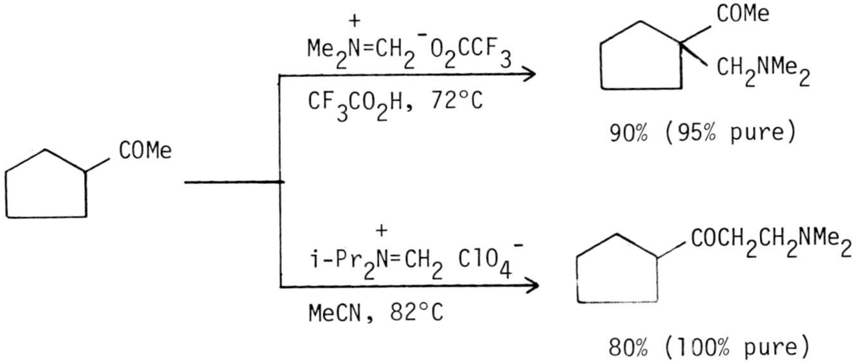

Other ketones studied to determine the regioselectivity in the formation of the Mannich bases.

I.A.7.a.1a-14 J. P. Albarella, J. Org. Chem., 42, 2009 (1977).

$$RCH_2CN \xrightarrow[\text{THF, } -78°]{\text{1) 2.3 eq. LDA}} RCH(CN)CO_2Et$$
$$\text{2) (EtO)}_2CO$$

56-86%

I.A.7.a.1a-15 S. A. DiBiase and G. W. Gokel, Synthesis, 1977, 629.

$$R^1COR^2 \xrightarrow[\text{MeCN}]{\text{KOH (pellets)}} R^1R^2C=CHCN$$

R^1	R^2	% Yield
Et	Et	35
n-Bu	n-Bu	65
Ph	Me	15
$-(CH_2)_6-$		78
$-(CH_2)_5-$		70

I.A.7.a.1a-16 K. Takahashi, T. Okamoto, K. Yomada, and H. Iida, Synthesis, 1977, 58.

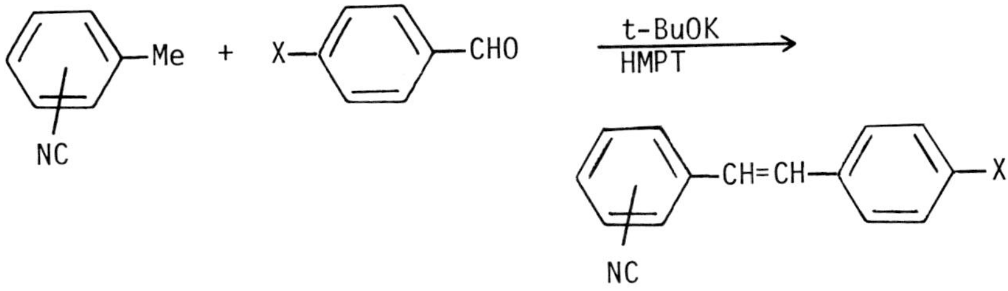

$$\text{C}_6\text{H}_4(\text{NC})\text{-Me} + \text{X-C}_6\text{H}_4\text{-CHO} \xrightarrow{\text{t-BuOK}/\text{HMPT}} \text{C}_6\text{H}_4(\text{NC})\text{-CH=CH-C}_6\text{H}_4\text{-X}$$

CN position	X	% Yield
4	OMe	58
2	OMe	84
2	NMe_2	58
2	Cl	35

I.A.7.a.1a-17 F. M. Hauser and R. Rhee, Synthesis, 1977, 245.

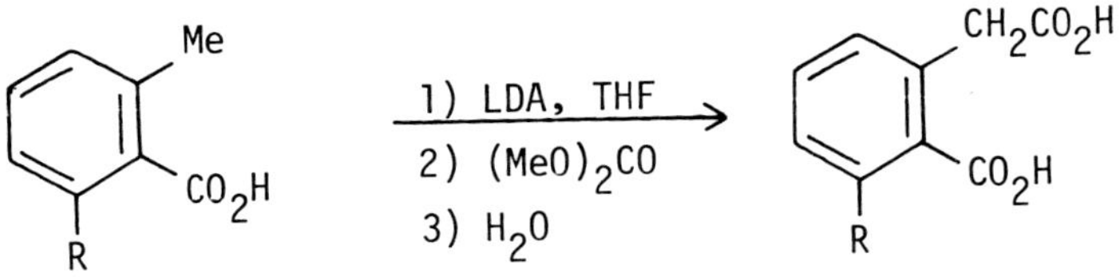

$$\text{o-Me-C}_6\text{H}_3(\text{R})\text{-CO}_2\text{H} \xrightarrow[\text{2) (MeO)}_2\text{CO}]{\text{1) LDA, THF}} \text{o-(CH}_2\text{CO}_2\text{H})\text{-C}_6\text{H}_3(\text{R})\text{-CO}_2\text{H}$$
3) H_2O

R = H, 85%

R = OMe, 90%

I.A.7.a.1b-1 D. Seebach, M. S. Hoekstra, and G. Protschuk, Angew. Chem. Int. Ed. Engl., 16, 321 (1977).

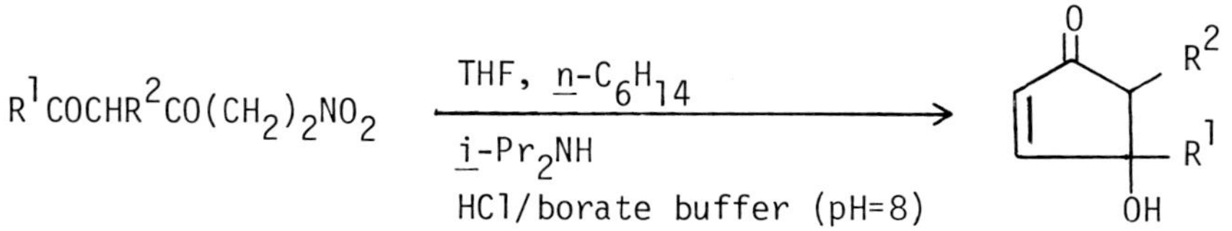

$$R^1COCHR^2CO(CH_2)_2NO_2 \xrightarrow[\text{HCl/borate buffer (pH=8)}]{\text{THF, } \underline{n}-C_6H_{14}, \underline{i}-Pr_2NH}$$

Upon prolonged treatment the initial product isomerizes to

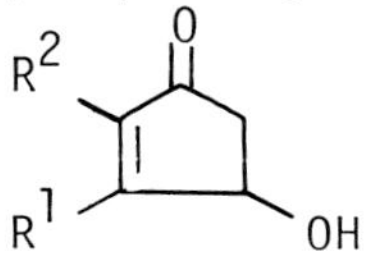

I.A.7.a.1b-2 J. Sraga and H. Hrnciar, Coll. Czech. Chem. Commun., 42, 998 (1977).

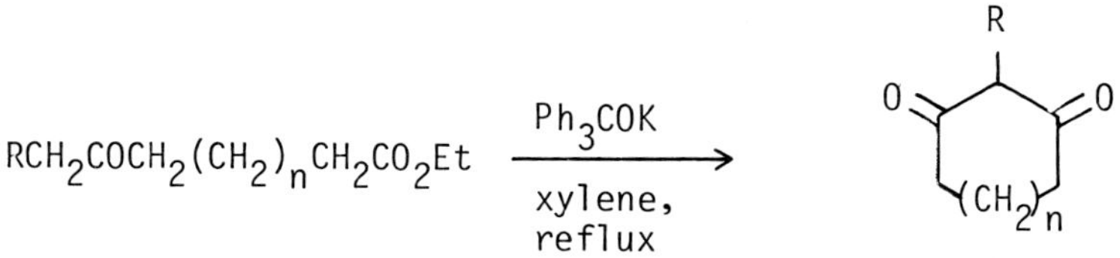

$$RCH_2COCH_2(CH_2)_nCH_2CO_2Et \xrightarrow[\text{xylene, reflux}]{Ph_3COK} A \text{ or } B$$

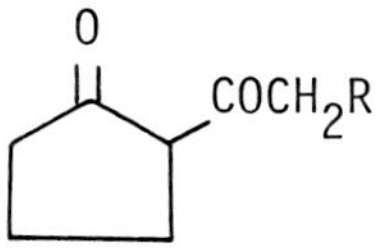

R	n	% Yield A	% Yield B
H	0	60	-
H	1	100	-
H	2	-	70
Me	2	-	80

CARBON–CARBON BOND FORMING REACTIONS

I.A.7.a.1b-3 R. B. Mitra, G. H. Kulkarni, and P. N. Khanna, Synthesis, **1977**, 415.

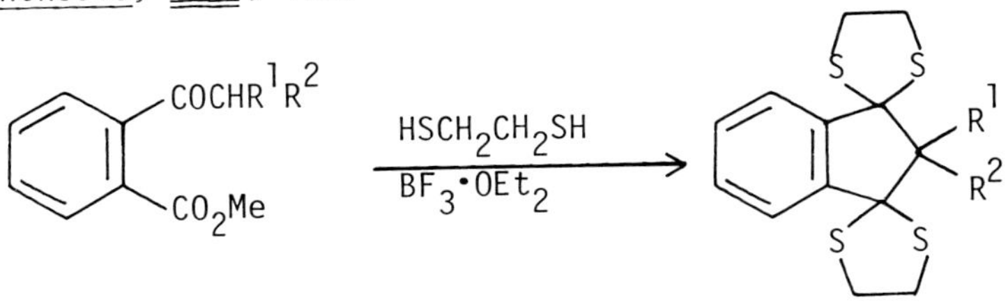

R^1	R^2	% Yield
H	H	60
H	Et	85
H	Ph	50
Me	Me	40

The products can be desulfurized with Raney Ni to give 2-substituted indans.

I.A.7.a.1b-4 K.-F. Cheng and P.-C. Li, Synth. Commun., **7**, 423 (1977).

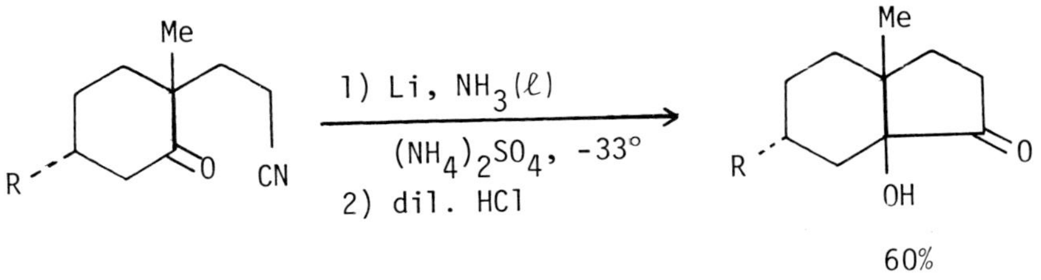

R = H or CMe=CH$_2$

I.A.7.a.2-1 R. Epsztein and F. Mercier, Synthesis, 1977, 183.

$$R^1C\equiv CCH_2NMe_2 \xrightarrow[\text{3) }R^2COR^3]{\text{1) n-BuLi} \atop \text{2) }ZnI_2} R^1C\equiv CCH(NMe_2)C(OH)R^2R^3$$

R^1	R^2	R^3	% Yield
Me_3Si	H	H	50
Me_3Si	Ph	H	79
$\underline{n}$-C_5H_{11}	Me	Me	50
Me_3Si	$-(CH_2)_5-$		73

I.A.7.a.2-2 D. Seebach, D. Enders, and B. Renger, Chem. Ber., 110, 1852 (1977); see also: B. Renger, H.-O. Kalinowski, and D. Seebach, ibid, 1866 (1977); B. Renger and D. Seebach, ibid, 2334 (1977).

$$R^1N(NO)CH_2R^2 \xrightarrow[\text{2) }R^3COR^4]{\text{1) LDA, THF, -78°C}} R^1N(NO)CHR^2C(OH)R^3R^4$$

R^1	R^2	R^3	R^4	% Yield
Et	Me	Ph	H	90
$-(CH_2)_2-$		Ph	Ph	65
$-(CH_2)_4-$		Me	H	85
$-(CH_2)_4-$		$-(CH_2)_5-$		63

CARBON—CARBON BOND FORMING REACTIONS

I.A.7.a.2-3 S. E. Davis, L. M. Schaffer, N. L. Shealy, K. D. Shealy, and C. F. Beam, Synth. Commun., $\underline{7}$, 261 (1977).

$$p\text{-}XC_6H_4C(Me)=NCH_2Ph \xrightarrow[\text{2. p-Y-C}_6\text{H}_4\text{COPh}]{\text{1. n-BuLi}} p\text{-}X\text{-}C_6H_4C(Me)=NCHPhC(OH)Ph$$
$$\underset{p\text{-}YH_4C_6}{|}$$

37-42%

The product is obtained if 1, 2, or 3 equivalents of $\underline{n}$-BuLi are employed.

I.A.7.a.2-4 T. Kauffmann, D. Berger, B. Scheerer, and A. Woltermann, Chem. Ber., $\underline{110}$, 3034; see also: T. Kauffmann, H. Berg, E. Köppelmann, and D. Kuhlmann, ibid, 2659 (1977).

$$PhN=NMe \xrightarrow[\text{2) R}^1\text{COR}^2]{\text{1) n-BuLi, THF, -70°C}} PhNHN=CHC(OH)R^1R^2$$

R^1	R^2	% Yield
H	$\underline{n}$-Pr	47
H	Ph	53
$-(CH_2)_4-$		77

I.A.7.a.2-5 U. Schollkopf, D. Stafforst, and R. Jentsch, Justus Liebigs Ann. Chem., $\underline{1977}$, 1167.

$$PhCH=CHNC \xrightarrow[\text{2) RCOCl}]{\substack{\text{1) n-BuLi, -110°C} \\ \text{Trapp mixture}}} PhCH=C\underset{\diagdown COR}{\overset{\diagup NC}{}}$$

R = OEt, 70%; R = Ph, 94%

The alkenyl isocyanides can act as acyl anion synthons.

I.A.7.a.2-6 A. S. Fletcher, K. Smith, and K. Swaminathan, J. Chem. Soc., Perkin Trans. I, 1977, 1881.

$$\underline{i}\text{-Pr}_2\text{NCHO} \xrightarrow[\text{2) R}^1\text{COR}^2]{\text{1) t-BuLi}} \underline{i}\text{-Pr}_2\text{NCOC(OH)R}^1\text{R}^2$$

62-85%

The carbomoyl-lithium species could be formed directly with t-BuLi under appropriate solvent and temperature conditions (Et$_2$O/THF (80:20), -78°C or Trapp solution, -95°C).

I.A.7.a.2-7 R. Schlecker and D. Seebach, Helv. Chim. Acta, 60, 1459 (1977).

 Ph-substituted bicyclic succinimide N-Me $\xrightarrow[\text{2) R}^1\text{COR}^2]{\text{1) sec-BuLi, -100°C, THF, HMPT}}$ N-CH$_2$C(OH)R^{1}R^2 product

R^1	R^2	% Yield
Ph	H	68
Ph	Ph	54
-(CH$_2$)$_5$-		91

Sterically less hindered succinimides gave mainly dimers as products.

CARBON—CARBON BOND FORMING REACTIONS

I.A.7.a.2-8 P. Beak, B. G. McKinnie, and D. B. Reitz, Tetrahedron Lett., 1839 (1977).

$$\text{ArCOYMe} \xrightarrow[\text{2) PhCHO}]{\text{1) RLi, THF, -78°C}} \text{ArCOYCH}_2\text{CH(OH)Ph}$$

Y = NMe, 75%

Y = S, 80%

I.A.7.a.2-9 K.-H. Geiss, D. Seebach, and B. Seuring, Chem. Ber., 110, 1833 (1977).

$$\text{PhCH}_2\text{SH} \xrightarrow[\substack{\text{2) R}^1\text{COR}^2 \\ \text{3) MeI}}]{\text{1) 2 eq. n-BuLi} \\ \text{THF, TMEDA}} \text{PhCH(SMe)C(OH)R}^1\text{R}^2$$

$R^1, R^2 = \text{-}(CH_2)_5\text{-}$, 85%

$R^1 = Ph, R^2 = H$, 48%

$$R^1R^2C(OH)CH_2CH=CHSMe$$
A

$$\text{CH}_2\text{=CHCH}_2\text{SH} \xrightarrow[\substack{\text{2) R}^1\text{COR}^2 \\ \text{3) MeI}}]{\text{1) 2 eq. n-BuLi} \\ \text{THF, TMEDA}}$$

+

$$\text{CH}_2\text{=CHCH(SMe)C(OH)R}^1\text{R}^2$$
B

$R^1, R^2 = \text{-}(CH_2)_4\text{-}$, 73%, A/B = 72/28

$R^1 = R^2 = Ph$, 88%, A/B = 77/23

I.A.7.a.2-10 B. M. Trost, D. E. Keeley, H. C. Arndt, J. H. Rigby, and M. J. Bogdanowicz, J. Am. Chem. Soc., 99, 3080 (1977).

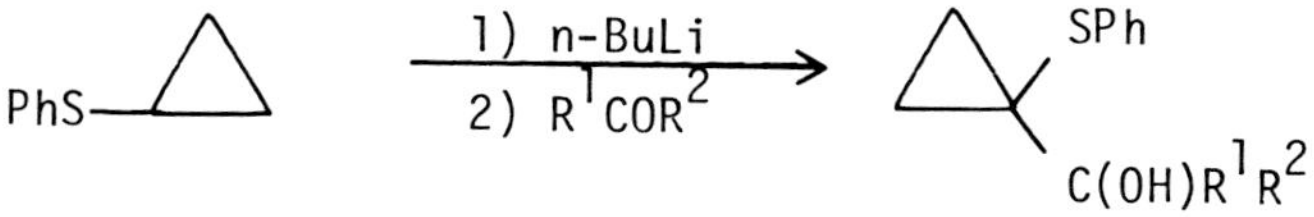

PhS—△ $\xrightarrow[\text{2) R}^1\text{COR}^2]{\text{1) n-BuLi}}$ △⟨SPh / C(OH)R^{1}R^2

A full paper describing the stereochemistry, regiochemistry, and chemoselectivity of the addition is presented.

I.A.7.a.2-11 R. D. Balanson, V. M. Kobal, and R. R. Schumaker, J. Org. Chem., 42, 393 (1977).

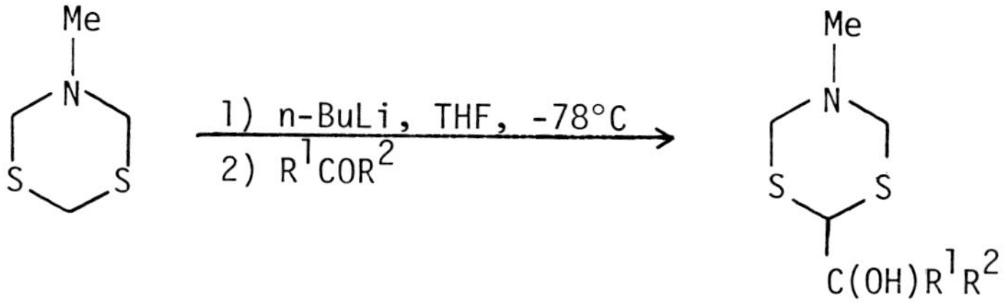

$\xrightarrow[\text{2) R}^1\text{COR}^2]{\text{1) n-BuLi, THF, -78°C}}$

R^1	R^2	% Yield
Ph	H	100
n-Pr	H	85
Ph	Me	90

The products hydrolyze to the corresponding aldehydes more easily than the 1,3-dithane analogues.

CARBON – CARBON BOND FORMING REACTIONS

I.A.7.a.2-12   U. Klein and W. Sucrow, Chem. Ber., 110, 1611 (1977).

$$MeSO_2NMe_2 \quad \xrightarrow[\text{2) } R^1COR^2]{\text{1) LDA, THF}} \quad R^1R^2C(OH)CH_2SO_2NMe_2$$

$$R^1, R^2 = -(CH_2)_5, \; 79\%$$

I.A.7.a.2-13   W. Dumont, M. Sevrin, and A. Krief, Angew. Chem. Int. Ed. Engl., 16, 541 (1977).

$$R^1R^2C(Br)SePh \quad \xrightarrow[\text{2) } R^3R^4CO]{\text{1) n-BuLi, THF, -78°C}} \quad R^1R^2C(SePh)C(OH)R^3R^4$$

R^1	R^2	R^3	R^4	% Yield
$\underline{n}\text{-}C_6H_{15}$	H	H	$\underline{n}\text{-}C_6H_{13}$	63
Me	H	Me	Me	62
H	H	Me	Me	75
Me	Me	H	$\underline{n}\text{-}C_6H_{13}$	62

I.A.7.a.2-14   B.-T. Gröbel and D. Seebach, Chem. Ber., 110, 867 (1977).

$$H_2C=C(Br)SiMe_3 \quad \xrightarrow[\text{2) PhCHO}]{\text{1) t-BuLi, THF, -78°C}} \quad H_2C=C(SiMe_3)CHOHPh$$

$$64\%$$

$$PhCH=C\begin{smallmatrix}\diagup MPh \\ \diagdown SnMe_3\end{smallmatrix} \quad \xrightarrow[\text{2) PhCOBr}]{\text{1) n-BuLi, THF, -78°C}} \quad PhCH=C(MPh)COPh$$

$$M = S, \; 52\%$$
$$M = Se, \; 55\%$$

I.A.7.a.3-1 K. Abe, T. Sato, N. Nakamura, and T. Sakan, Chem. Lett., 1977, 645.

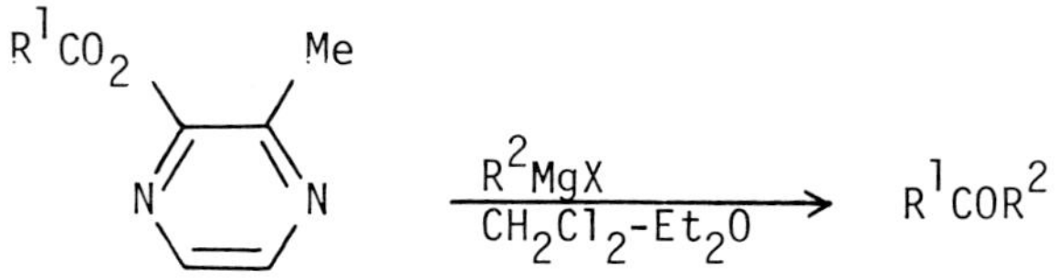

$$R^1CO_2\text{-pyrazine-Me} \xrightarrow[CH_2Cl_2-Et_2O]{R^2MgX} R^1COR^2$$

R^1	R^2	% Yield
Ph	Me	80
Ph	n-Bu	90
Ph	Ph	92
Me	Ph	79

I.A.7.a.3-2 S. Masson, M. Saquet, and A. Thuillier, Tetrahedron, 33, 2949 (1977).

$$MeC(S)SMe \xrightarrow[2) MeI]{1) RMgBr, THF} RCMe(SMe)_2$$

$$R = CH_2=CHCH_2-, PhCH_2 \text{ or } CH_2=CH-$$

$$R^1C(S)SMe \xrightarrow[2) MeI]{1) R^2R^3C=CHCH_2MgX} CH_2=CHCR^2R^3CR^1(SMe)_2$$

I.A.7.a.3-3 R. Gauthier, C. P. Axiotis, and M. Chastrette, <u>J. Organomet. Chem.</u>, <u>140</u>, 245 (1977); see also: <u>ibid</u>, <u>Tetrahedron Lett.</u>, <u>1977</u>, 23.

$$R^1OCHR^2CN \xrightarrow[2) R^4M]{1) R^3MgX} R^1OCHR^2CR^3R^4NH_2 + R^1OCHR^2COR^3$$
$$\phantom{R^1OCHR^2CN \xrightarrow[2) R^4M]{1) R^3MgX}} \quad A B$$

R^1	R^2	R^3	R^4M	A:B
Me	H	Et	$\underline{n}$-BuLi	4:1
$-(CH_2)_4-$		Me	$CH_2=CHCH_2MgBr$	7:3
$-(CH_2)_3-$		$\underline{n}$-Bu	$\underline{n}$-BuLi	7:3

I.A.7.a.3-4 A. Sekiya and N. Ishikawa, <u>Chem. Lett.</u>, <u>1977</u>, 81.

$$\text{(succinic anhydride)} \xrightarrow[\text{pyridine, THF}]{(CF_3)_2CFZnI} (CF_3)_3CFCOCH_2CH_2CO_2H$$
$$ 74\%$$

$$PhCOCl \xrightarrow[\substack{\text{pyridine, THF} \\ ZnF_2}]{(CF_3)_2CFZnI} PhCOCF(CF_3)_2$$
$$ 89\%$$

I.A.7.a.3-5 G. Cahiez, D. Bernard, and J. F. Normant, Synthesis, 1977, 130; see also: G. Cahiez and J. F. Normant, Bull. Soc. Chim. Fr., 1977, 570.

$$R^2COCl \xrightarrow[Et_2O]{R^1MnI} R^1COR^2$$

$$60-92\%$$

The scope and limitations of this reaction are discussed.

I.A.7.a.3-6 G. Cahiez and J. F. Normant, Tetrahedron Lett., 3383 (1977).

$$R^1COR^2 \xrightarrow[2) H_3O^+]{1) n-BuMnI} R^1R^2C(OH)Bu$$

R^1	R^2	% Yield
n-Bu	H	89
Et	Et	86
i-Pr	i-Pr	88

I.A.7.a.3-7 Y. Okude, S. Hirano, T. Hiyama, and H. Nozaki, J. Am. Chem. Soc., 99, 3179 (1977).

$$R^1R^2C=CHCH_2Br \xrightarrow[2) R^3COR^4]{1) "Cr(II)"} H_2C=CHR^1R^2C(OH)R^3R^4$$

R^1	R^2	R^3	R^4	% Yield
H	H	H	MeO$_2$C(CH$_2$)$_4$	75
H	H	H	Me(CH$_2$)$_3$CO(CH$_2$)$_4$	66
Me	H	H	Ph	96
Me	Me	H	n-C$_6$H$_{13}$	83
H	H		-(CH$_2$)$_5$-	78

"Cr(II)" is either commercial chromous chloride or a reagent prepared from chromic chloride by reduction with LiAlH$_4$.

I.A.7.a.3-8   N. Collignon, Bull. Soc. Chim. Fr., 1977, 120.

$$HC \equiv CH \xrightarrow[\substack{\text{1) Cs, THF} \\ \text{2) } R^1COR^2 \\ \text{3) } H_3O^+}]{} R^1R^2C(OH)C \equiv CH$$

$R^1, R^2 = -(CH_2)_5-$, 60%

$$RC \equiv CH \xrightarrow[\substack{\text{1) Cs, THF-heptane} \\ \text{2) } CO_2 \\ \text{3) } H_3O^+}]{} RC \equiv CCO_2H$$

R = Ph, 69%
R = n-Bu, 91%

I.A.7.a.3-9   H. Taguchi, H. Yamamoto, and H. Nozaki, Bull. Chem. Soc. Japan, 50, 1588 (1977).

$$R^1COR^2 + CHXY_2 \xrightarrow{LiNR^3_2} R^1R^2C(OH)CXY_2$$

R^1	R^2	X	Y	% Yield
C_8H_{17}	H	H	Cl	73
C_8H_{17}	H	H	I	79
$-(CH_2)_5-$		H	Cl	89-100
$-(CH_2)_{11}-$		H	Br	81
$-(CH_2)_5-$		Br	Br	91

I.A.7.a.3-10   J. Villieras, P. Periot, and J. F. Normant, Bull. Soc. Chim. Fr., 1977, 765.

$$R^1CCl_2H \xrightarrow[\text{2) } R^2COR^3]{\text{1) n-BuLi, TMEDA} \atop Et_2O, THF, -90°C} R^1CCl_2R^4$$

R^1	R^2	R^3	R^4	% Yield
Me	i-Pr	i-PrCO$_2$	i-PrCO	86
PhCH(OSiMe$_3$)	Me	MeCO$_2$	MeCO	75
Me	n-Bu	H	n-BuCH(OH)	62
Me	Ph	H	PhCH(OH)	83

I.A.7.a.3-11   A. Merz and R. Tomahogh, Chem. Ber., 110, 96 (1977).

$$R^1COR^2 \xrightarrow[PhCH_2\overset{+}{N}Et_3Cl^-]{50\% \text{ NaOH, CHCl}_3} R^1R^2C(OH)CCl_3$$

R^1	R^2	% Yield
Ph	H	80
i-Pr	H	34
Me	Me	69
-(CH$_2$)$_4$-		33

I.A.7.a.3-12 J. Ficini, S. Falou, A.-M. Touzin, and J. d'Angelo, Tetrahedron Lett., 3589 (1977).

$$EtOCH=CHSnBu_3 \xrightarrow[\substack{2)\ R^1COR^2 \\ 3)\ H_3O^+}]{1)\ BuLi,\ THF,\ -70°C} R^1R^2C=CHCHO$$

R^1	R^2	% Yield
Ph	H	80
C_6H_{13}	H	80
$\text{-}(CH_2)_5\text{-}$		50

I.A.7.a.3-13 R. H. Wollenberg, K. F. Albizati, and R. Peries, J. Am. Chem. Soc., 99, 7365 (1977).

$$Bu_3SnCH=CHOEt \xrightarrow[\substack{2)\ R^1COR^2 \\ 3)\ Chromatograph \\ (Florisil\ or\ silca\ gel)}]{1)\ n\text{-}BuLi,\ THF,\ -78°C} R^1R^2C=CHCHO$$

R^1	R^2	% Yield
Ph	H	76
Me	1-cyclohexenyl	84
$\text{-}(CH_2)_5\text{-}$		97
$\text{-}(CH_2)_3CH=CH\text{-}$		66

I.A.7.a.3-14 R. K. Boeckman, Jr. and K. J. Bruza, Tetrahedron Lett., 4187 (1977).

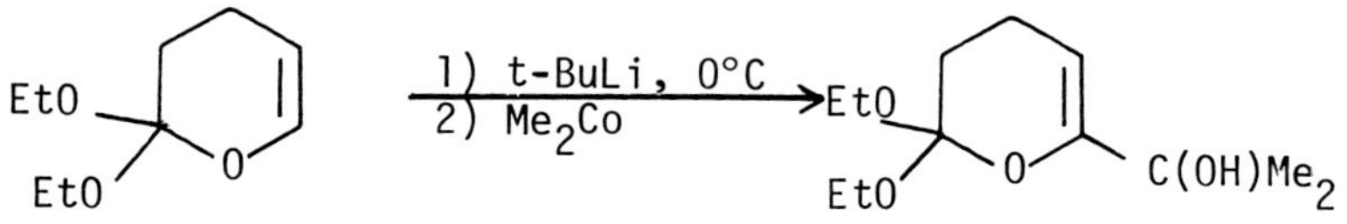

EtO-(ring with O, =)-EtO $\xrightarrow{\text{1) t-BuLi, 0°C} \atop \text{2) Me}_2\text{Co}}$ EtO-(ring with O)-EtO — C(OH)Me$_2$

53%

I.A.7.a.3-15 J.-M. Dollat, J.-L. Luche, and P. Crabbe', J. Chem. Soc., Chem. Commun., 1977, 761.

$$R^1R^2C(OAc)C\equiv CH \xrightarrow[\text{2) }(R^3CO)_2O]{\text{1) Me}_2\text{CuLi}} R^1R^2\overset{\overset{COR^3}{|}}{C}C\equiv CH$$

R^1	R^2	R^3	% Yield
C_5H_{11}	H	Me	50
C_5H_{11}	H	Ph	60
$-(CH_2)_5-$		Me	45

CARBON—CARBON BOND FORMING REACTIONS

I.A.7.a.3-16   M. Pohmakotr and D. Seebach, Angew. Chem. Int. Ed. Engl., 16, 320 (1977).

$$PhCOCH_2CH_2CH=CH_2 \xrightarrow[\substack{2)\ \text{sec-BuLi, TMEDA} \\ -78° \text{ to } 0°C \\ 3)\ R^1COR^2}]{1)\ KH,\ THF,\ 0°C} PhCOCH_2CH=CHCH_2C(OH)R^1R^2$$

R^1	R^2	% Yield	Z:E
Et	H	20-35	100:0
Ph	H	38-53	0:100
Ph	Ph	40-65	0:100
$-(CH_2)_5-$		25	100:0

I.A.7.a.3-17   D. Ayalon-Chass, E. Ehlinger, and P. Magnus, J. Chem. Soc., Chem. Commun., 1977, 772.

$$Me_3SiCH_2CH=CH_2 \xrightarrow[2)\ R^1COR^2]{1)\ n\text{-BuLi, THF}} R^1R^2\overset{OH}{\underset{|}{C}}CH_2CH=CHSiMe_3$$

$$R^1, R^2 = -(CH_2)_5-,\ 73\%$$

$$R^1, R^2 = -(CH_2)_4-,\ 77\%$$

The allylic anion from allyltrimethylsilane can act as a α-acylcarbanion synthon by further elaboration of the initial adduct.

I.A.7.a.3-18 F. Cooke and P. Magnus, <u>J. Chem. Soc., Chem. Commun.</u>, <u>1977</u>, 513.

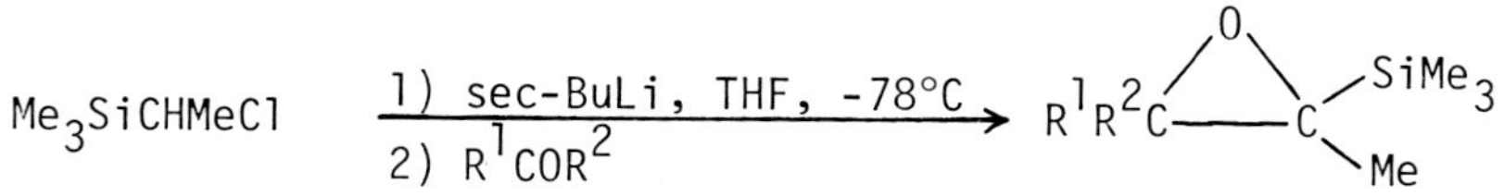

$$Me_3SiCHMeCl \xrightarrow[2)\ R^1COR^2]{1)\ sec\text{-}BuLi,\ THF,\ -78°C} R^1R^2C\overset{O}{\underset{}{\triangle}}C(SiMe_3)Me$$

R^1	R^2	% Yield
Pr	H	53
Ph	Ph	78
4-MeC$_6$H$_4$	H	82
$-(CH_2)_5-$		82

I.A.7.a.3-19 K. Tanaka, T. Nakai, and H. Ishikawa, <u>Chem. Lett.</u>, <u>1977</u>, 1379.

$$CF_3CH_2OPh \xrightarrow[2)\ R^1COR^2]{1)\ LDA\ (2\ eq.)\ THF,\ -60°C} R^1R^2C=C(OPh)CF_2OH$$
$$A$$

$$\xrightarrow[100°]{5\%\ aq.\ TFA} R^1R^2C=C(OPh)CO_2H$$
$$B$$

		% Yield	
R^1	R^2	A	B
Me	Me	60	76
Me	Pr	51	53
H	C$_6$H$_{13}$	62	78
$-(CH_2)_5-$		70	87

CARBON—CARBON BOND FORMING REACTIONS

I.A.7.a.3-20 P. Beak and B. G. McKinnie, J. Am. Chem. Soc., **99**, 5213 (1977).

$$ArCO_2Me \xrightarrow[\text{2) R}^1\text{COR}^2]{\text{1) sec-BuLi, TMEDA, THF}} ArCO_2CH_2C(OH)R^1R^2$$

$R^1 = R^2 = Ph, 69\%$

$R^1 = Ph, R^2 = Et, 63\%$

I.A.7.a.3-21 T. Kauffman, H. Fischer, and A. Woltermann, Angew. Chem. Int. Ed. Engl., **16**, 53 (1977); see also: T. Kauffman, R. Joussen, and A. Woltermann, ibid, 709 (1977).

$$Ph_2\overset{O}{\underset{\|}{As}}Me \xrightarrow[\text{2) R}^1\text{COR}^2]{\text{1) LDA, THF, -40°}} Ph_2\overset{O}{\underset{\|}{As}}CH_2C(OH)R^1R^2$$

$R^1 = H, R^2 = \underline{n}\text{-Bu}, 60\%$

$R^1 = R^2 = Ph, 81\%$

I.A.7.a.3-22 F. A. Davis and P. A. Mancinelli, J. Org. Chem., **42**, 398 (1977).

$$PhSN=CR^1R^2 \xrightarrow[\substack{\text{2) aq. HCl} \\ \text{3) aq. NaOH}}]{\text{1) R}^3\text{Li, Et}_2\text{O}} R^1R^2R^3CNH_2$$

R^1	R^2	R^3	% Yield
Ph	H	Me	79
Ph	H	Ph	59
Me	Me	Ph	61
Me	Me	$\underline{t}$-Bu	43
$-(CH_2)_5-$		Me	68

I.A.7.a.3-23   E. Vedejs and W. T. Stolle, Tetrahedron Lett., 135 (1977).

$$R^1CH=NNHTs \xrightarrow{R^2Li \atop THF \text{ or } TMEDA} R^1CH_2R^2$$

R^1	R^2	% Yield
$PhCH_2CH_2$	$\underline{n}$-Bu	49
$PhCH_2CH_2$	$\underline{sec}$-Bu	40
$PhCH_2CH_2$	$\underline{t}$-Bu	61
Ph	$\underline{t}$-Bu	38
PhCH=CH	$\underline{t}$-Bu	58

I.A.7.a.3-24   S. T. Srisethnil and S. S. Hall, J. Org. Chem., 42, 4266 (1977).

$$R^1CO_2R^2 \xrightarrow[2)\ Li,\ NH_3(\ell),\ NH_4Cl]{1)\ PhLi,\ Et_2O} R^1CHPh_2$$

R^1 = Me, R^2 = Et, 70%

R^1 = Ph, R^2 = Me, 84%

$$\underset{(CH_2)_n}{\overset{O\diagdown\diagup O}{\bigcirc}} \xrightarrow[2)\ Li,\ NH_3(\ell),\ NH_4Cl]{1)\ PhLi,\ Et_2O} Ph_2CH(CH_2)_nCH_2CH_2OH$$

n = 1, 69%

n = 2, 53%

CARBON—CARBON BOND FORMING REACTIONS

I.A.7.a.3-25 T. Shono, I. Nishiguchi, and H. Ohmizu, Chem. Lett., 1977, 1021.

$$PhCHR^1Cl + R^2COCl \xrightarrow[MeCN,\ Et_4NOTs]{electrolysis} PhCHR^1COR^2$$

R^1	R^2	% Yield
H	Et	57
H	t-Bu	34
Me	i-Pr	47
Me	$(CH_2)_2CO_2Me$	44

I.A.7.a.3-26 H. Lund, Acta Chem. Scand., B31, 424 (1977).

anthracene $\xrightarrow[\text{TBAI, electrolysis}]{Ac_2O,\ DMF}$ 9-(1-acetoxyethyl)-9,10-dihydroanthracene

66-75%

I.A.7.a.3-27 J. M. McIntosh, Can. J. Chem., 55, 4200 (1977).

$$R^1CHO + R^2CH=CHCH_2Br \xrightarrow[CH_2Cl_2,\ TEBAC]{KCN\ (aq.)} R^1CH(CN)OCH_2CH=CHR^2$$

R^1	R^2	% Yield
i-Pr	H	44
Ph	H	70
Ph	Me	72

Cyanohydrin acetates can be made by this phase transfer procedure utilizing acetic anhydride instead of $R^2CH=CHCH_2Br$. Ketones do not give similar products.

I.A.7.b.1-1   G. A. Kraus and H. Sugimoto, <u>Tetrahedron Lett.</u>, 3929 (1977).

$$CH_2=CHCH=CHOSiMe_3 \xrightarrow[\substack{\text{2) } R^1R^3C=CR^2COR^4 \\ \text{3) } Me_3SiCl}]{\text{1) n-BuLi, THF}}$$

[cyclohexene with Me_3SiO, COR^4, R^2, R^3, R^1 substituents]

R^1	R^2	R^3	R^4	% Yield
H	H	H	OMe	60
Me	H	H	Me	44
Me	CO_2Me	H	OMe	70

I.A.7.b.1-2   I. Belsky, <u>J. Chem. Soc., Chem. Commun.</u>, <u>1977</u>, 237.

$$MeNO_2 + PhCH=CHCOPh \xrightarrow[\text{18-crown-6, 81°C}]{\text{KF, MeCN}} O_2NCH_2CHPhCH_2COPh$$

(excess)   94%

$$CH_2(CN)_2 + CH_2=CHCN \xrightarrow{\text{same as above}} (NC)_2C(CH_2CH_2CN)_2$$

                      (excess)   69%

I.A.7.b.1-3 F. G. Cowherd, M.-C. Doria, E. Galeazzi, and J. M. Muchowski, Can. J. Chem., **55**, 2919 (1977).

$$\text{Phthalimide-NCH}_2\text{NO}_2 \xrightarrow[\text{2) R}^1\text{CH=CHCOR}^2]{\text{1) NaH}} \text{Phthalimide-N-CH(NO}_2)\text{CHR}^1\text{CH}_2\text{COR}^2$$

R^1	R^2	% Yield
H	Me	43-63
Ph	Me	8
H	OMe	43-64
$-(CH_2)_3-$		50-67

N-Nitromethylphthalimide acts as a formyl anion equivalent.

I.A.7.b.1-4 P. A. Grieco and Y. Yokoyama, J. Am. Chem. Soc., **99**, 5210 (1977).

$$C_6H_{13}CH(CN)SePh \xrightarrow{\text{1) LDA, THF, -78°C}} \text{cyclohexenone} \rightarrow \text{product, 91\%}$$

(product: 3-substituted cyclohexanone with $-C(CN)(C_6H_{13})(SePh)$ group)

Treatment of the aryl selenocyanates with H_2O_2 gives unsaturated nitriles.

I.A.7.b.2-1   S.-H. Liu, J. Org. Chem., 42, 3209 (1977).

$$CH_2=CHCO_2Et \xrightarrow[Et_2O,\ -20°\ to\ -50°C]{RMgX} RCH_2CHCO_2Et$$

R	% Yield
Ph	54
$\underline{n}$-C$_5$H$_{11}$	80
PhCH$_2$	69
cyclohexyl	68

I.A.7.b.2-2   U. Schöllkopf and R. Meyer, Justus Liebigs Ann. Chem., 1977, 1174.

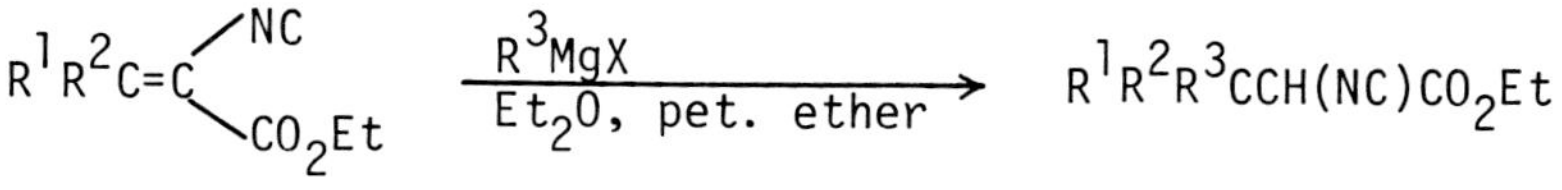

$$R^1R^2C=C\begin{smallmatrix}NC\\CO_2Et\end{smallmatrix} \xrightarrow[Et_2O,\ pet.\ ether]{R^3MgX} R^1R^2R^3CCH(NC)CO_2Et$$

R^1	R^2	R^3	% Yield
Me	Me	Et	63
Ph	Me	Ph	87
Ph	H	Et	78
$-(CH_2)_5-$		Ph	74

I.A.7.b.2-3 A. Marfat, P. R. McGuirk, R. Kramer, and P. Helquist, J. Am. Chem. Soc., 99, 253 (1977).

R^3MgBr $\xrightarrow{\begin{array}{l}1.\ Me_2S \cdot CuBr\\ 2.\ R^4C\equiv CH\\ 3.\ R^1CH=CHCOR^2\\ 4.\ aq.\ NH_4Cl\end{array}}$ $\underset{R^4}{\overset{R^3}{>}}C=C\underset{H}{\overset{CHR^1CH_2COR^2}{<}}$

R^1	R^2	R^3	R^4	% Yield
Ph	Ph	Et	$\underline{n}$-C$_6$H$_{13}$	52
Ph	Me	Et	$\underline{n}$-C$_6$H$_{13}$	35
-(CH$_2$)$_3$-		$\underline{n}$-C$_6$H$_{13}$	Et	68

I.A.7.b.2-4 M. Isobe, S. Kondo, N. Nagasawa, and T. Goto, Chem. Lett., 1977, 679.

1) R$_3$ZnLi·2 LiCl, THF
2) NH$_4$Cl (aq.)

R	% Yield
Me	92
$\underline{n}$-Bu	92
$\underline{sec}$-Bu	66
$\underline{t}$-Bu	58
Ph	15

I.A.7.b.2-5 E. C. Ashby, J. J. Lin, and J. J. Watkins, J. Org. Chem., 42, 1099 (1977).

$$R^1R^2C=CR^3COR^4 \xrightarrow[\text{THF or Et}_2O]{Li_xCu_y(Me)_z} R^1R^2CMeCHR^3COR^4$$

Comparison of reactions of $LiCu_2(Me)_3$, $Li_2Cu_3(Me)_5$, and $Li_2Cu(Me)_3$ with $LiCu(Me)_2$.

I.A.7.b.2-6 P. A. Wender and S. L. Eck, Tetrahedron Lett., 1245 (1977).

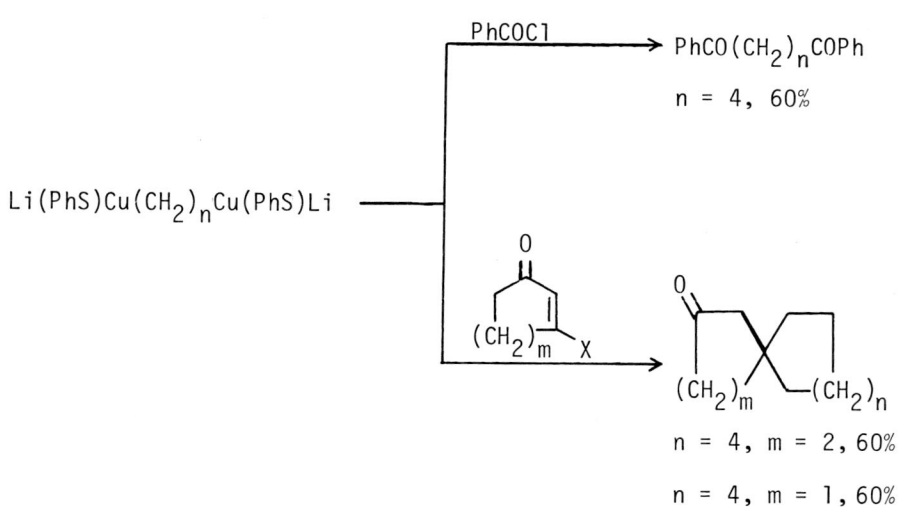

I.A.7.b.2-7   M. P. Cooke, Jr. and R. Goswami, J. Am. Chem. Soc., **99**, 642 (1977).

$$R^1R^2C=CR^3COC(Ph_3P)CO_2Et \quad \xrightarrow[\text{2) E, -78°C}]{\text{1) } R^4Li, \text{ THF}\atop -78°C \text{ to } 25°C}$$

$$R^1R^2R^4CCR^5R^3COC(Ph_3P)CO_2Et$$

R^1	R^2	R^3	R^4Li	R^5	E	% Yield
H	H	H	MeLi	n-Bu	n-BuI	83
H	H	H	t-BuO$_2$CCH$_2$Li	Me	MeI	72
Me	H	H	PhLi	PhCHOH	PhCHO	92
Me	Me	H	MeLi	Me	MeI	82
H	-(CH$_2$)$_4$-		MeLi	H	H$_2$O	85

I.A.7.b.2-8   T. M. Dolak and T. A. Bryson, Tetrahedron Lett., 1961 (1977).

$$i\text{-PrCH}_2SPh \quad \xrightarrow[\text{2) } R^2CH=CR^1COR^3]{\text{1) Li, HMPA}\atop \text{THF, -78°C}} \quad i\text{-PrCH(SPh)CHR}^2\text{CHR}^1\text{COR}^3$$

$R^1 = H, \; R^2,R^3 = -(CH_2)_3-, \; 56\%$

$R^1 = Me, \; R^2,R^3 = -(CH_2)_2-, \; 57\%$

I.A.7.b.2-9 P. C. Ostrowski and V. V. Kane, Tetrahedron Lett., 3549 (1977).

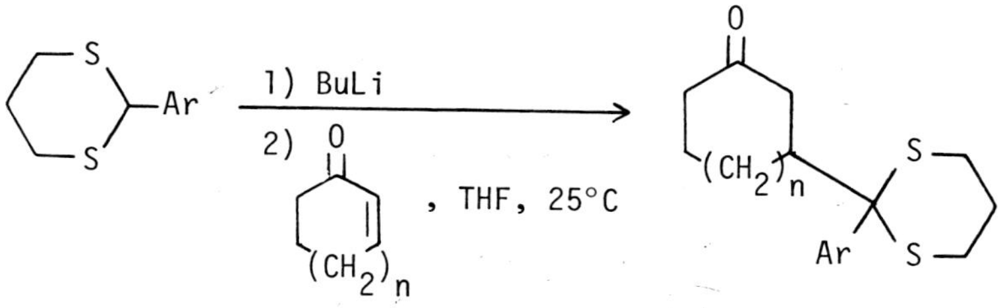

Ar	n	% Yield
Ph	2	86
Ph	1	93
Ph	0	81
1-naphthyl	1	96
2-thienyl	1	90

I.A.7.b.2-10 G. A. Kraus and H. Sugimoto, Synth. Commun., 7, 505 (1977).

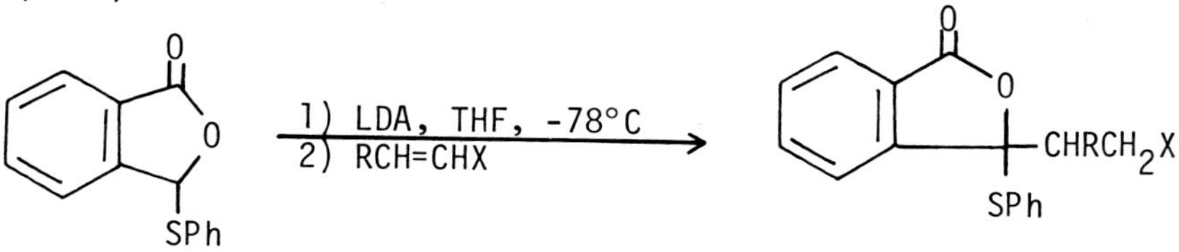

R	X	% Yield
Me	CO_2Et	90
H	CN	66
Me	COMe	51
$-(CH_2)_3CO-$		75

I.A.7.b.2-11 R. Bürstinghaus and D. Seebach, Chem. Ber., 110, 841 (1977).

$Me_3MCH(SMe)_2$ $\xrightarrow[\text{2)}]{\text{1) LDA, THF; HMPTA, -78°C}}$ cyclo-$(CH_2)_n$-C(=O)-CH_2-C(SMe)_2MMe_3

M	n	% Yield
Si	2	73
Si	3	78
Si	4	57
Sn	2	82
Sn	3	81

The intermediate enolates can alkylated.

I.A.7.b.2-12 B. Giese and J. Meister, Chem. Ber., 110, 2588 (1977).

$R^1HgOAc + R^2CH=CR^3R^4 \xrightarrow{NaBH_4} R^1CHR^2CHR^3R^4$

R^1	R^2	R^3	R^4	% Yield
cyclohexyl	H	H	CN	61
cyclohexyl	H	H	CO_2Me	62
cyclohexyl	H	Me	CO_2Me	84
cyclohexyl	CO_2Me	H	CO_2Me	34
$\underline{n}$-C_6H_{13}	H	H	COMe	51
$\underline{t}$-Bu	H	H	COMe	69

I.A.7.b.2-13 A. Hosomi and H. Sakurai, J. Am. Chem. Soc., 99, 1673 (1977).

$$Me_3SiCH_2CH=CR_2^1 + R^2R^3C=CHCOR^4 \xrightarrow[CH_2Cl_2]{TiCl_4} CH_2=CHCR_2^1CR^2R^3CH_2COR^4$$

R^1	R^2	R^3	R^4	% Yield
H	H	H	Me	59
Me	H	H	Me	79
H	Me	Me	Me	87
H	Ph	H	Ph	96
H	Me	$-CH_2CMe_2CH_2-$		76

I.A.7.b.3-1 T. Shono, I. Nishiguchi, and H. Ohmizu, J. Am. Chem. Soc., 99, 7396 (1977).

$$R^1R^2C=CR^3Y + (R^4CO)_2O \xrightarrow[Et_4NOTs, MeCN]{electrolysis} R^4COCR^1R^2CHR^3Y$$

R^1	R^2	R^3	Y	R^4	% Yield
H	H	H	CO_2Me	Me	62
Me	H	H	CO_2Me	i-Pr	74
Ph	H	H	CN	Me	76
H	$-(CH_2)_4-$		CO_2Et	Me	74

I.A.7.b.3-2 B. Fraser-Reid, N. L. Holder, D. R. Hicks, and D. L. Walker, Can. J. Chem., 55, 3978 (1977).

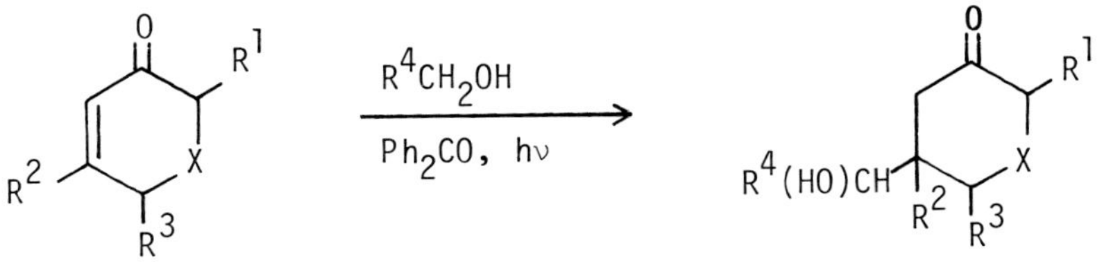

R^1	R^2	R^3	R^4	X	% Yield
CH_2OH	H	OEt	H	O	66
CH_2OAc	H	OEt	Me	O	50
CH_2OCPh_3	Me	OMe	H	O	61
H	H	H	H	CH_2	33
H	H	H	Me	CH_2	33

I.A.7.b.3-3 B. Fraser-Reid, R. C. Anderson, D. R. Hicks, and D. L. Walker, Can. J. Chem., 55, 3986 (1977).

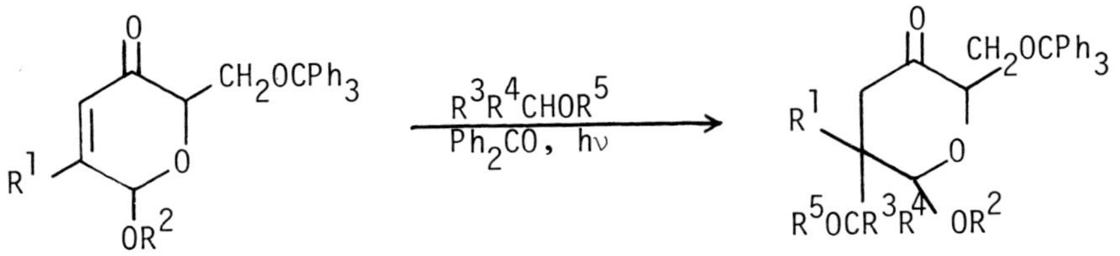

R^1	R^2	R^3	R^4	R^5	% Yield
H	Et	H	CH_2OH	H	79
H	Et	n-Pr	$-OCH_2CH_2-$		62
Me	Me	CH_2OH	$-OCH_2CH_2-$		46

I.A.7.b.3-4 S. F. Martin and S. R. Desai, J. Org. Chem., 42, 1664 (1977).

$$\text{MeC(OEt)=CHPPh}_3\text{I}^- \xrightarrow[\text{3) H}_3\text{O}^+]{\text{1) n-BuLi} \atop \text{2) R}^1\text{COCR}^2\text{=CR}^3\text{R}^4}$$

R^1	R^2	R^3	R^4	% Yield
Me	H	H	Me	48
Me	H	Me	Me	61
Me	CO$_2$Me	Me	Me	26
-(CH$_2$)$_4$-		H	Me	36

I.A.7.b.3-5 H.-J. Bestmann, M. Ettlinger, and R. W. Saalfrank, Justus Liebigs Ann. Chem., 1977, 276.

$$\text{Ph}_3\text{P=C=C(OEt)}_2 + \text{R}^1\text{CH}_2\text{R}^2 \longrightarrow \text{Ph}_3\text{P=CHC(OEt)=CR}^1\text{R}^2$$

R^1	R^2	% Yield
H	NO$_2$	49
H	CN	44
H	SO$_2$Me	61
-CH$_2$=CH-CH=CH$_2$-		39
-CO$_2$CH$_2$CH$_2$-		63

I.A.7.b.3-6   B. B. Snider, L. A. Brown, R. S. Eichen Conn, and
T. A. Killinger, Tetrahedron Lett., 2831 (1977).

$$R^1CH_2CR^2=CR^3R^4 + MeCOC\equiv CH \xrightarrow[CH_2Cl_2]{ZnX} R^1CH=CR^2CR^3R^4CH=CHCOMe \text{ (A)} + R^1CH_2CR^2=CR^3CH=CHCOMe \text{ (B)}$$

R^1	R^2	R^3	R^4	% Yield A	% Yield B
Me	Et	H	H	54	3
H	Me	Me	H	38	14
H	Me	Me	Me	45	-

Same amounts of 2 + 2 cycloaddition products are also formed.

I.A.7.b.3-7   H. Stetter, P. H. Schmitz, and M. Schreckenberg,
Chem. Ber., 110, 1971 (1977)

$$R^1CHO + Me_2N(CH_2)_2COR^2 \xrightarrow{catalyst} R^1COCH_2CH_2COR^2$$

R^1	R^2	catalyst	% Yield
Ph	Me	$^-$CN	45
Ph	t-Bu	$^-$CN	60
n-Pr	CH=CMe$_2$	I	53
Ph	1-cyclohexen-1-yl	II	25

catalyst I = 3-benzyl-5-(2-hydroxyethyl)-4-methylthiazolium chloride.

catalyst II = 3-ethyl-5-(2-hydroxyethyl)-4-methylthiazonium bromide.

I.A.7.b.3-8 R. K. Freidlina and F. K. Velichko, Synthesis, 1977, 145.

Review: "Synthetic Applications of Homolytic Addition and Telomerisation Reactions of Bromine-Containing Addends with Unsaturated Compounds Containing Electron-Withdrawing Substituents."

I.A.8-1 H.M.R. Hoffmann and T. Tsushima, J. Am. Chem. Soc., 99, 6008 (1977).

$$R^1CH_2CR^2=CHR^3 \xrightarrow[\text{(cyclohexyl)}_2\text{NH, CH}_2\text{Cl}_2]{\text{MeCOSbCl}_6^-} R^1CH=CR^2CHR^3COMe$$

R^1	R^2	R^3	% Yield
H	Me	H	>90
$-(CH_2)_5-$		H	73

I.A.8-2 A. S. Schegolev, W. A. Smit, S. A. Khurshudyan, V. A. Chertkov, and V. F. Kucherov, Synthesis, 1977, 324.

$$R^1C \equiv CH \xrightarrow[\text{ArH, CH}_2\text{Cl}_2]{R^2COBF_4} ArCR^1=CHCOR^2$$

R^1	R^2	Ar	% Yield
Me	Me	Ph	40
Me	t-Bu	MeC$_6$H$_4$	56
4-ClC$_6$H$_4$	t-Bu	Ph	79

I.A.8-3 A. Kh. Khusid, G. V. Kryshtal, V. F. Kucherov, and L. A. Yanovskaya, Tetrahedron Lett., 907 (1977).

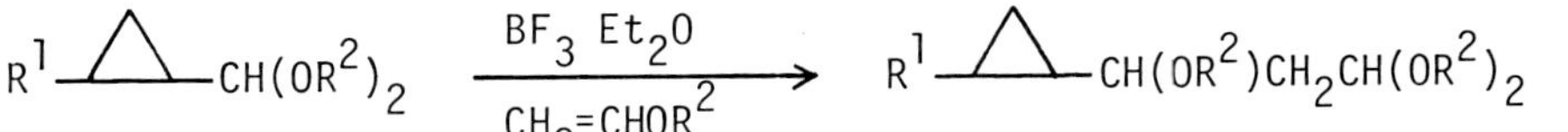

$$R^1-\triangle-CH(OR^2)_2 \xrightarrow[CH_2=CHOR^2]{BF_3\ Et_2O} R^1-\triangle-CH(OR^2)CH_2CH(OR^2)_2$$

$$R^1 = H,\ R^2 = Et,\ 45\%$$
$$R^1 = (MeO)_2CHCH_2,\ R^2 = Me,\ 70\%$$

I.A.8-4 E. Nakamura, K. Hashimoto, and I. Kuwajima, J. Org. Chem., **42**, 4166 (1977).

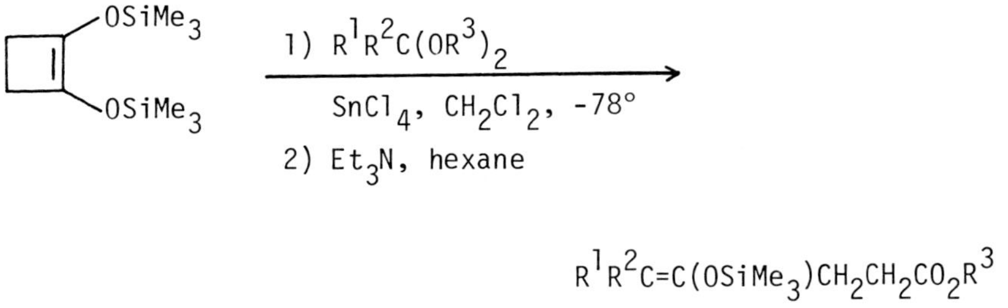

[cyclobutane with OSiMe$_3$ and OSiMe$_3$ substituents]

1) $R^1R^2C(OR^3)_2$
   $SnCl_4$, CH_2Cl_2, $-78°$
2) Et_3N, hexane

$$R^1R^2C=C(OSiMe_3)CH_2CH_2CO_2R^3$$

R^1	R^2	R^3	% Yield
Et	Et	Me	87
$-(CH_2)_5-$		Et	93
$-(CH_2)_{11}-$		Me	91

I.A.8-5 T. Sato, M. Arai, and I. Kuwajima, J. Am. Chem. Soc., **99**, 5827 (1977).

$$R^1CH=C(SiMe_3)OSiMe_3 \xrightarrow[BF_3 \cdot Et_2O]{R^2CH(OR^3)_2} R^2CH(OR^3)CHR^1COSiMe_3$$

R^1	R^2	R^3	% Yield
Me	n-Pr	Me	87
Me	Ph	Et	92
PhCH$_2$	PhCH=CH	Et	87

Products can be reacted with Bu_4NOH to give α,β-unsaturated aldehydes.

I.A.8-6 J. Barluenga, S. Fustero, V. Rubio, and V. Gotor, Synthesis, **1977**, 780.

$$R^1COCH_2R^2 \xrightarrow[AlCl_3, 80°]{R^3C\equiv N} R^1COCHR^2CR^3=NH$$

R^1	R^2	R^3	% Yield
Ph	Me	Ph	65
Ph	Et	Me	55
Et	Me	Ph	60

I.A.8-7 J. A. Marshall and P.G.M. Wuts, Synth. Commun., 7, 233 (1977).

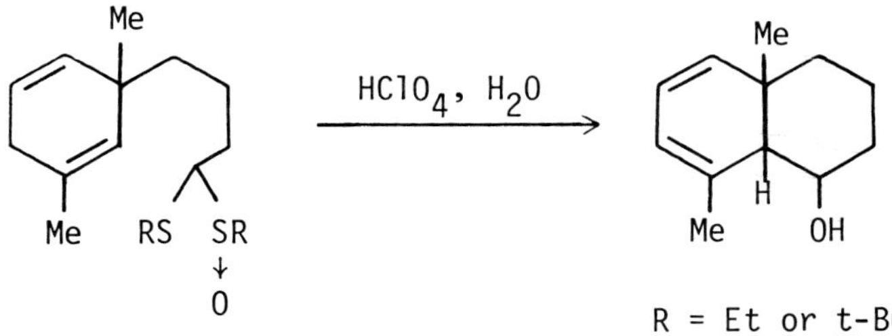

R = Et or t-Bu

"high yield"

I.A.8-8 W. L. Mock and M. E. Hartman, J. Org. Chem., 42, 459 (1977).

$$R^1COR^2 + N_2CHCO_2R^3 \xrightarrow[CH_2Cl_2,\ 0^\circ C]{Et_3O^+BF_4^-} R^1COCH(CO_2R^3)R^2$$

A full paper describing the synthetic scope of this reaction is presented.

I.A.8-9 D. M. Orere and C. B. Reese, J. Chem. Soc., Chem. Commun., 1977, 280.

$$R^1COR^2 \xrightarrow[\text{2) KCN, MeOH reflux}]{\text{1) ArSO}_2\text{NHNH}_2} R^1R^2CHCN$$

R^1	R^2	% Yield
t-Bu	Me	60
Pr	Pr	74
C_7H_{15}	H	72
$-(CH_2)_4-$		47

I.A.8-10 G. Stork and P. G. Willard, J. Am. Chem. Soc., 99, 7067 (1977).

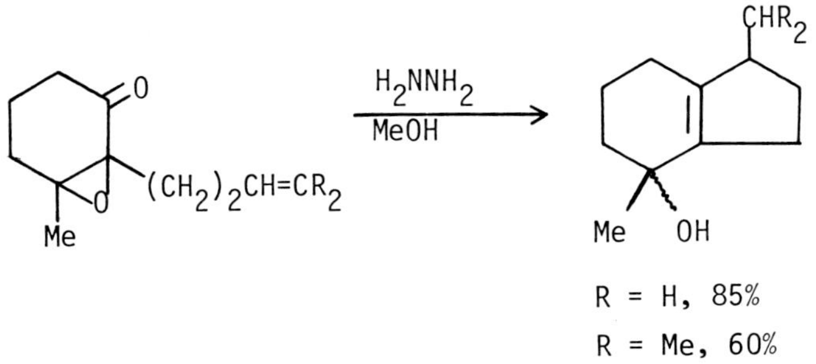

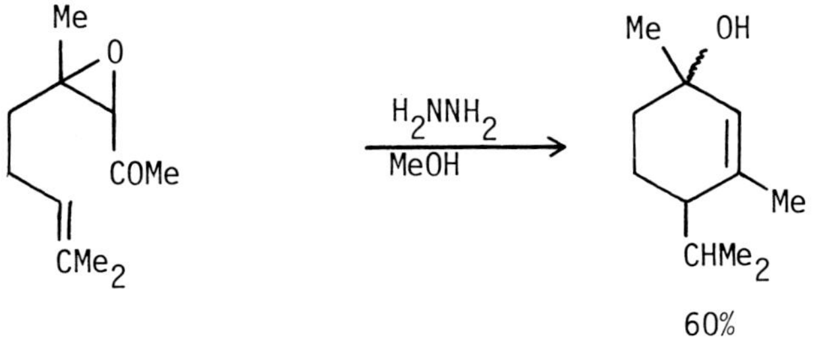

R = H, 85%
R = Me, 60%

60%

I.A.8-11 P. Gölitz and A. de Meijere, <u>Angew. Chem. Int. Ed. Engl.</u>, <u>16</u>, 854 (1977).

$$RNH_2 \xrightarrow[\text{MeOH, } -70°]{CF_3NO} RN=NCF_3 \xrightarrow{h\nu} RCF_3$$

R = alkyl, 30-69%

I.A.8-12 J. A. Ors and R. Srinivasan, <u>J. Org. Chem.</u>, <u>42</u>, 1321 (1977).

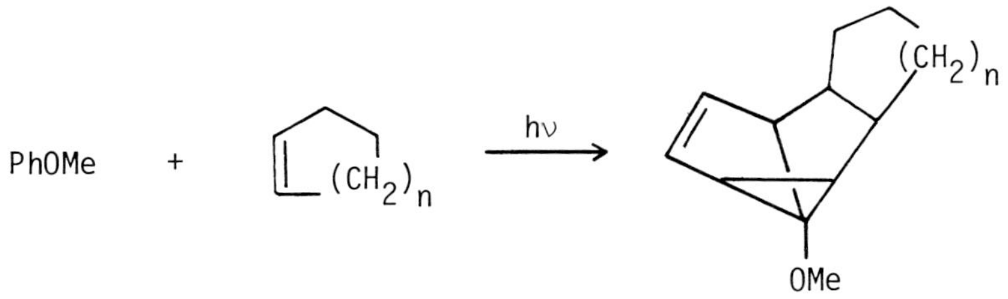

PhOMe + cyclic olefin-(CH$_2$)$_n$ $\xrightarrow{h\nu}$ [product with OMe, (CH$_2$)$_n$]

n = 1, 85%
n = 3, 62%

Other cyclic olefins also studied.

I.A.8-13 H. Hart and R. Willer, <u>Tetrahedron Lett.</u>, 2310 (1977).

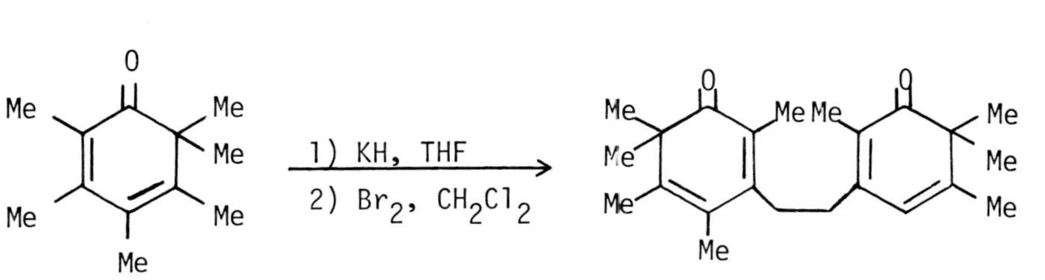

[hexamethyl cyclohexadienone] $\xrightarrow{\text{1) KH, THF} \atop \text{2) Br}_2\text{, CH}_2\text{Cl}_2}$ [dimer product]

90%

I.A.8-14 H. J. Günther, V. Jäger, and P. S. Skell, <u>Tetrahedron Lett.</u>, 2539 (1977).

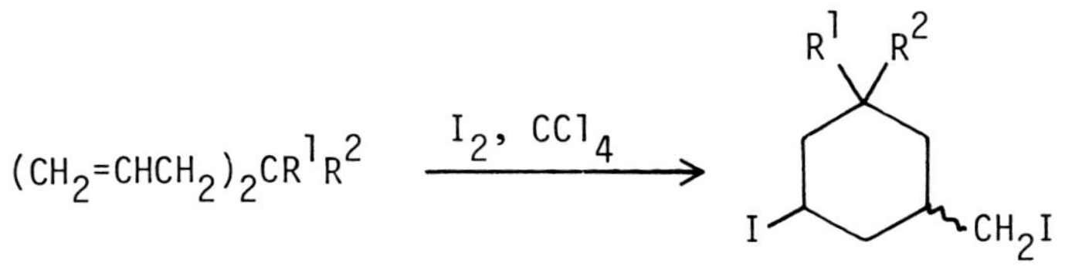

$$(CH_2=CHCH_2)_2CR^1R^2 \xrightarrow{I_2, CCl_4}$$

$R^1 = R^2 = Me$, 85%
$R^1 = Me$, $R^2 = Et$, 99%

If $R^1=R^2=H$ then only open chain diiodo- and tetraiodo-compounds are isolated. With $R^1=R^2=CO_2Et$, iodolactonization occurs.

I.A.8-15 A. Murai, M. Ono, and T. Masamune, <u>J. Chem. Soc., Chem. Commun.</u>, <u>1977</u>, 573.

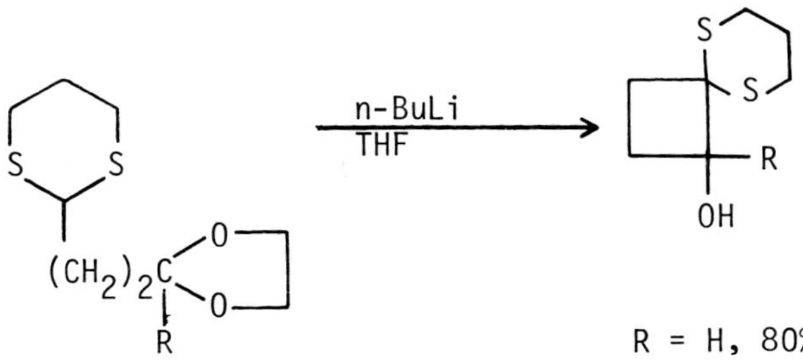

$$\xrightarrow{\text{n-BuLi}}_{\text{THF}}$$

R = H, 80%
R = Me, 60%

I.A.8-16 N. Engl, B. Kübel, and W. Steglich, Angew. Chem. Int. Ed. Engl., 16, 394 (1977).

$$R^1CONHCHPhCO_2CH_2CH=CR^2R^3 \xrightarrow[\text{Et}_3N, \text{MeCN}]{\text{COCl}_2 \text{ or } Ph_3P/CCl_4}$$

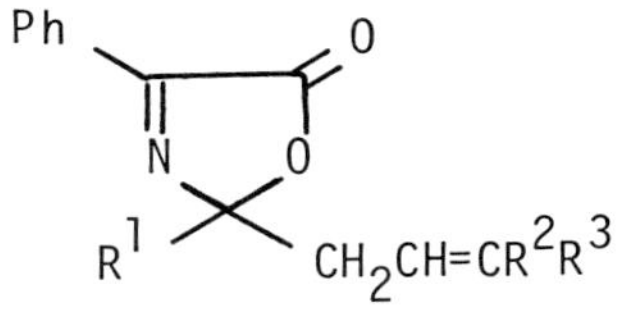

<img: oxazolone with Ph, N, O, R¹, CH₂CH=CR²R³ substituents>

R^1	R^2	R^3	% Yield
Ph	Me	Me	90
PhCH=CH	Me	Me	73
p-ClC$_6$H$_4$	Ph	H	87

I.A.8-17 T.B.R.A. Chen, J. J. Burger, and E. R. de Waard, Tetrahedron Lett., 4527 (1977).

$$CH_2=CHCH=CHSO_2CH_2Cl \xrightarrow{RSO_2Na} CH_2=CHCH=CHCH_2SO_2R$$

R = Ph or CH=CHCH=CH$_2$

I.B.1-1 K. B. Becker, Helv. Chim. Acta, 60, 68 (1977); see also: ibid, 81 (1977).

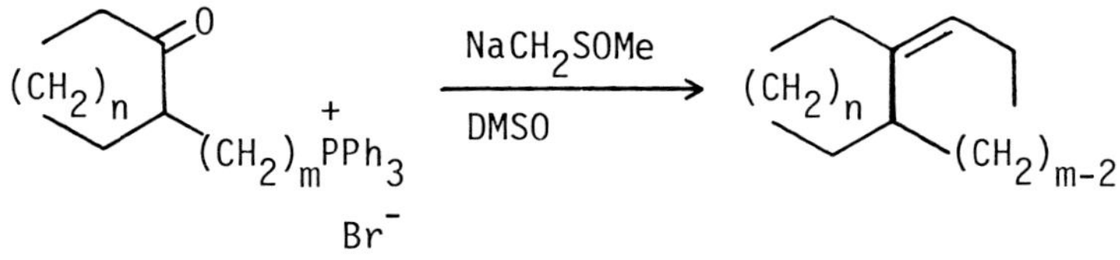

n = 1, 2, or 3
m = 3, 4, or 5
50-65%

I.B.1-2 A. Turcant and M. LeCorre, Tetrahedron Lett., 789 (1977).

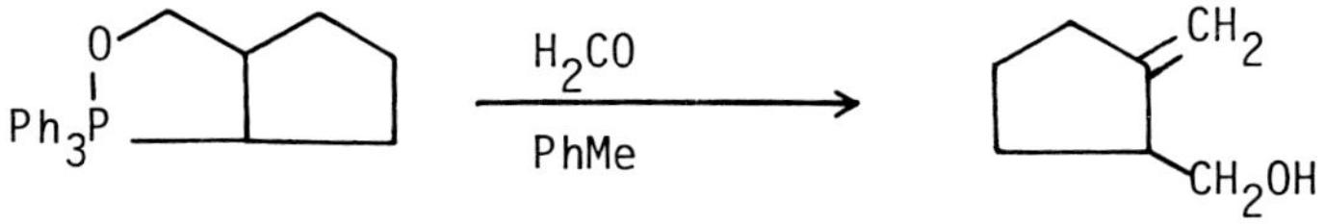

85%

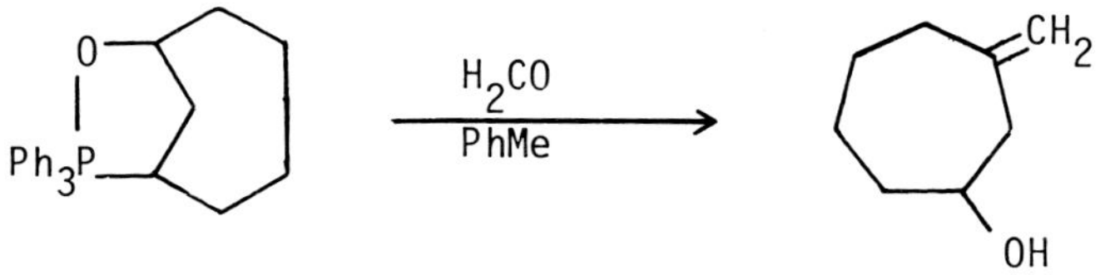

54%

I.B.1-3 M. T. Reetz and F. Eibach, Justus Liebigs Ann. Chem., 1977, 242.

$Ph_2PCH_2CO_2Et$ $\xrightarrow{\begin{array}{l}1)\ LDA\\2)\ PhCHO\\3)\ MeI\end{array}}$ $PhCH=CHCO_2Et$

70%

$Ph_2PCH_2COR^1$ $\xrightarrow{\begin{array}{l}1)\ R^2MgX\\2)\ MeI\\3)\ \underline{t}\text{-BuOK, HMPT}\end{array}}$ $CH_2=CR^1R^2$

$R^1 = Me,\ R^2 = Ph;\ 31\%$

$Ph_2PCH_2CO_2Et$ $\xrightarrow{\begin{array}{l}1)\ RMgX\\2)\ MeI\\3)\ \underline{t}\text{-BuOK, HMPT}\end{array}}$ $CH_2=CR_2$

$R = PhCH_2,\ 65\%$

I.B.1-4 A. Gossauer, R.-P. Hinze, and H. Zilch, Angew. Chem., Int. Ed. Engl., 16, 418 (1977).

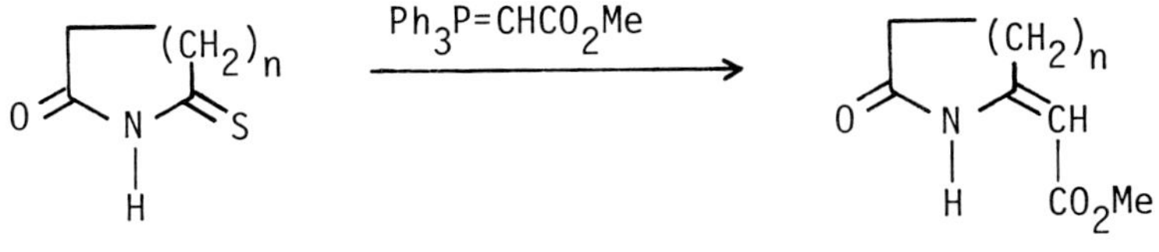

<chemical structure: 2-thioxo lactam with (CH$_2$)$_n$ ring> $\xrightarrow{Ph_3P=CHCO_2Me}$ <chemical structure: =CHCO$_2$Me product>

n = 1-3, 45-66%

I.B.1-5 D. Seyferth, K. R. Wursthorn, and R. E. Mammarella, J. Org. Chem., 42, 3104 (1977).

$$Ph_3\overset{+}{P}CHR^1CH_2MMe_3 \xrightarrow[\text{2) }R^2COR^3]{\text{1) LDA or MeLi}} R^2R^3C=CR^1CH_2MMe_3$$

M	R^1	R^2	R^3	% Yield
Sn	H	$\underline{n}$-C_6H_{13}	H	98
Sn	H		-(CH$_2$)$_5$-	79
Si	H		-(CH$_2$)$_5$-	85
Si	Me	Ph	H	72

I.B.1-6 W. G. Salmond, Tetrahedron Lett., 1239 (1977).

$$RCHO \xrightarrow{(Me_2N)_3P=CCl_2} RCH=CCl_2$$

85-94%

I.B.1-7 C. G. Kruse, N.L.J.M. Broekhof, A. Wijsman, and A. van der Gen, Tetrahedron Lett., 885 (1977).

[dithiane]-$\overset{+}{P}Ph_3$ Cl$^-$ $\xrightarrow[\text{2) }R^1COR^2]{\text{1) KOH or NaH}}$ R^1R^2C=[dithiane]

55-96%

[dithiane]-$P(O)(OEt)_2$ $\xrightarrow[\text{2) }R^1COR^2]{\text{1) n-BuLi}}$ 85-96%

I.B.1-8 J. I. Grayson and S. Warren, J. Chem. Soc., Perkin Trans. I, 1977, 2263.

$$Ph_2P(O)CHR^1SR^2 \xrightarrow[2)\ R^3COR^4]{1)\ n\text{-BuLi},\ -78°C} R^2SCR^1=CR^3R^4$$

R^1	R^2	R^3	R^4	% Yield
Et	Ph	Ph	H	93
i-Pr	Ph	PhCH=CH	H	93
H	Ph	Ph	Me	90
H	Ph	PhSCHMe	H	76
Et	Me	$-(CH_2)_5-$		63

The α-phenylthioalkylphosphine oxides, acylanion equivalents, are easily prepared.

I.B.1-9 A. Loupy, K. Sogadji, J. Seyden-Penne, Synthesis, 1977, 126.

$$R^1COR^2 + Ph_2P(O)CH_2CN \xrightarrow{\text{t-BuOK}}_{\text{THF or DMF}} R^1R^2C=CHCN$$

R^1	R^2	% Yield
Ph	H	95
i-Pr	H	94
Ph	Me	82

I.B.1-10 C. Earnshaw, C. J. Wallis, and S. Warren, J. Chem. Soc., Chem. Commun., 1977, 314.

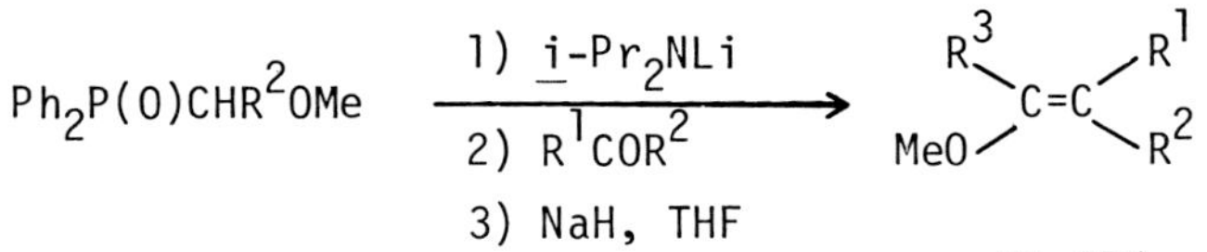

$$Ph_2P(O)CHR^2OMe \xrightarrow[\substack{\text{2) } R^1COR^2 \\ \text{3) NaH, THF}}]{\text{1) } i\text{-}Pr_2NLi} \begin{array}{c} R^3 \\ MeO \end{array} C=C \begin{array}{c} R^1 \\ R^2 \end{array}$$

50-65%

Decomposition of the initial adducts proceeds to give single geometrical isomers of vinyl ethers.

I.B.1-11 S. F. Martin and P. J. Garrison, Tetrahedron Lett., 3875 (1977).

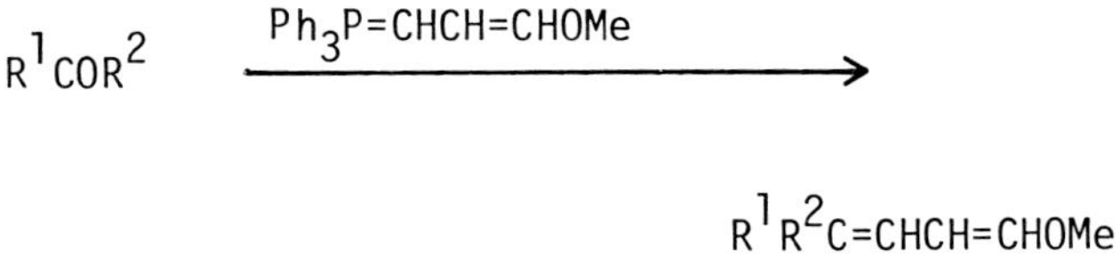

$$R^1COR^2 \xrightarrow{Ph_3P=CHCH=CHOMe} R^1R^2C=CHCH=CHOMe$$

R^1	R^2	% Yield
Ph	H	75
$Me_2C=CH$	Me	41
$\text{--}(CH_2)_5\text{--}$		75

Products can either be isolated or directly hydrolyzed to $R^1R^2CHCH=CHCHO$.

CARBON—CARBON BOND FORMING REACTIONS

I.B.1-12 P. Dalla Croce and A. Zaniboni, Synthesis, 1977, 552.

$$\underset{ClO_4^-}{ArNHN=CHPPh_3^+} \xrightarrow[Et_3N]{RCHO} ArN=NCH=CHR$$

Ar	R	% Yield
Ph	Ph	45
2-ClC$_6$H$_4$	Ph	55
2-ClC$_6$H$_4$	CCl$_3$	70
Ph	PhCO	50

I.B.1-13 H. Yoshida, T. Ogata, and S. Inokawa, Synthesis, 1977, 626; see also: ibid, Bull. Chem. Soc. Japan, 50, 3315 (1977).

$$Ph_3P=CHCR^2=NR^1 \xrightarrow{R^3CHO} R^3CH=CHCR^2=NR^1$$

R^1	R^2	R^3	% Yield
Ph	Ph	Ph	94
Ph	OMe	Me	82
p-MeC$_6$H$_4$	Ph	Me	87
Ph	SMe	Ph	91

I.B.1-14 A. Dehnel, J. P. Finet, and G. Lavielle, Synthesis, 1977, 474.

$$(EtO)_2P(O)CHR^1N=CHR^2 \xrightarrow[\text{2) } R^3COR^4]{\text{1) NaH, THF}} R^3R^4C=CR^1N=CHR^2$$

R^1	R^2	R^3	R^4	% Yield
H	Ph	H	Ph	>95
Ph	Ph	H	Ph	80
Ph	Ph	H	$\underline{i}$-Pr	>95
Me	Ph	$-(CH_2)_5-$		>95

I.B.1-15 T. Kauffmann, U. Koch, F. Steinseifer, and A. Vahrenhorst, Tetrahedron Lett., 3341 (1977).

$$Ph_2C=NCH_2G \xrightarrow[\text{2) } R^1COR^2]{\text{1) LDA or n-BuLi}} R^1R^2C=CHN=CPh_2$$

R^1	R^2	G	% Yield
Ph	Ph	$SiPh_3$	81
Ph	Ph	$P(O)Ph_2$	92
Me	Me	$SiPh_3$	72
$\underline{n}$-Pr	H	$P(O)Ph_3$	56

I.B.1-16 A. Sekiguchi and W. Ando, Chem. Lett., 1977, 1293.

$$R^1COR^2 \xrightarrow[Ph_3P, 150°C]{Me_3SiCH_2Cl} R^1R^2C=CH_2$$

R^1	R^2	% Yield
Ph	Ph	78
Me_3Si	Ph	70
t-Bu	Ph	57

If ketone has α H's then significant amounts of silyl enol ether is formed along with the olefin.

I.B.1-17 D. Seyferth, J. L. Lefferts, and R. L. Lambert, Jr., J. Organomet. Chem., 142, 39 (1977).

$$RCHO \xrightarrow[-115°C, THF]{(Me_3Si)_2CBrLi} RCH=C(Br)SiMe_3$$

$$52-73\%$$

I.B.1-18 B.-T. Gröbel and D. Seebach, Chem. Ber., 110, 852 (1977).

$$Me_3SiC(X)(Y)Li \xrightarrow{R^1COR^2} XCY=CR^1R^2$$

X	Y	R^1	R^2	% Yield
MeS	MeS	n-C_5H_{11}	H	82
MeS	MeS	Ph	Me	57
PhSe	PhSe	Ph	H	75
Me_3Si	H	Ph	H	70
Me_3Si	Me_3Si	Ph	H	71
Me_3Si	MeS	Ph	H	84

I.B.1-19 D. Seebach, R. Bürstinghaus, B.-T. Gröbel, and M. Kolb, Justus Liebigs Ann. Chem., 1977, 830.

$$R^1CH=C(SiMe_3)X \xrightarrow[\text{2) } R^3CHO]{\text{1) } R^2Li} R^1R^2CX=CHR^3$$

$$X = SiMe_3 \text{ or } SMe$$

I.B.1-20 D. H. Lucast and J. Wemple, Tetrahedron Lett., 1103 (1977).

$$Me_3SiCH_2C(O)SR^1 \xrightarrow[\text{2) } R^2COR^3, -78° \text{ to } 25°C]{\text{1) LDA, THF, } -78°C}$$

$$R^2R^3C=CHC(O)SR^1$$

R^1	R^2	R^3	% Yield
t-Bu	Ph	H	73
i-Pr	i-Pr	H	51
t-Bu	Ph	Me	52
PhCH$_2$	$(CH_2)_5$		49

I.B.1-21 T. Kauffmann, R. Kriegesmann, and A. Woltermann, Angew. Chem. Int.Ed. Engl., 16, 862 (1977).

$$Ph_3SnCH_2I \xrightarrow[\text{2) } R^1COR^2]{\text{1) n-BuLi, Et}_2\text{O, -50°C}} Ph_3SnCH_2C(OH)R^1R^2$$

A

$$\xrightarrow{110-175°C} CH_2=CR^1R^2$$

B

| | | % Yield | |
R^1	R^2	A	B
Ph	H	65	95
n-Pr	H	31	100
Me	Ph	41	70
-(CH$_2$)$_5$-		37	96

I.B.1-22 H. Pommer, Angew. Chem. Int. Ed., Engl, 16, 423 (1977).

Review: "The Wittig Reaction in Industrial Practice"

I.B.2.a-1 J. Mulzer and G. Brüntrup, Angew. Chem. Int. Ed. Engl., 16, 255 (1977).

$$R^3R^4C(OH)CR^1R^2CO_2H \xrightarrow[Ph_3P, THF]{EtO_2CN=NCO_2Et} R^3R^4C=CR^1R^2$$

55-75%

I.B.2.a-2 H. Sliwa and A. Tartar, <u>Tetrahedron</u>, <u>33</u>, 3111 (1977).

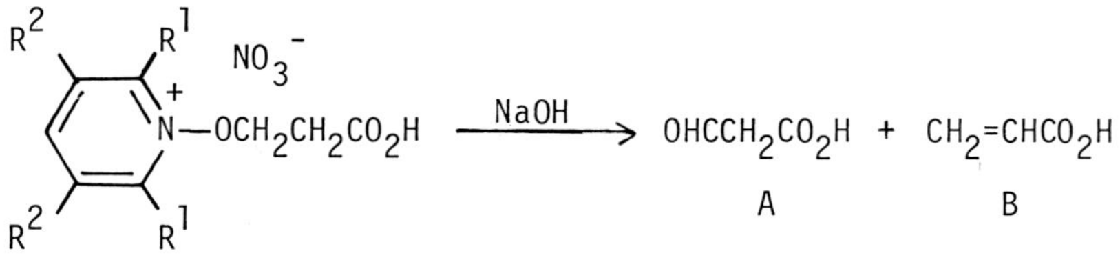

$$\underset{R^2\ \ \ \ R^1}{\underset{|}{\overset{R^2\ \ \ \ R^1}{\bigcirc}}}\overset{+}{N}-OCH_2CH_2CO_2H\ \ NO_3^- \xrightarrow{NaOH} \underset{A}{OHCCH_2CO_2H} + \underset{B}{CH_2=CHCO_2H}$$

Major product is A when R^1 = Me, R^2 = H. Major product is B when R^1 = H, R^2 = Me or $R^1 = R^2$ = H.

I.B.2.a-3 T. Sato, M. Arai, and I. Kuwajima, <u>J. Am. Chem. Soc.</u>, <u>99</u>, 5827 (1977).

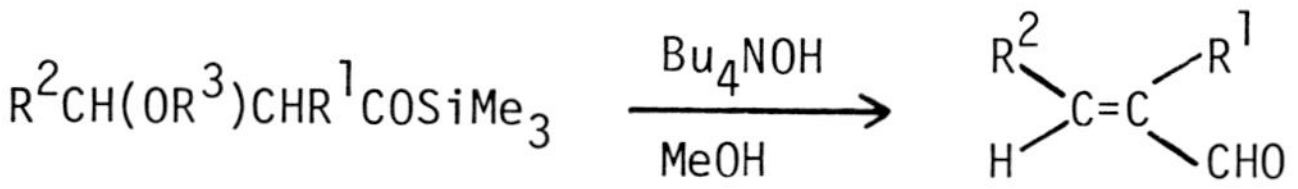

$$R^2CH(OR^3)CHR^1COSiMe_3 \xrightarrow[MeOH]{Bu_4NOH} \underset{H}{\overset{R^2}{>}}C=C\underset{CHO}{\overset{R^1}{<}}$$

R^1	R^2	R^3	% Yield
Me	n-Pr	Me	91
Me	Ph	Et	94
PhCH$_2$	Ph	Et	93

I.B.2.a-4 A. H. Davidson, I. Fleming, J. I. Grayson, A. Pearce, R. L. Snowden, and S. Warren, J. Chem. Soc., Perkin Trans. I, 1977, 550; see also: A. H. Davidson, C. Earnshaw, J. I. Grayson, and S. Warren, ibid, 1452.

$$\text{PhCH(OH)}\underset{\underset{\text{Et}}{|}}{\overset{\overset{\text{Me}}{|}}{\text{C}}}\text{-P(O)Ph}_2 \xrightarrow{\text{TFA}} \text{Ph}_2\text{P(O)CHPhCMe=CHMe}$$

90%, only E-isomer

$$\text{PhCH(OH)}\underset{\underset{\text{Et}}{|}}{\overset{\overset{\text{CH}_2\text{SiMe}_3}{|}}{\text{C}}}\text{-P(O)Ph}_2 \xrightarrow{\text{TFA}} \text{Ph}_2\text{P(O)CHPhC(Et)=CH}_2$$

95%

I.B.2.a-5 J. A. Marshall and M. E. Lewellyn, J. Org. Chem., 42, 1311 (1977).

Diol $\xrightarrow{\text{1) Cl}_2\text{POX}}_{\text{2) Li, NH}_3(\ell)}$ A + B

n	Stereochemistry of diol	X	% Yield	A:B
6	cis	NMe$_2$	81	90:10
6	cis	OEt	57	87:13
6	trans	OEt	55	19:81
8	cis	NMe$_2$	86	90:10
8	trans	OEt	86	8:92

I.B.2.a-6 G. Goto, Bull. Chem. Soc. Japan, 50, 186 (1977).

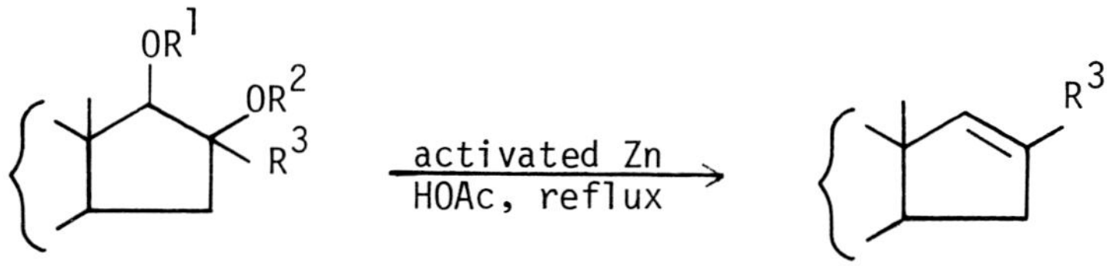

$$\text{cyclopentane with } OR^1, OR^2, R^3 \xrightarrow[\text{HOAc, reflux}]{\text{activated Zn}} \text{cyclopentene with } R^3$$

R^1	R^2	R^3	% Yield
Ac	H	Me	85
H	Ac	H	88
H	Ac	H	7
Ac	H	Ph	87
EtCO	H	Et	91
-EtCCOMe-		Et	92

I.B.2.a-7 A.G.M. Barrett, D.H.R. Barton, R. Bielski, and S. W. McCombie, J. Chem. Soc., Chem. Commun., 1977, 866.

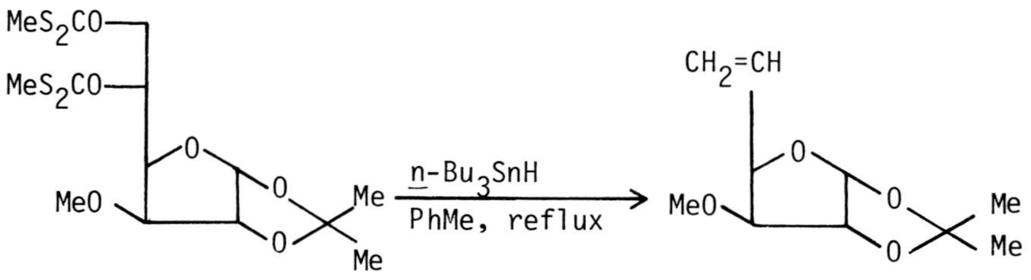

$$\text{sugar diol bis(methylxanthate)} \xrightarrow[\text{PhMe, reflux}]{\text{n-Bu}_3\text{SnH}} \text{alkene product}$$

62%

Other diols derived from sugars were also studied.

I.B.2.a-8 P. F. Hudrlik, A. M. Hudrlik, R. J. Rona, R. N. Misra, and G. P. Withers, J. Am. Chem. Soc., 99, 1993 (1977).

$$\underset{RCH-CHSiMe_3}{\triangle^O} \xrightarrow{HX} RCH(OH)CH(X)SiMe_3 \xrightarrow[base]{acid\ or} RCH=CHX$$

X = Br, OAc, OMe or NHAc

The reactions are stereospecific and choice of conditions allow the formation of either the (E)- or (Z)-isomer.

I.B.2.a-9 T. Inoue, T. Uchimaru, and T. Mukaiyama, Chem. Lett., 1977, 1215.

$$R^1CH_2-\overset{O}{\triangle}\overset{Me}{\underset{Me}{<}} \xrightarrow[\substack{2,6-dimethylpyridine \\ Et_2O}]{9-BBNOTf} R^1CH_2CHOHCMe=CH_2$$

51-100%

I.B.2.a-10 D.L.J. Clive and S. M. Menchen, J. Chem. Soc., Chem. Commun., 1977, 658.

$$R^1-\overset{O}{\triangle}-R^2 \xrightarrow[EtOH]{(EtO)_2P(O)Te^-Na^+} R^1CH=CHR^2$$

R^1	R^2	% Yield
H	C_6H_{13}	73
H	C_8H_{17}	70
$-(CH_2)_4-$		89

"Non-terminal epoxides react slowly and, therefore, selective deoxygenations are possible...."

I.B.2.a-11 V. Reutrakul and W. Kanghae, <u>Tetrahedron Lett.</u>, 1377 (1977).

$$R^1CH_2\underset{R^2}{\diagdown}\!\!\!\triangle\!\!\!-S(O)Ph \xrightarrow{\Delta} R^1CH=CR^2CHO$$

R^1	R^2	% Yield
$\underline{n}$-C_5H_{11}	H	33
$-(CH_2)_2-$		40
$-(CH_2)_5-$		77
$-(CH_2)_6-$		90

I.B.2.a-12 D. Hoppe and R. Follmann, <u>Angew. Chem. Int. Ed. Engl.</u>, <u>16</u>, 462 (1977).

$$\underset{H}{\overset{R^1}{\diagdown}}\!\!\!\underset{S\diagup\diagdown O}{\overset{+NMe_3}{\|}}\!\!\!\underset{R^3}{\diagup\overset{R^2}{}} \xrightarrow[80°C]{MeNO_2} Me_2N\overset{O}{\overset{\|}{C}}-S\underset{R^1}{\diagdown}C=C\underset{R^3}{\diagup R^2}$$

R^1	R^2	R^3	% Yield
H	Me	Me	64
H	Me	Ph	81
Ph	Me	Me	78

I.B.2.a-13 M. Kim and J. D. White, J. Am. Chem. Soc., 99, 1172 (1977).

$$R^1R^2C(O)(O)N(N=S(O)Me_2)CR^3R^4 \xrightarrow{110°-140°C} R^1R^2C=CR^3R^4$$

R^1	R^2	R^3	R^4	% Yield
H	H	H	H	92
Ph	H	H	H	94
Ph	H	Ph	H	97
‑(CH$_2$)$_4$‑		H	H	98
H	‑(CH$_2$)$_6$‑		H	61

Also useful for making highly strained olefins like bicyclo[3.3.1]non-1-ene.

I.B.2.b-1 M. Schlosser and C. Tarchini, Helv. Chim. Acta., 60, 3060 (1977).

$$C_8H_{17}Cl(Br) \xrightarrow[\text{TGME}]{\text{KOH, Me}_3\text{N}} C_6H_{13}CH=CH + (C_8H_{17})_2O$$
$$\phantom{C_8H_{17}Cl(Br) \xrightarrow[\text{TGME}]{\text{KOH, Me}_3\text{N}} } 8 \quad : \quad 1$$

A complete study on the factors influencing elimination vs substitution were carried out.

I.B.2.b-2   P. F. King and L. A. Paquette, Synthesis, 1977, 279.

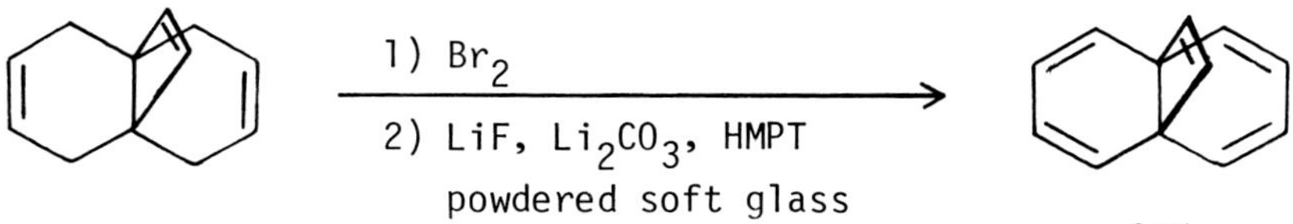

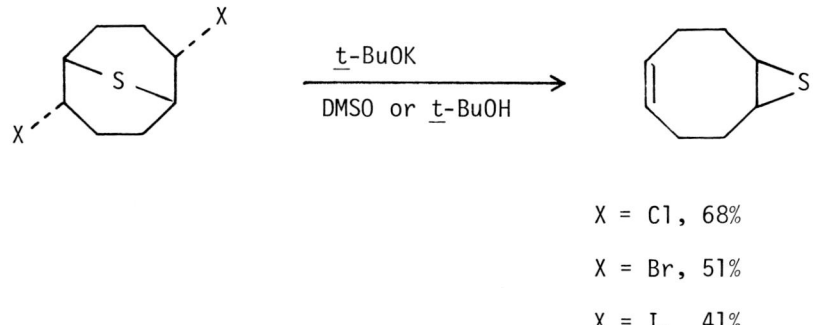

I.B.2.b-3   P. H. McCabe, C. M. Livingston, and A. Stewart, J. Chem. Soc., Chem. Commun., 1977, 661.

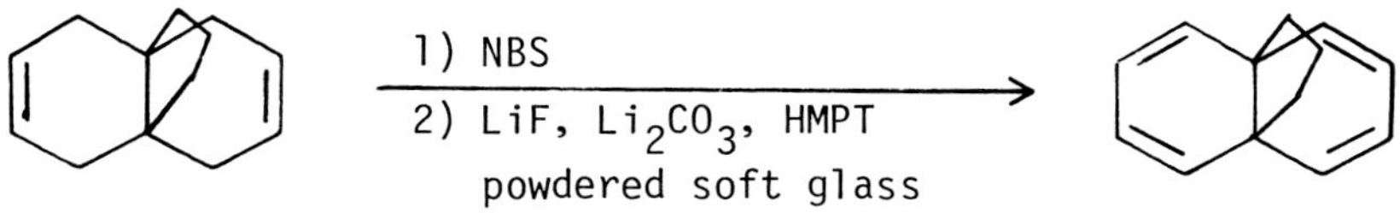

X = Cl, 68%

X = Br, 51%

X = I, 41%

I.B.2.b-4   T.-L. Ho and G. A. Olah, Synthesis, 1977, 170.

$$R^1CHBrCHBrR^2 \xrightarrow[\text{[V(II)]}]{VCl_3/LiAlH_4} R^1CH=CHR^2$$

$$R^1 = R^2 = Ph, 97\%$$

$$R^1, R^2 = -(CH_2)_6-, 68\%$$

I.B.2.b-5   R. B. Miller and G. McGarvey, Synth. Commun., 7, 475 (1977).

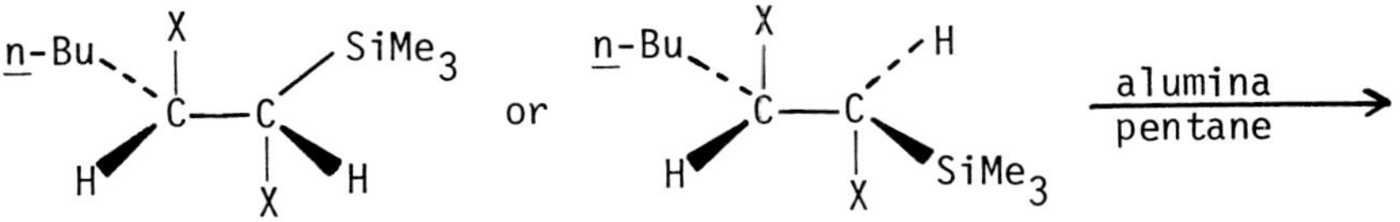

n-BuCH=CHX

C

starting material	X	% Yield C	C cis:trans
A	Cl	65	23:77
A	Br	84	5:95
B	Cl	62	99:1
B	Br	68	99:1

I.B.2.c-1 T. Hayashi, A. Sakurai, and T. Oishi, Chem. Lett., 1977, 1483.

$$R^1CHCH_2R^2 \xrightarrow[Li_2CO_3, LiF]{MeI, HMPA} \underset{H}{\overset{R^1}{>}}C=C\underset{R^2}{\overset{H}{<}}$$
$$|$$
$$S-C(S)-N\square$$

72-95%

The S substituent was at an allylic or benzylic position.

I.B.2.c-2 M. F. Semmelhack and J. C. Tomesch, J. Org. Chem., $\underline{42}$, 2657 (1977).

$$(CH_2)_n \begin{matrix} CH(SPh)_2 \\ OH \end{matrix} \xrightarrow[THF, CuOTf]{n-BuLi} OHC(CH_2)_{n+2}CH=CHSPh$$

n = 1, 93%
n = 2, 92%

I.B.2.c-3 R. Tanikaga, H. Sugihara, K. Tanaka, and A. Kaji, Synthesis, 1977, 299.

$$XCH_2CH_2S(O)Ar \xrightarrow{heat} XCH=CH_2$$
Ar = 4-ClC$_6$H$_4$

X = O$_2$NCMe$_2$, 85%
X = (EtO$_2$C)$_2$CEt, 93%
X = PhS, 86%

$$\underset{ArSCH_2CH_2CX_2CH_2CH_2SAr}{\overset{O \quad\quad\quad O}{\|\quad\quad\quad\|}} \xrightarrow{heat} CH_2=CHCX_2C=CH_2$$

X = NC, 64%

Starting materials are prepared by X-H addition to CH$_2$=CHS(O)Ar.

I.B.2.c-4 M. Sevrin, W. Dumont, and A. Krief, <u>Tetrahedron Lett.</u>, 3835 (1977).

$$R^1R^2CHCR^3(SeR^4)_2 \xrightarrow{\text{MeI}}[\text{DMF}] R^1R^2C=CR^3SeR^4$$

50-65%

I.B.2.c-5 E. Vedejs and D. A. Engler, <u>Tetrahedron Lett.</u>, 1241 (1977).

$$R^3R^2CHCR^1(CN)\overset{+}{N}MeCH_2CO_2Et \xrightarrow{\text{DBU}} R^3R^2C=CR^1CN$$

$$CF_3SO_3^-$$

R^1	R^2	R^3	% Yield
Ph	Me	H	95
$\underline{n}$-C_5H_{11}	H	H	73
$-(CH_2)_4-$		H	57*

*Also 20% of

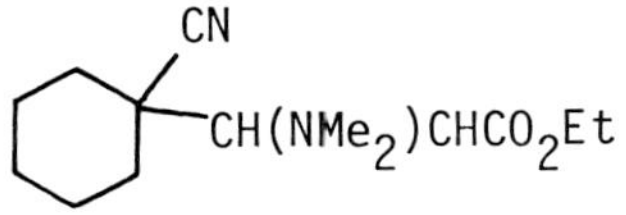

$$\underset{}{\text{cyclohexyl}}\begin{array}{c}CN\\|\\-CH(NMe_2)CHCO_2Et\end{array}$$

I.B.2.c-6   N. Kornblum and L. Cheng, J. Org. Chem., 42, 2944 (1977).

$$R^1R^2C(NO_2)C(NO_2)R^3R^4 \xrightarrow[HMPA]{Ca(Hg)} R^1R^2C=CR^3R^4$$

R^1	R^2	R^3	R^4	% Yield
Me	$(CH_2)_2CN$	Me	$(CH_2)_2CN$	87
Me	$(CH_2)_2COMe$	Me	$(CH_2)_2COMe$	77
$-(CH_2)_5-$		Me	$(CH_2)_2CO_2Me$	66
$-(CH_2)_5-$		$-(CH_2)_5-$		86

I.B.2.c-7   T.-H. Chan, Acc. Chem. Res., 10, 442 (1977).

Review: "Alkene Synthesis via β-Functionalized Organosilicon Compounds."

I.B.3-1   R. T. Taylor, C. R. Degenhardt, W. P. Melega, and L. A. Paquette, Tetrahedron Lett., 159 (1977).

$$R^1CH_2CR^2=NNHTs \xrightarrow[\text{2) Me}_3MCl]{\text{1) 4 eq. n-BuLi, TMEDA, -78°C}} R^1CH=C(MMe_3)R^2$$

M = Si, 15-71%

M = Ge, 47-64%

M = Sn, 43-56%

I.B.3-2   B. M. Trost and W. J. Frazee, J. Am. Chem. Soc., 99, 6124 (1977).

[structure: cyclopropane fused to cyclobutanone with MeO$_2$C, CO$_2$Me, Me substituents] $\xrightarrow{\text{NaOMe} \atop \text{MeOH, reflux}}$ (MeO$_2$C)$_2$CHCH$_2$CH=CMeCH$_2$CH$_2$CO$_2$Me

80-84%

The olefin formation is stereospecific.

I.B.3-3   G. M. Rubottom, R. Marrero, D. S. Krueger, and J. L. Schreiner, Tetrahedron Lett., 4013 (1977).

(CH$_2$)$_n$ [cyclopropane with OSiMe$_3$] $\xrightarrow{\text{Pb(OAc)}_4 \atop \text{HOAc}}$ CH$_2$=CH(CH$_2$)$_n$CO$_2$H

n = 3, 62%
n = 4, 83%
n = 5, 92%

I.B.3-4   P. Gygax and A. Eschenmoser, Helv. Chim. Acta., 60, 507 (1977).

[bicyclic structure with N-cyclohexyl, N-O, and two C=O groups, fused to cyclohexene]

1) Me$_3$O$^+$PF$_6^-$, CH$_2$Cl$_2$, 83°C
2) i-Pr$_2$NEt, CH$_2$Cl$_2$, 36°C
3) 2 $\underline{N}$ H$_2$SO$_4$, CCl$_4$ 60%

1) R$_2$NLi, THF, -40°C
2) Me$_3$SiCl, THF, 0°C   64%
3) 2 $\underline{N}$ H$_2$SO$_4$, 0°C
4) CH$_2$N$_2$, Et$_2$O

→ [benzene ring with CO$_2$Me]

I.B.3-5 W. Walter and H.-W. Lüke, Angew. Chem. Int. Ed. Engl., 16, 535 (1977).

$$HC(S)N(SiMe_3)_2 \xrightarrow[\text{2) Me}_3\text{SiCl, rt}]{\text{1) RCH}_2\text{Li}, \text{ n-C}_5\text{H}_{12}, -45°C} \begin{array}{c} R \\ H \end{array} C=C \begin{array}{c} H \\ N(SiMe_3)_2 \end{array}$$

R = H, Me, Pr

I.B.3-6 E. Friedrich and W. Lutz, Angew. Chem. Int. Ed. Engl., 16, 413 (1977).

[cyclohexene with OSiMe$_3$ and R substituent] $\xrightarrow[\text{2) Ph}_3\text{P, MeOH}]{\text{1) }^1\text{O}_2}$ [cyclohexenone with R substituent]

22-83%

Particularly useful for synthesis of chirally substituted cyclohexenones at C-6.

I.B.3-7 J. A. Hyatt and J. J. Krutak, J. Org. Chem., 42, 169 (1977).

$$R^1R^2NCH_2CH_2SO_2F \xrightarrow[\text{CHCl}_3]{\text{MnO}_2} R^1R^2NCH=CHSO_2F$$

33-96%

I.B.3-8   M. T. Reetz and W. Stephan, Angew. Chem. Int. Ed. Engl., **16**, 44 (1977).

$$R^1CH_2CR^2R^3M \xrightarrow{\text{Hydride acceptor}} R^1CH=CR^2R^3$$

R^1	R^2	R^3	M	Hydride acceptor	% Yield
n-Bu	Ph	Ph	Li	Ph_3CBF_4	65
H	Me	Me	Li	sec-Bu_3B	95
n-Bu	Ph	Ph	Li	sec-Bu_3B	80
$-(CH_2)_4-$		H	MgCl	Ph_3CBF_4	75

I.B.3-9   M. T. Reetz and C. Weis, Synthesis, **1977**, 135.

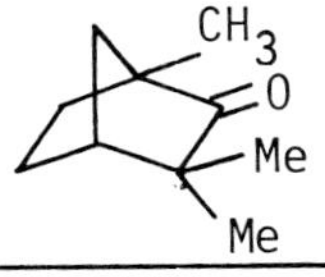

$$R^1R^2CHCHR^3MgX \xrightarrow[\text{Et}_2\text{O, reflux}]{} R^1R^2C=CHR^3$$

(reagent: camphor-derived ketone)

R^1	R^2	R^3	% Yield
C_6H_{13}	H	H	78
Et	H	n-Pr	91
$-(CH_2)_5-$		H	91
H	$-(CH_2)_4-$		93

I.B.3-10   S.E.J. Glue and I. T. Kay, Synthesis, 1977, 607.

$$ArN=C=S \xrightarrow[\text{2) } R^1O_2CCHClCOR^2]{\text{1) KCN/MeOH}} ArNHC(CN)-C(COR^2)CO_2R^1$$

27-64%

R^1 = Et, R^2 = Me and
R^1 = Me, R^2 = OMe

I.B.3-11   A. Nürrenbach, J. Paust, H. Pommer, J. Schneider, and B. Schulz, Justus Liebigs Ann. Chem., 1977, 1146.

$$RCMe=CHCH_2\overset{+}{P}Ph_3 \; HSO_4^- \xrightarrow[\text{aq. } Na_2CO_3]{30\% \; H_2O_2} RCMe=CHCH=CHCH=CMeR$$

80%

R =

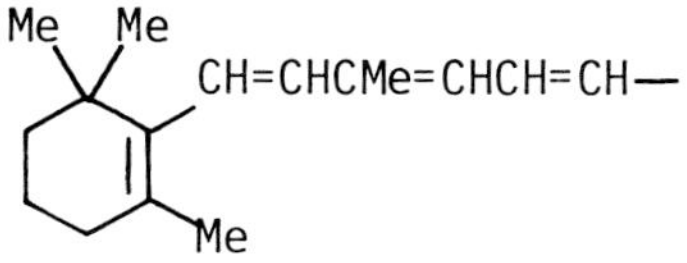

CH=CHCMe=CHCH=CH—

I.B.3-12   T. Severin, I. Bräutigam, and K.-H. Bräutigam, Chem. Ber., 110, 1669 (1977).

$$XYC=CHCH=NO_2^- \xrightarrow{AgNO_3} XYC=CHCH=CHCH=CXY$$

X	Y	% Yield
CO_2Me	CO_2Me	74
MeCO	CO_2Et	56
EtCO	Me	46
MeCO	H	35

CARBON—CARBON BOND FORMING REACTIONS

I.B.3-13   H. Teichmann and T. Am, Z. Chem., 17, 93 (1977).

$$R^1_2CHCONHR^2 \xrightarrow[\text{2) Et}_3\text{N}]{\text{1) Ph}_3\text{P, CCl}_4\text{, CHCl}_3} R^1_2C=C=NR^2$$

R^1	R^2	% Yield
Me	Ph	77
Ph	4-ClC$_6$H$_4$	88
4-ClC$_6$H$_4$	4-MeC$_6$H$_4$	77

I.B.3-14   L. Y. Kryukova, L. N. Kryukov, T. D. Truskanova, V. L. Isaev, R. N. Sterlin, and I. L. Knunyants, Dokl. Chem. (Eng. Transl.), 232, 91 (1977).

$$(RO_2C)_2CHCF_3 \xrightarrow[180-200°C]{P_2O_5} \underset{CF_3}{\overset{RO_2C}{>}}C=C=O$$

41-66%

I.B.3-15   G. Seybold and C. Heibl, Chem. Ber., 110, 1225 (1977).

$$\underset{R^2}{\overset{R^1}{\diagdown}}\!\!\!\diagup\!\!\!\underset{S}{\overset{N\!\!\diagdown\!\!N}{|}} \xrightarrow{500-600°C} R^1R^2C=C=S$$

60-75%

I.B.4-1 Y. Okude, T. Hiyama, and H. Nozaki, Tetrahedron Lett., 3829 (1977).

$$R^1 \underset{Br}{\overset{Br}{\triangle}} R^2 \xrightarrow[DMF]{CrCl_3-LiAlH_4} R^1CH=C=CHR^2$$

$R^1, R^2 = {-(CH_2)_6-}$, 100%

$R^1 = Ph$, $R^2 = H$, 62%

I.B.4-2 D. G. Oelberg and M. D. Schiavelli, J. Org. Chem., **42**, 1804 (1977).

$$R^1R^2C(X)C\equiv CR^3 \xrightarrow[CH_2Cl_2,\ 35°C]{AgClO_4} R^1R^2C=C=C(X)R^3$$

X = OAc, 46-70%

X = $(EtO)_2P(O)O$, 43-62%

X = $ArCO_2$, 35-46%

I.B.4-3 K. A. Parker and J. J. Petraitis, Tetrahedron Lett., 4561 (1977).

$$R^1R^2C(OH)C\equiv CR^3 \xrightarrow[\substack{xylene\ or\ o-Cl_2C_6H_4 \\ 140-170°C}]{Et_2NCH(OEt)_2} R^1R^2C=C=CR^3CONEt_2$$

R^1	R^2	R^3	% Yield
Me	Me	Ph	84
H	$n\text{-}C_8H_{17}$	Me	93
$-(CH_2)_5-$		H	50
$-(CH_2)_5-$		Me	92

I.B.4-4 B. Cazes and S. Julia, Synth. Commun., 7, 273 (1977).

$$R^1C{\equiv}CCH_2OCHR^2CN \xrightarrow[\text{THF, }-78°C]{\text{excess LDA}} CH_2{=}C{=}CR^1COR^2$$

R^1	R^2	% Yield
H	n-Bu	49
H	i-Bu	51
Et	n-Bu	61

I.B.4-5 P. Cresson, Comptes. rendus, Ser. C, 284, 247 (1977).

$$R^2C{\equiv}CCH_2\overset{+}{N}Me_2CHR^1CN \xrightarrow[\text{THF, }-40°C]{\text{t-BuOK}} CH_2{=}C{=}CR^2CR^1(CN)NMe_2$$
$$\sim 100\%$$

The products are hydrolyzed (Cu^{+2}, H_2O) to α-allenic ketones and aldehydes.

I.B.4-6 L.-I. Olsson, Acta Chem. Scand., B31, 639 (1977).

$$R^1C{\equiv}CCHR^2SC(S)R^3 \xrightarrow[\text{2) MeI}]{\text{1) EtMgBr, THF, }-30°C} R^2CH{=}C{=}CR^1\overset{\displaystyle SEt}{\underset{\displaystyle SMe}{C}}R^3$$

R^1	R^2	R^3	% Yield
H	H	Et	64
Me	H	C_6H_{13}	74
Ph	H	Et	58

I.B.4-7 T. Kunieda and T. Takizawa, Chem. Pharm. Bull., 25, 1809 (1977).

$$Ar_2CHCX_3 \text{ or } ArC{=}CX_2 \xrightarrow[45°C]{\text{Cu powder}} Ar_2C{=}C{=}C{=}CAr_2$$

Ar	X	% Yield
Ph	Br	64
Ph	Cl	48
p-ClC_6H_4	Cl	51

I.C-1 B. Beijer and H. Suhr, Justus Liebigs Ann. Chem., 1977, 1614.

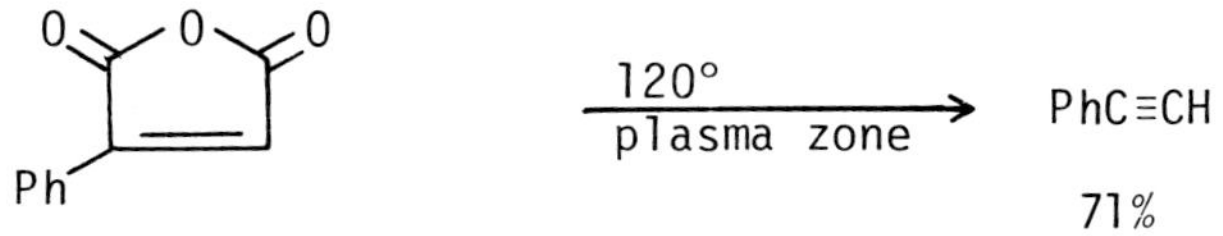

O
O
O
Ph
120°
plasma zone
PhC≡CH
71%

I.C-2 J. L. Coke, H. J. Williams, and S. Natarajan, J. Org. Chem., 42, 2380 (1977).

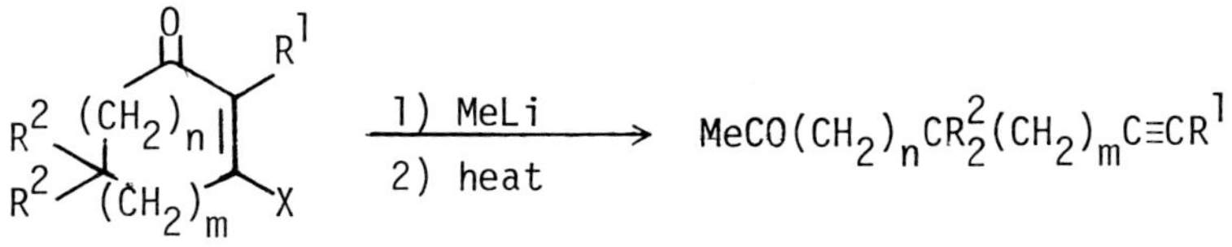

O
R1
R2
R2
(CH2)n
(CH2)m
X
1) MeLi
2) heat
MeCO(CH2)nCR22(CH2)mC≡CR1

X = Cl or Br

n	m	R^1	R^2	% Yield
1	1	H	Me	70-75
1	0	Me	H	32
1	1	Et	H	44

I.C-3 T. Nakai, K. Tanaka, H. Setoi, and N. Ishikawa, Bull. Chem. Soc. Japan, 50, 3069 (1977).

$$R^1SCH_2CF_3 \xrightarrow[Et_2O,\ -78°C]{3\ eq.\ LiNR^2_2} R^1SC\equiv CNR^2_2$$

R^1	R^2	% Yield
Ph	Et	91
Ph	i-Pr	70
Et	Et	83

I.C-4 E. W. Colvin and B. J. Hamill, J. Chem. Soc., Perkin Trans. I, 1977, 869.

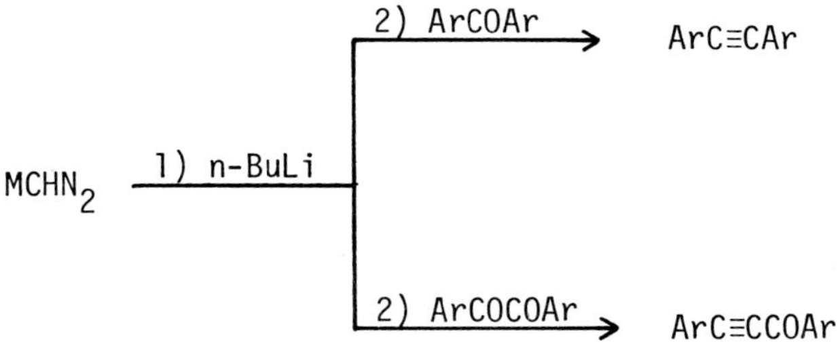

$$MCHN_2 \xrightarrow{1)\ n\text{-}BuLi} \begin{array}{l} \xrightarrow{2)\ ArCOAr} ArC\equiv CAr \\ \\ \xrightarrow{2)\ ArCOCOAr} ArC\equiv CCOAr \end{array}$$

$$M = Me_3Si,\ (RO)_2\overset{O}{\overset{\|}{P}},\ Ph_2\overset{O}{\overset{\|}{P}}$$

I.C-5 V. I. Sorokin, V. N. Drozd, N. P. Akimova, and I. I. Grandberg, J. Org. Chem. USSR, 13, 673 (1977).

$$\text{Me-cyclopropene(Me,Me)} \xrightarrow[\text{2) CuCl, Et}_2\text{O, -70°C}]{\text{1) }\underline{n}\text{-BuLi, Et}_2\text{O, 0°C}} \text{Me}_2\text{C=CMeC}\equiv\text{CCMe=CMe}_2$$

49%

I.D.1-1 F. Fedoryński, Synthesis, 1977, 783.

$$R^1R^2C=CR^3R^4 + HCBr_2Cl \xrightarrow[\text{Dibenzo-18-crown-6}]{50\% \text{ NaOH}} R^1R^2\triangle(Br,Cl)R^3R^4$$

43-76%

I.D.1-2 D. J. Burton and J. L. Hahnfeld, J. Org. Chem., 42, 828 (1977).

$$R^1R^2C=CR^3R^4 \xrightarrow[\substack{\text{THF-hexane} \\ -116°\text{C}}]{CFX_3, \underline{n}\text{-BuLi}} R^1R^2\triangle(F,X)R^3R^4$$

R^1	R^2	R^3	R^4	X	% Yield
Me	Me	Me	Me	Cl	49
Me	H	Me	Me	Cl	32
Me	Me	Me	Me	Br	73
H	H	Me	Me	Br	60
H	$-(CH_2)_4-$		H	Br	57

I.D.1-3 K. Nanjo, K. Suzuki, and M. Sekiya, Chem. Lett., 1977, 553.

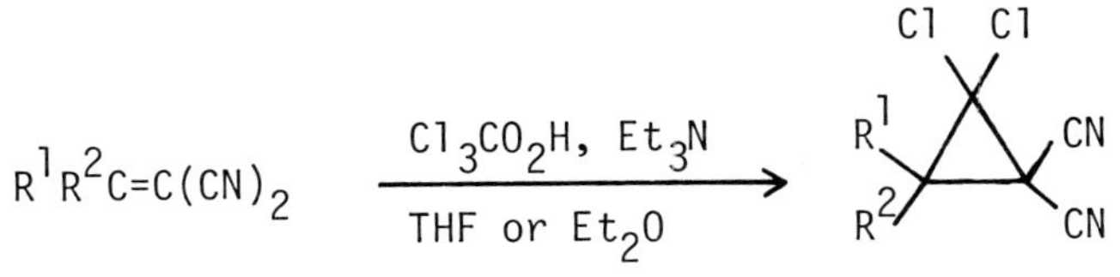
R1R2C=C(CN)2
Cl3CO2H, Et3N
THF or Et2O
Cl
Cl
R1
R2
CN
CN

R^1	R^2	% Yield
Ph	$PhCH_2$	46
Et	Et	53
H	i-Pr	44
$-(CH_2)_5-$		74

I.D.1-4 P. J. Stang and D. P. Fox, J. Org. Chem., 42, 1667 (1977).

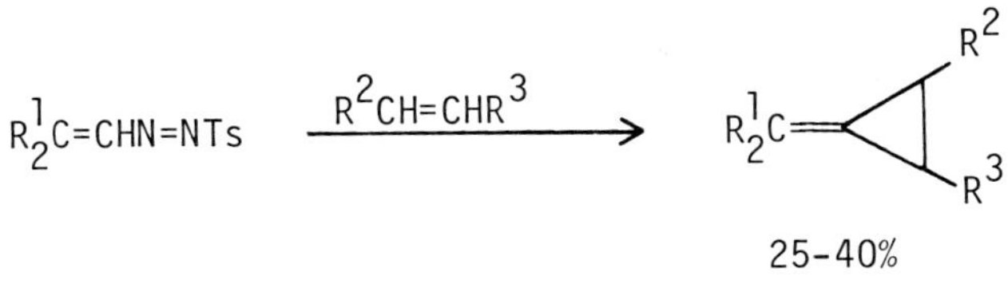
R12C=CHN=NTs
R2CH=CHR3
R12C
R2
R3
25-40%

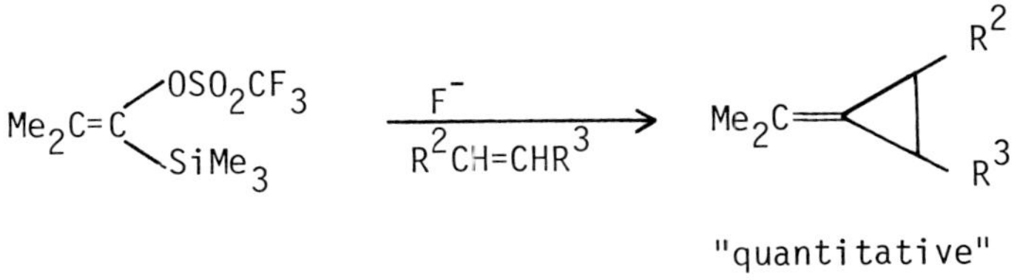
Me2C=C
OSO2CF3
SiMe3
F-
R2CH=CHR3
Me2C
R2
R3
"quantitative"

I.D.1-5 C. P. Casey and S. W. Polichnowski, J. Am. Chem. Soc., 99, 6097.

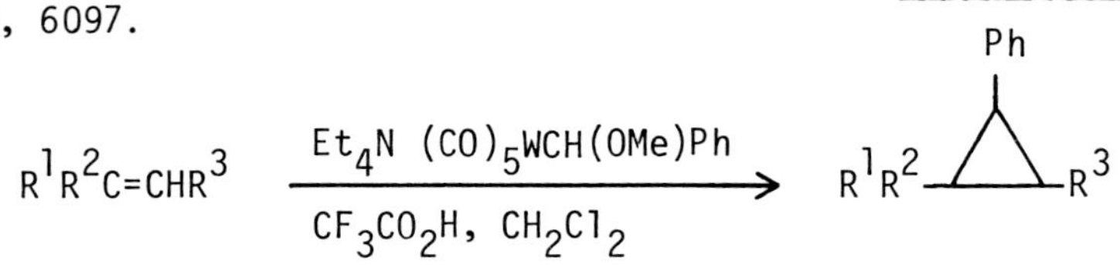

$$R^1R^2C=CHR^3 \xrightarrow[CF_3CO_2H, CH_2Cl_2]{Et_4N\ (CO)_5WCH(OMe)Ph} R^1R^2\triangle R^3\ (Ph)$$

R^1	R^2	R^3	% Yield
Me	Me	H	98
Me	H	H	80
t-Bu	H	H	69
Me	H	Me	82

I.D.1-6 S. Miyano, Y. Izumi, H. Fujii, and H. Hashimoto, Synthesis, 1977, 700.

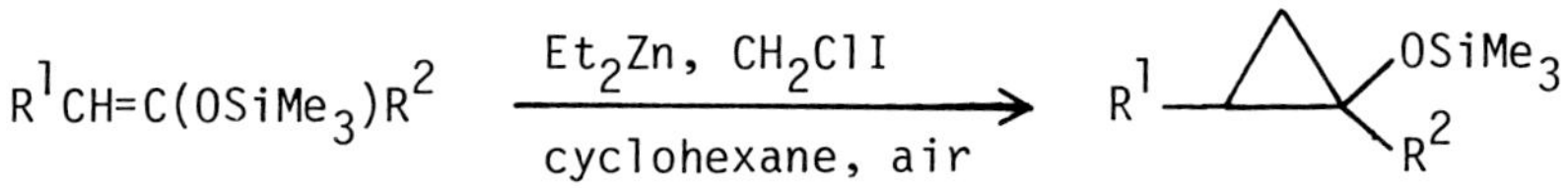

$$R^1CH=C(OSiMe_3)R^2 \xrightarrow[cyclohexane,\ air]{Et_2Zn,\ CH_2ClI} R^1\triangle{OSiMe_3,R^2}$$

R^1	R^2	% Yield
Et	H	69
Me	Et	51
Ph	H	88
-(CH$_2$)$_5$-		73

I.D.1-7 I. Ryu, S. Murai, and N. Sonoda, Tetrahedron Lett., 4611 (1977).

$$\text{(CH}_2)_n\text{-cyclohexenyl-OSiMe}_3 \xrightarrow[\text{2) CH}_2\text{I}_2]{\text{1) ZnEt}_2,\ \text{CH}_2\text{I}_2,\ \text{PhH, 20°C}} \text{(CH}_2)_n\text{-cyclopropanated-OSiMe}_3$$

n = 4, 66%

n = 3, 63%

I.D.2-1 M. Breugelmans and M.J.O. Anteunis, Bull. Soc. Chim. Belg., 86, 809 (1977).

$$R^1\text{CHBrCH}_2\text{CHBrR}^2 \xrightarrow[\text{2) NH}_4\text{Cl}]{\text{1) Na, NH}_3(\ell),\ \text{THF}} R^1\triangle R^2 + R^1(CH_2)_3R^2$$
                                                                    A   B

Overall yields were 78-91% with A usually >80% of the product.

I.D.2-2 A. A. Kamyshova, E. T. Chukovskaya, and R. K. Freidlina, Dokl. Chem. (Eng. Transl.), 233, 112 (1977).

$$Cl_3CCH_2CHClR \xrightarrow[\text{EtOH}]{Zn(HCl)} \triangle\text{-R}$$

82-87%

I.D.2-3   L. Fitjer, Synthesis, 1977, 189.

$X^1CH_2CH_2CHX^2COMe$ $\xrightarrow{\text{KF, 110°C}}_{\text{diethylene glycol}}$ △ X^2 COMe

X^1	X^2	% Yield
Br	Br	62
Br	Cl	91
Br	F	40
Cl	Cl	68

I.D.2-4   A. L. Baumstark, C. J. McCloskey, T. J. Tolson, and G. T. Syripoulas, Tetrahedron Lett., 3003 (1977).

PhCHOHCH$_2$CHOHPh $\xrightarrow[\text{reflux}]{\text{TiCl}_3\text{-LiAlH}_4}_{\text{THF or glyme}}$ Ph—△—Ph

∿ 40%

I.D.2-5   J. Masson, P. Metzner, and J. Vialle, Tetrahedron, 33, 3089 (1977).

MeCOCMe$_2$C(S)R^1 $\xrightarrow{\text{1) R}^2\text{MgBr, THF}}_{\text{2) H}_2\text{O}}$

Me, OH / Me, SR^2 / Me, R^1

R^{2}S and OH groups are cis

R^1	R^2	% Yield
Me	Me	68
Me	t-Bu	59
SMe	Et	85
SEt	i-Pr	75

I.D.2-6 R. Verhe', D. DeKimpe, L. DeBuyck, D. Courtheyn, and N. Schamp, Bull. Soc. Chim. Belg., 86, 55 (1977).

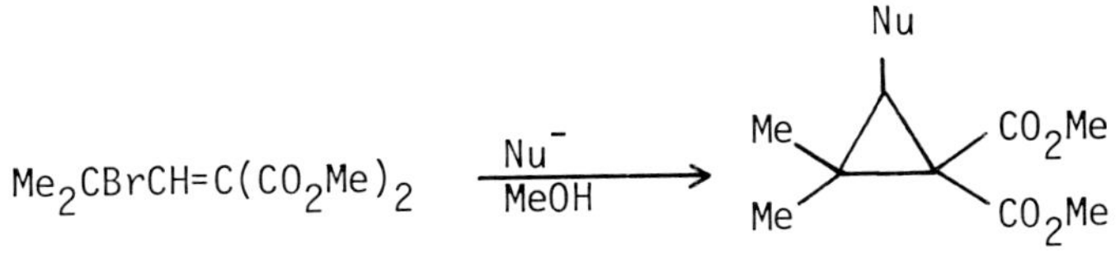

$Me_2CBrCH=C(CO_2Me)_2 \xrightarrow{Nu^-/MeOH}$ cyclopropane with Me, Me, CO_2Me, CO_2Me substituents

66-85%

Nu = MeO, CN, PhS, n-BuS, HS, H

Amines give direct substitution products.

I.D.2-7 F. Hammerschmidt and E. Zbiral, Justus Liebigs Ann. Chem., 1977, 1026.

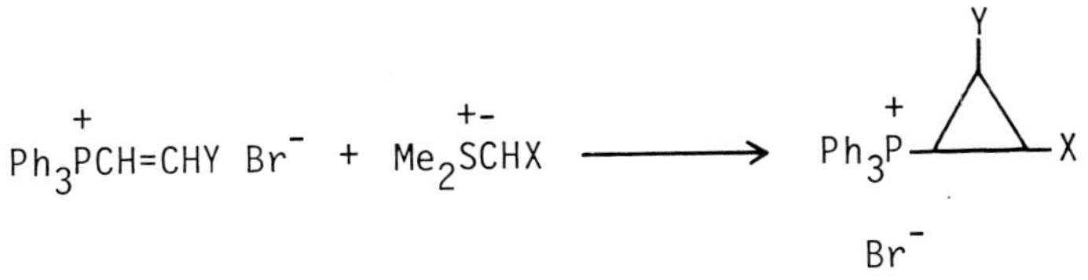

$Ph_3\overset{+}{P}CH=CHY\ Br^- + Me_2\overset{+}{S}\overset{-}{C}HX \longrightarrow Ph_3\overset{+}{P}$—cyclopropane(Y,X) Br^-

Y	X	% Yield
COPh	COPh	91
CO_2Et	COMe	82
CO_2Et	CN	66

I.D.2-8 K. Takaki and T. Agawa, J. Org. Chem., 42, 3303 (1977).

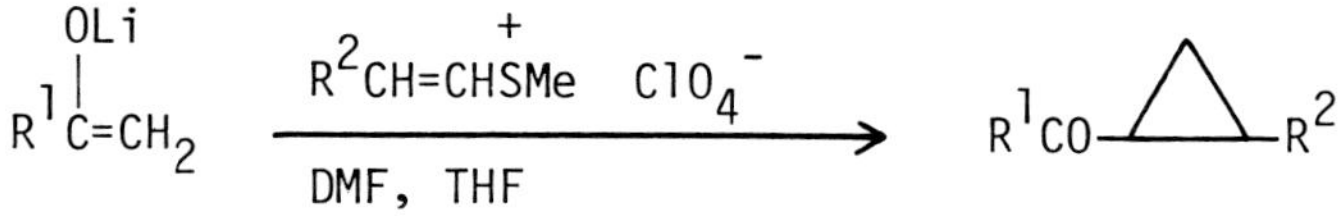

$$R^1\underset{C=CH_2}{\overset{OLi}{|}} \quad \xrightarrow[DMF, THF]{R^2CH=CHSMe \;\; ClO_4^-} \quad R^1CO\triangle R^2$$

R^1	R^2	% Yield
Ph	Ph	70
Ph	i-Pr	80
p-ClC$_6$H$_5$	i-Pr	67

I.D.2-9 N. N. Magdesieva, L. N. Ngn, and N. M. Koloskova, J. Org. Chem. USSR, 13, 928 (1977).

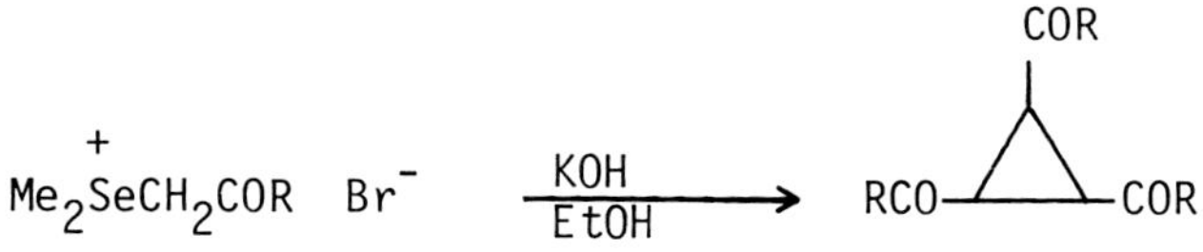

$$Me_2\overset{+}{S}eCH_2COR \;\; Br^- \xrightarrow{\text{KOH} \atop \text{EtOH}} RCO\triangle COR$$

R = 2-thienyl, 70%

$$Me_2\overset{+}{S}e\overset{-}{C}HCOAr^1 \xrightarrow{Ar^2CH=CHCOAr^3 \atop EtOH} Ar^2\triangle COAr^3 \text{ (COAr}^1\text{)}$$

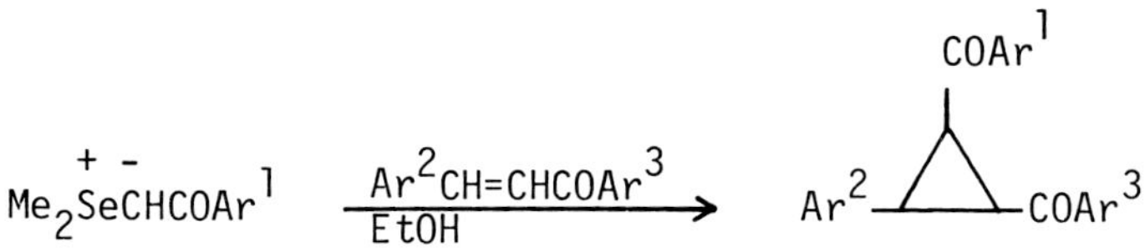

76-82%

I.D.2-10 D. A. White, Synth. Commun., 7, 559 (1977).

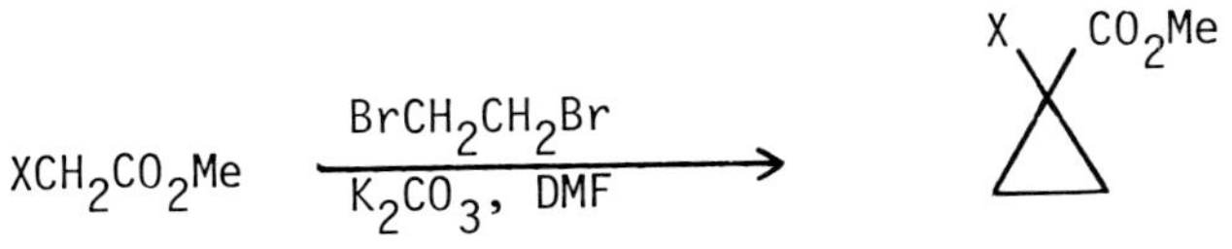

$$XCH_2CO_2Me \xrightarrow[K_2CO_3, \ DMF]{BrCH_2CH_2Br}$$ [cyclopropane with X and CO$_2$Me]

X = CO$_2$Me, 73%

X = CN, 70%

$$NCCHBrCO_2Et \xrightarrow[DMF]{K_2CO_3}$$ [cyclopropane with NC, CO$_2$Et, NC, CN, EtO$_2$C, CO$_2$Et]

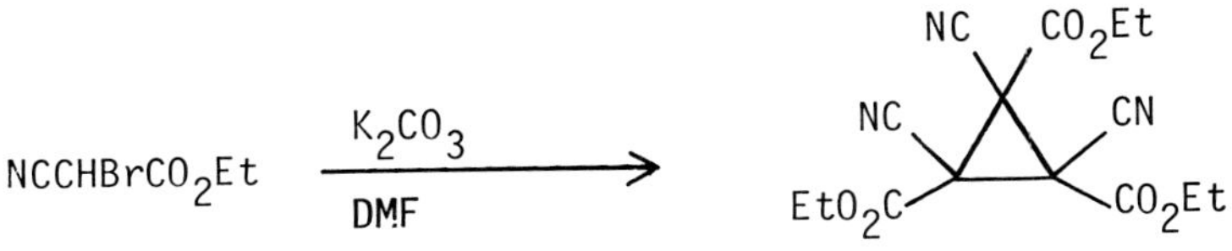

67%

I.D.2-11 H. Sakurai, T. Imai, and A. Hosomi, Tetrahedron Lett., 4045 (1977).

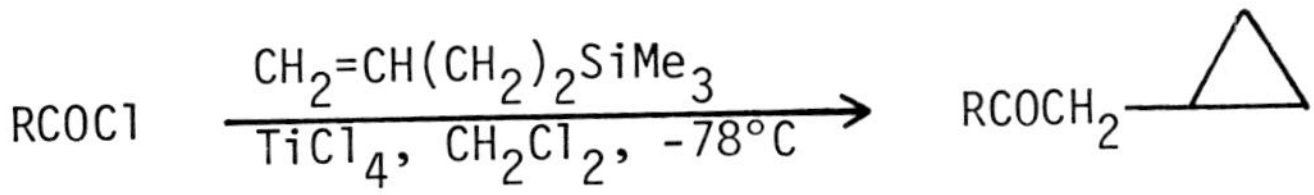

$$RCOCl \xrightarrow[TiCl_4, \ CH_2Cl_2, \ -78°C]{CH_2=CH(CH_2)_2SiMe_3} RCOCH_2-\triangle$$

38-65%

I.D.2-12 G. C. Crockett and T. H. Koch, J. Org. Chem., 42, 2721 (1977).

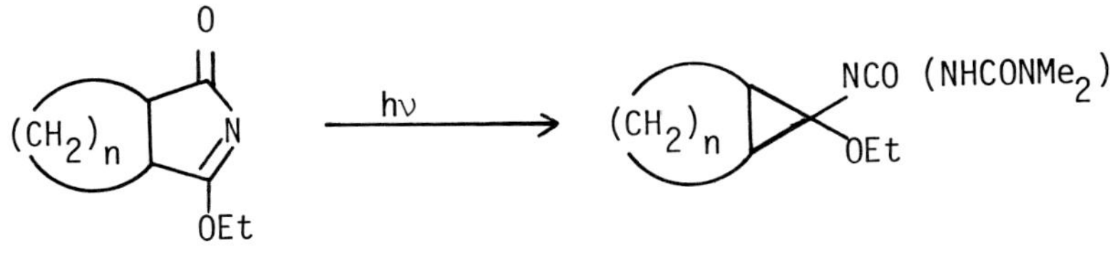

[bicyclic structure with (CH$_2$)$_n$, O, N, OEt] $\xrightarrow{h\nu}$ [bicyclic cyclopropane with (CH$_2$)$_n$, NCO (NHCONMe$_2$), OEt]

n = 2-4, 43-76%

I.D.2-13 M. Franck-Neumann and J.-J. Lohmann, <u>Angew. Chem. Int. Ed. Engl.</u>, <u>16</u>, 323 (1977).

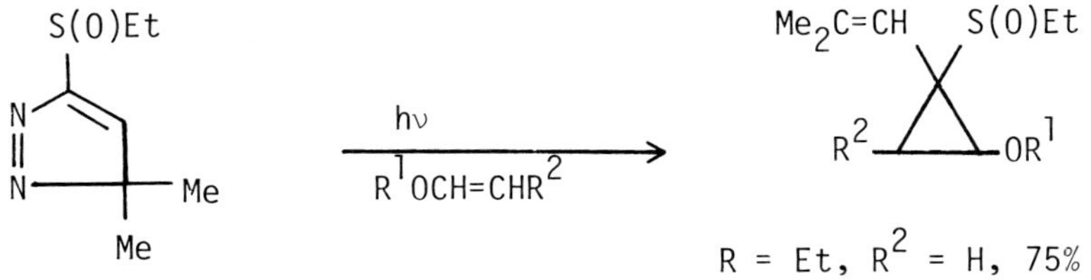

$$\text{pyrazoline}(S(O)Et, Me, Me) \xrightarrow[R^1OCH=CHR^2]{h\nu} \text{cyclopropane}(Me_2C=CH, S(O)Et, R^2, OR^1)$$

R = Et, R^2 = H, 75%

R^1, R^2 = -CH=CH-, 95%

I.D.2-14 C. Dietrich-Buchecker and M. Franck-Neumann, <u>Tetrahedron</u>, <u>33</u>, 751 (1977).

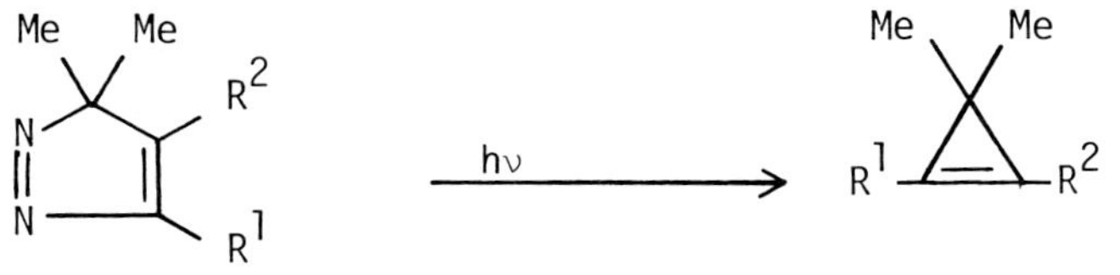

$$\text{pyrazoline}(Me, Me, R^2, R^1) \xrightarrow{h\nu} \text{cyclopropene}(Me, Me, R^1, R^2)$$

R^1	R^2	% Yield
CO_2Me	H	75
CO_2Me	CO_2Me	90
CO_2Me	Ph	87
$\underline{n}$-C_5H_{11}	CO_2Me	91

I.D.2-15 M. Ikeda, S. Matsugashita, F. Tabusa, and Y. Tamura, J. Chem. Soc., Perkin Trans. I, 1977, 1166.

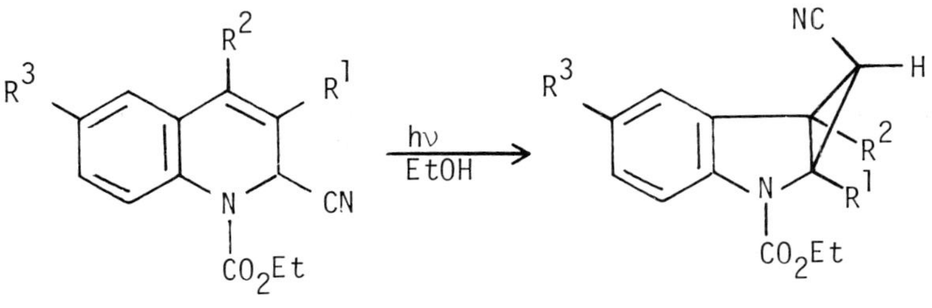

$R^1 = R^3 = H, R^2 = Me, 61\%$

$R^1 = H, R^2 = R^3 = Me, 63\%$

$R^1 = H, R^2 = Me, R^3 = OMe, 59\%$

I.E.1-1 W. G. Dauben and H. O. Krabbenhoft, J. Org. Chem., 42, 282 (1977).

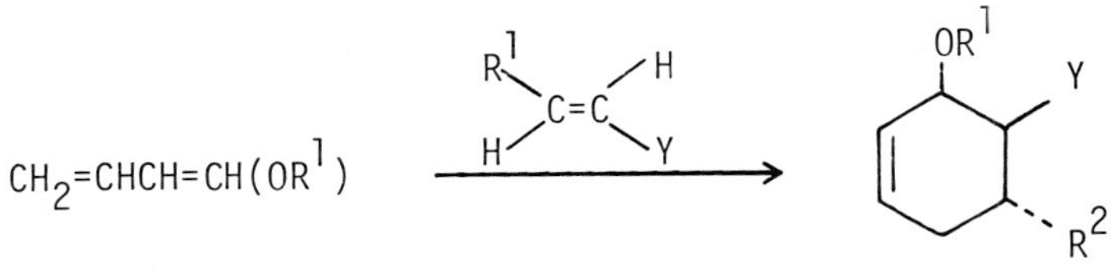

A study of the Diels-Alder reaction of various systems under high pressure (15,000 atm) was carried out.

I.E.1-2 S. Tanimoto, Y. Matsumura, T. Sugimoto, and M. Okano, Tetrahedron Lett., 2899 (1977); see also: S. Tanimoto, ibid, 2903 (1977).

$$CH_2=CHCH=CH_2X \xrightarrow{YC\equiv CZ, \; Et_2O, \; reflux}$$ [benzene ring with Y and X substituents ortho]

X	Y	Z	% Yield
NEt_2	CO_2Et	CO_2Et	70
NEt_2	H	CO_2Me	31
OEt	CO_2Me	CO_2Me	44

I.E.1-3 S. Danishevsky, R. K. Singh, and T. Harayama, J. Am. Chem. Soc., 99, 5810 (1977); see also: S. Danishefsky, C. F. Yan, and P. M. McCurry, Jr., J. Org. Chem., 42, 1819 (1977).

[cyclopentenone/cyclohexenone with S(O)Ph substituent and $(CH_2)_n$ tether]
1) $CH_2=C(OTMS)CH=CHOMe$, heat
2) H^+
→ [hydroxy-substituted bicyclic ketone with $(CH_2)_n$]

n = 1, 68%
n = 2, 80%

$MeO_2CCMe=CHS(O)Ph$
1) $CH_2=C(OTMS)CH=CHOMe$, heat
2) H^+
→ [cyclohexadienone with Me and CO_2Me substituents]

83%

I.E.1-4 R. Bonjouklian and R. A. Ruden, J. Org. Chem., 42, 4095 (1977).

$R^1CH=CR^2CR^3=CHR^4$ $\xrightarrow[\text{MeCN, 145°C}]{\overset{+}{CH_2}=CHPPh_3 \ Br^-}$ [cyclohexene product with R^1, R^2, R^3, R^4 substituents and $\overset{+}{P}Ph_3 \ Br^-$ group]

90-96%

By reaction of the phosphorane from the product with an aldehyde one has an overall allene equivalent in the Diels-Alder reaction using vinyltriphenylphosphonium bromide.

I.E.1-5 G. Pitacco, A. Risaliti, M. L. Trevisan, and E. Valentin, Tetrahedron, 33, 3145 (1977).

[morpholine enamine of 1-acetyl-cyclohexene] $\xrightarrow{PhCH=CHNO_2}$ [octahydronaphthalene with Ph, NO_2, and morpholine-N substituents]

44%

$\xrightarrow{H_3O^+}$ [decahydronaphthalene with Ph, O_2N, H substituents]

Both products arise from kinetic control.

I.E.1-6 Y. Wakatsuki and H. Yamazaki, J. Organomet. Chem., 139, 169 (1977).

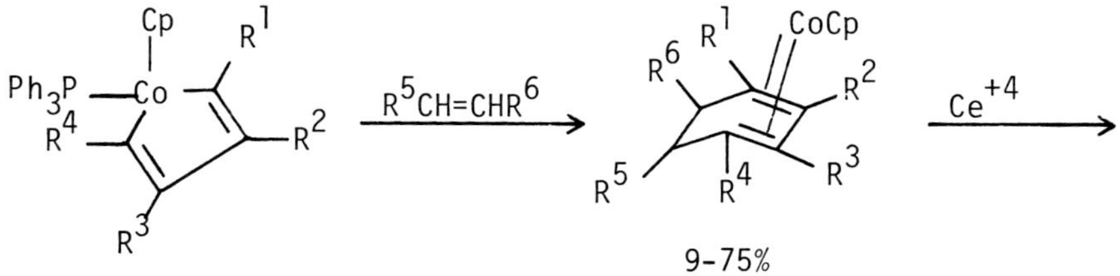

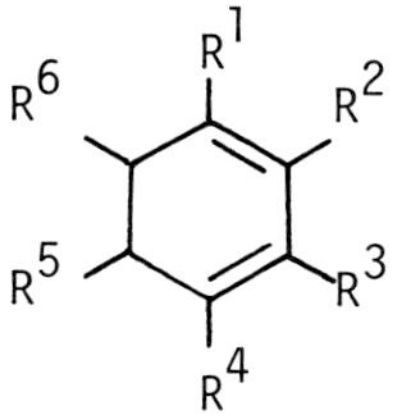

I.E.1-7 W. B. Manning, J. E. Tomaszewski, G. M. Muschik, and R. I. Sato, J. Org. Chem., 42, 3465 (1977); see also: J. E. Tomaszewski, W. B. Manning, and G. M. Muschik, Tetrahedron Lett., 971 (1977).

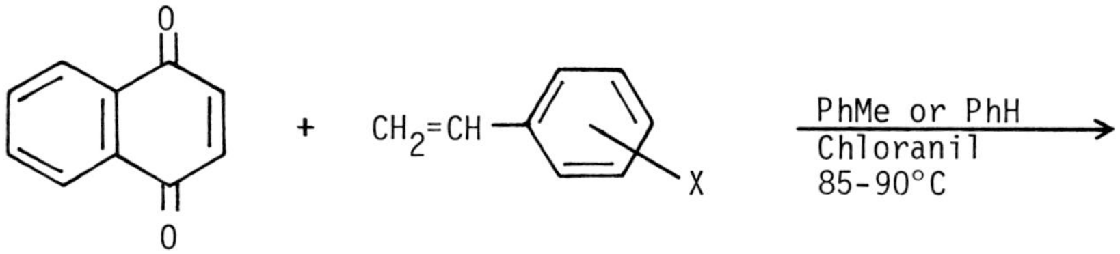

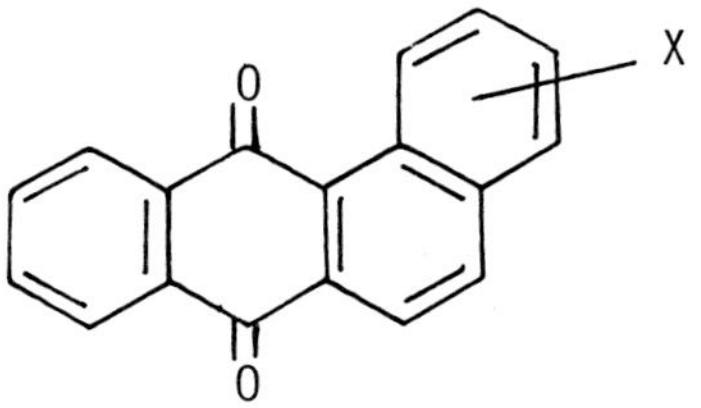

I.E.1-8 B. I. Rosen and W. P. Weber, J. Org. Chem., 42, 3465 (1977).

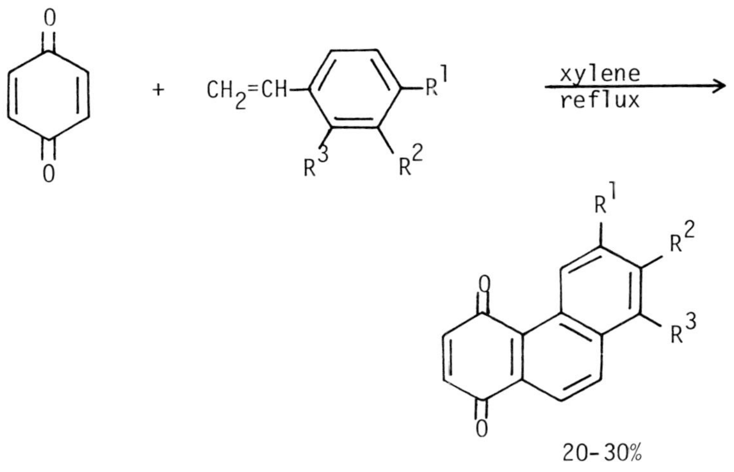
xylene
reflux
20-30%

I.E.1-9 A. P. Kozikowski, W. C. Floyd, and M. P. Kuniak, J. Chem. Soc., Chem. Commun., 1977, 582.

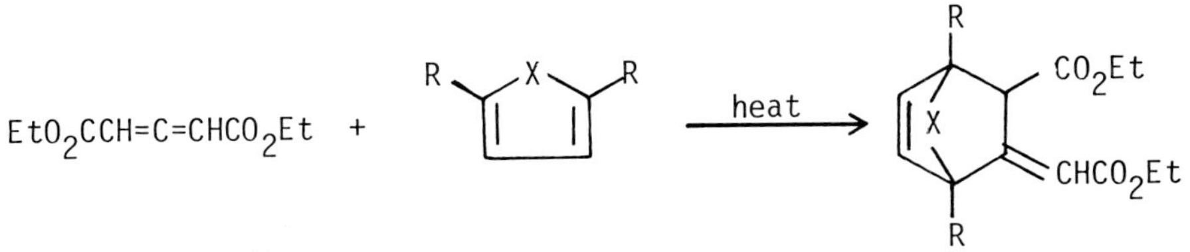
$EtO_2CCH{=}C{=}CHCO_2Et$ +
heat
CO_2Et
$CHCO_2Et$

R	X	Temp. (°C)	% Yield
H	O	40	87
Me	O	40	64
H	NCO_2Me	80	60
Me	NCO_2Me	80	66
H	CH_2	80	80

I.E.1-10 M. M. Radcliffe and W. P. Weber, J. Org. Chem., **42**, 297 (1977).

[Reaction scheme: ortho-disubstituted benzene with $CR^1=CHCH=CH_2$ substituent, X^1 and X^2 substituents, heated at ~460°C to give a bicyclic (dihydronaphthalene) product with X^2, X^1, and R substituents.]

R	X^1	X^2	% Yield
H	OMe	H	68
H	H	OMe	26
Me	H	OMe	96

I.E.1-11 L. G. Kozar, R. D. Clark, and C. H. Heathcock, J. Org. Chem., **42**, 1386 (1977).

[Reaction scheme: cyclohexenol with Me and $CH_2C\equiv CMe$ substituents, treated with HCO_2H at 25°C to give bicyclic enone formate with Me, Me, R substituents.]

R = H or Me, 82-84%

[Reaction scheme: cyclohexene with Me_2SiO, Me, and $CH_2C\equiv CSiMe_3$ substituents; 1) HCO_2H, 25°C; 2) KOH, MeOH, H_2O — gives bicyclic ketone with Me substituent.]

> 76%

I.E.1-12 S. Sarel and M. Langbeheim, J. Chem. Soc., Chem. Commun., 1977, 593.

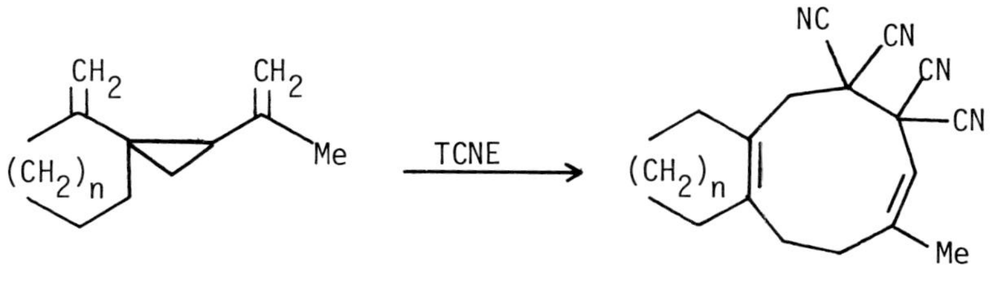

n = 1, 2, or 3, 88%

I.E.1-13 R. N. Warrener, R. A. Russell, and T. S. Lee, Tetrahedron Lett., 49 (1977); see also: W. R. Dolbier, Jr., K. Matsui, J. Michl, and D. V. Horak, J. Am. Chem. Soc., 99, 3876 (1977).

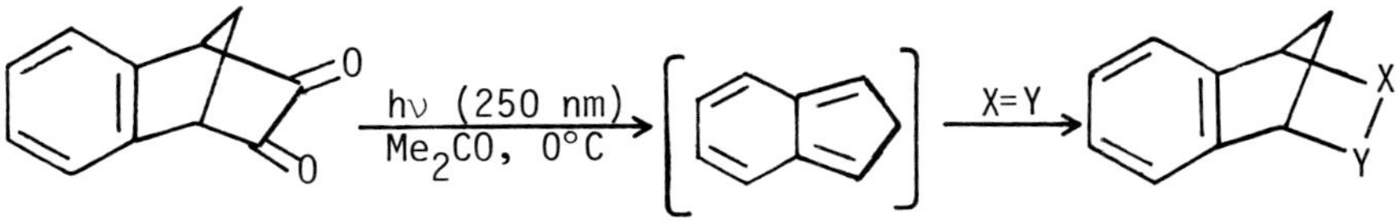

The intermediate was trapped with N-methyl maleimide or dimethyl azodicarboxylate; If no trapping agent is present then a 70% yield of indene is obtained.

I.E.1-14 K. Yokoyama, M. Kato, and R. Noyori, Bull. Chem. Soc. Japan, 50, 2201 (1977).

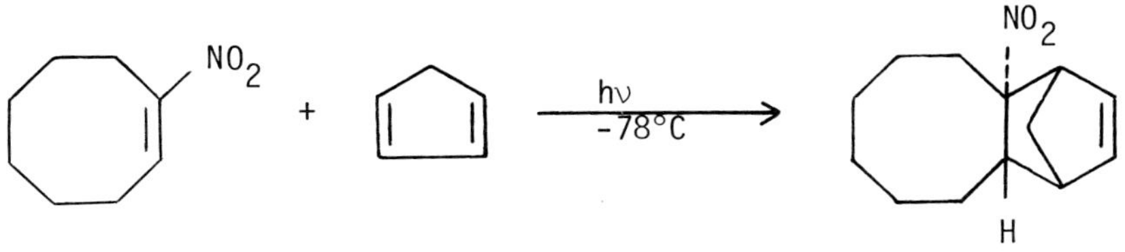

40%

I.E.1-15 H. E. Zimmerman and S. M. Aasen, J. Am. Chem. Soc., 99, 2342 (1977); see also: A. Padwa, T. J. Blacklock, D. Getman, and N. Hatanaka, ibid, 2344 (1977); H. E. Zimmerman, Pure and Appl. Chem., 49, 389 (1977).

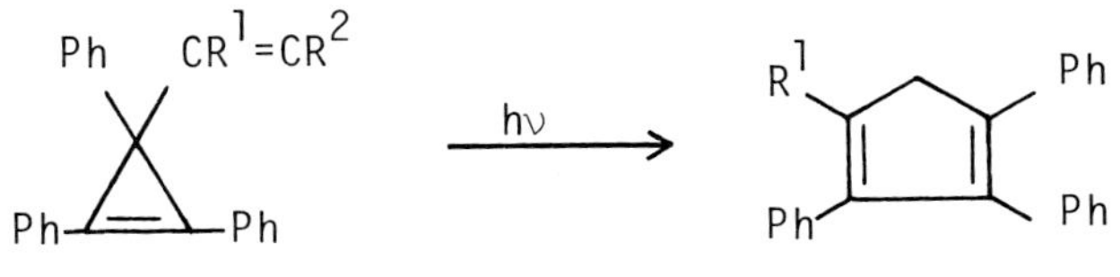

$$\text{Ph} \quad CR^1=CR^2 \xrightarrow{h\nu} $$

$R^1 = H$, 87%

I.E.1-16 W. Oppolzer, Angew. Chem. Int. Ed. Engl., 16, 10 (1977).

Review: "Intramolecular [4 + 2] and [3 + 2] Cycloadditions in Organic Synthesis."

I.E.2-1 G. B. Bennett, Synthesis, 1977, 589.

Review: "The Claisen Rearrangement in Organic Synthesis, 1967 to January 1977."

I.E.2-2 F. E. Ziegler, Acc. Chem. Res., 10, 227 (1977).

Review: "Stereo- and Regiochemistry of the Claisen Rearrangement: Applications to Natural Product Synthesis."

I.E.2-3 S. Jolidon and H.-J. Hansen, Helv. Chim. Acta., 60, 978 (1977).

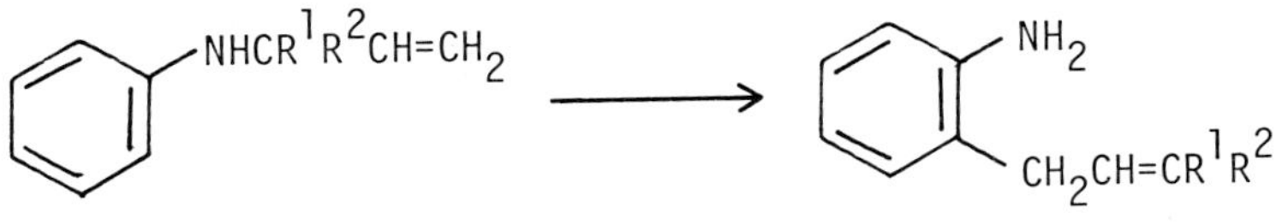

Ar-NHCR^{1}R^{2}CH=CH$_2$ ⟶ Ar(NH$_2$)-CH$_2$CH=CR^{1}R^2

A thorough study of the amino-Claisen rearrangement is reported.

I.E.2-4 R. M. Coates and I. M. Said, J. Am. Chem. Soc., 99, 2355 (1977).

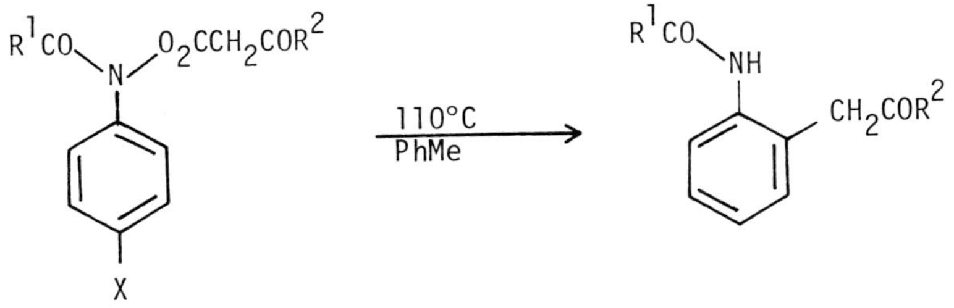

R^{1}CO-N(O$_2$CCH$_2$COR^2)-C$_6$H$_4$-X $\xrightarrow{\text{110°C, PhMe}}$ R^{1}CO-NH-C$_6$H$_4$-CH$_2$COR^2

R^1	R^2	X	% Yield
Me	Me	H	55
Ph	Me	Me	61
Me	Ph	Me	>50
Me	OEt	H	60
Ph	OEt	Me	80

I.E.2-5 W. C. Still and M. J. Schneider, J. Am. Chem. Soc., 99, 948 (1977).

R-cyclohexene with O$_2$CCH$_2$CH$_2$-N(pyrrolidine) substituent

1) LDA, THF, -45°C
2) Me$_3$SiCl
3) THF or PhMe, reflux
4) Me$_2$SO$_4$, MeOH, K$_2$CO$_3$

→ R-cyclohexene-CH(=CH$_2$)-CO$_2$Me

65-75%

I.E.2-6 J.L.C. Kachinski and R. G. Salomon, Tetrahedron Lett., 3235 (1977).

$$R^1COCH_2OCR^2R^3CR^4{=}CR^5R^6 \xrightarrow{\text{Me}_3\text{SiCl, Et}_3\text{N}}_{\text{DMF}} R^1\underset{\text{OSiMe}_3}{\overset{\text{CHO}}{C}}CR^5R^6CR^4{=}CR^2R^3$$

R^1	R^2	R^3	R^4	R^5	R^6	% Yield
Ph	H	H	H	H	H	70
Me	H	H	H	Me	Me	90
Ph	Me	Me	H	H	H	99
Me	H	H	-(CH$_2$)$_3$-		H	88

Product can be oxidatively cleaved to give allylic ketones.

I.E.2-7 V. A. Isidorov, B. V. Ioffe, and I. G. Zenkevich, Dokl. Chem. (Eng. Transl.), 230 584 (1976).

$$RMeC=NNMeCH_2CH=CH_2 \xrightarrow{140°-200°C} CH_2=CHCH_2CMeRN=NMe$$

R = Me, 94%
R = Et, 80%

I.E.2-8 E. Schaumann and F.-F. Grabley, Tetrahedron Lett., 4307 (1977).

$$R^1R^2C=CHCHR^3Br \xrightarrow[2) \ 50°C]{1) \ Me_2SiC\equiv CS^-} R^3CH=CHCR^1R^2C(SiMe_3)=C=S$$

R^1	R^2	R^3	% Yield
Me	H	H	57
Me	Me	H	80
H	$-(CH_2)_3-$		48

I.E.2-9 G. B. Gill and B. Wallace, J. Chem. Soc., Chem. Commun., 1977, 380.

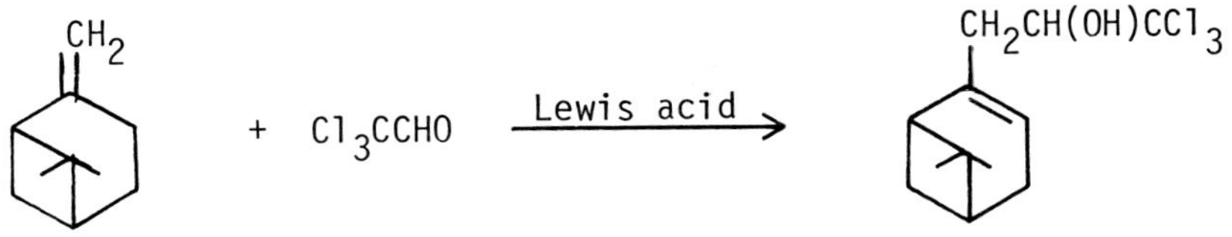

A study of the effect of various Lewis acids on the ene reaction was carried out.

I.E.2-10   B. Cazes and S. Julia, Bull. Soc. Chim. Fr., 1977, 925;
see also: ibid, 931.

$$R^1CH(CN)OCH_2CH=CR^2R^3 \xrightarrow[-78°C]{LDA, THF} R^1COCR^2R^3CH=CH_2$$

R^1	R^2	R^3	% Yield
Me	Me	Me	36
i-Pr	Me	Me	75
i-Pr	Me	$(CH_2)_2CH=CMe_2$	58
i-Pr	H	H	76 (9:1 mixture of β,γ- and α,β-)

I.E.2-11   M. T. Reetz, Chem. Ber., 110, 954 (1977); for fluoride
ion catalyzed rearrangements at 20°C, see: M. T. Reetz and N.
Greif, Chem. Ber., 110, 2958 (1977).

$$Me_3SiCR^1R^2OCH_2CH=CH_2 \xrightarrow{150°-190°C} CH_2=CHCH_2CR^1R^2OSiMe_3$$
$$97-99\%$$

I.E.2-12   J. M. Reuter and R. G. Salomon, J. Org. Chem., 42, 3360
(1977).

$$R^1R^2C=CR^3CR^4R^5OCH_2CH=CH_2 \xrightarrow[200°]{(Ph_3)P_3RuCl_2} R^4R^5C=CR^3CR^1R^2CHMeCHO$$

R^1	R^2	R^3	R^4	R^5	% Yield
H	H	Me	H	H	92
H	H	H	Me	Me	80
H	H	H	$-(CH_2)_5-$		78

CARBON – CARBON BOND FORMING REACTIONS

I.F.1-1   J. O. Morley, *J. Chem. Soc., Perkin Trans. II*, **1977**, 601.

$$ArH + PhCOCl \xrightarrow[150°C]{MO_x} ArCOPh$$

Ar	MO_x	% Yield
$3,4\text{-}Me_2C_6H_3$	ZnO	88
$2,4\text{-}Me_2C_6H_3$	Fe_2O_3	89
$2,4\text{-}Me_2C_6H_3$	SnO	91
$4\text{-}MeO\text{-}C_6H_5$	Fe_2O_3	85
$4\text{-}I\text{-}C_6H_5$	Fe_2O_3	76

I.F.1-2   J. O. Morley, *Synthesis*, **1977**, 54.

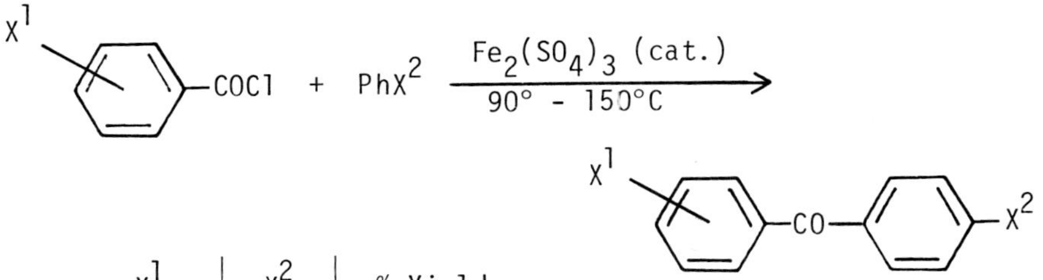

$$X^1\text{-}C_6H_{4-n}\text{-}COCl + PhX^2 \xrightarrow[90°-150°C]{Fe_2(SO_4)_3 \text{ (cat.)}} X^1\text{-}C_6H_4\text{-}CO\text{-}C_6H_4\text{-}X^2$$

X^1	X^2	% Yield
H	F	78
H	Br	78
$4\text{-}O_2N$	Cl	79
4-Br	Br	63

I.F.1-3 T. Keumi, H. Saga, R. Taniguchi, and H. Kitajima, Chem. Lett., 1977, 1099.

$$ArH + RCO_2H \xrightarrow[TFA]{\text{2-Py-OSO}_2CF_3} ArCOR$$

good yields

I.F.1-4 W. Ertel and K. Friedrich, Chem. Ber., 110, 86 (1977).

$$ArH + ClCH=CXY \xrightarrow[ClCH_2CH_2Cl]{AlCl_3} ArCH=CXY$$

Ar	X	Y	% Yield
Ph	CN	CO_2Et	30
4-MeOC$_6$H$_4$	CN	CO_2Et	60
4-Me$_2$NC$_6$H$_4$	CN	CO_2Et	62
Ph	CO_2Et	CO_2Et	44
4-MeC$_6$H$_4$	CO_2Et	CO_2Et	50

I.F.1-5 P. J. Stang and A. G. Anderson, Tetrahedron Lett., 1485 (1977).

$$R_2^2C=CR^1OSO_2CF_3 \xrightarrow[\text{4-Me-2,6-}(\underline{t}\text{-Bu})_2C_5H_2N]{ArH, \Delta} R_2^2C=CR^1Ar$$

R^1	R^2	Ar	% Yield
Ph	Ph	Ph	54
Ph	Me	ClC$_6$H$_9$	80
Ph	Me	MeOC$_6$H$_4$	79

I.F.1-6 D. Ben-Ishai and Z. Bernstein, Tetrahedron, 33, 3261 (1977).

$$R^1CH(OR^2)CO_2Me \xrightarrow[MeSO_3H]{PhH} PhCH(OR^2)CO_2Me + Ph_2CHCO_2Me$$
$$ A B$$

R^1	R^2	% Yield	A:B
MeO	Me	80	77:23
HO	Me	98	63:27
Ph	H	89	-
HO	H	91	35:65
Me$_2$NOCNH	Me	80	~11:1 (~15% was PhCH(NHCONMe$_2$)CO$_2$Me)

I.F.1-7 A. P. Krysin, N. V. Bodoev, and V. A. Kptyug, J. Org. Chem. USSR, 13, 1183 (1977).

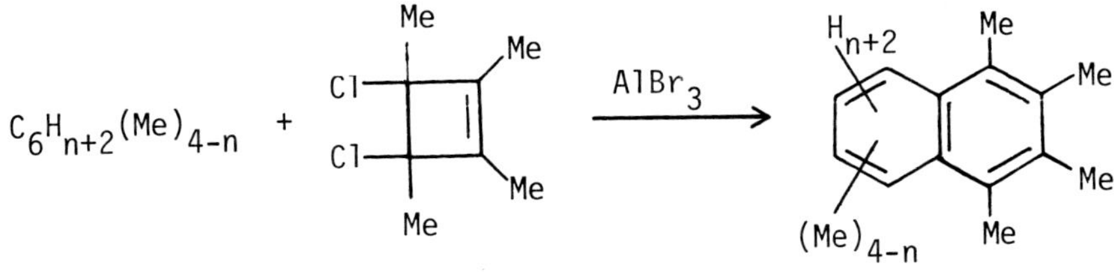

$$C_6H_{n+2}(Me)_{4-n} + \underset{\text{(dichlorotetramethylcyclobutene)}}{\text{Cl}_2\text{C}_4\text{Me}_4} \xrightarrow{AlBr_3} \text{(polymethylnaphthalene)}$$

18-60%

I.F.1-8 D. A. Evans, P. A. Cain, and R. Y. Wong, J. Am. Chem. Soc., 99, 7083 (1977).

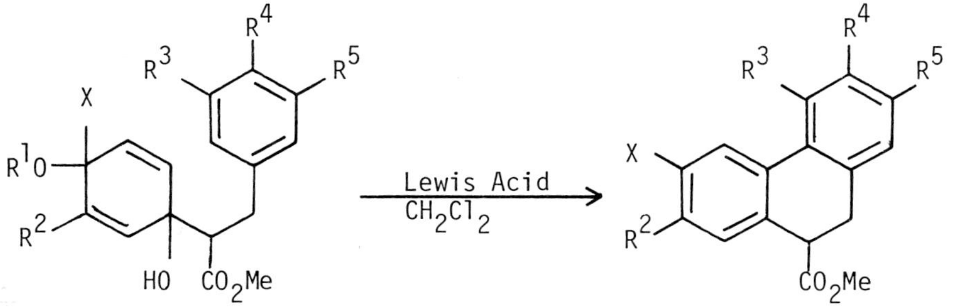

X	R^1	R^2	R^3	R^4	R^5	Lewis acid	% Yield
MeO	Me	H	MeO	MeO	MeO	P_2O_5-$MeSO_3H$	77
CN	Me_3Si	H	MeO	MeO	MeO	$SnCl_4$	61
MeO	Me	MeO	H	H	H	$BF_3 \cdot Et_2O$	64

I.F.2-1 G. van Koten, J.T.B.H. Jastrzebski, and J. G. Noltes, J. Chem. Soc., Chem. Commun., 1977, 203.

$$Ar_4Cu_4 \text{ or } Ar_nCu_n \xrightarrow{(CF_3SO_3)Cu} Ar\text{-}Ar$$

85-100%

Ar = $2\text{-}Me_2NCH_2C_6H_4$, $2\text{-}Me_2NC_6H_4$ or $4\text{-}MeC_6H_4$

I.F.2-2 W. Kalk, H.-S. Bien, and K.-H. Schündehütte, Justus Liebigs Ann. Chem., 1977, 329.

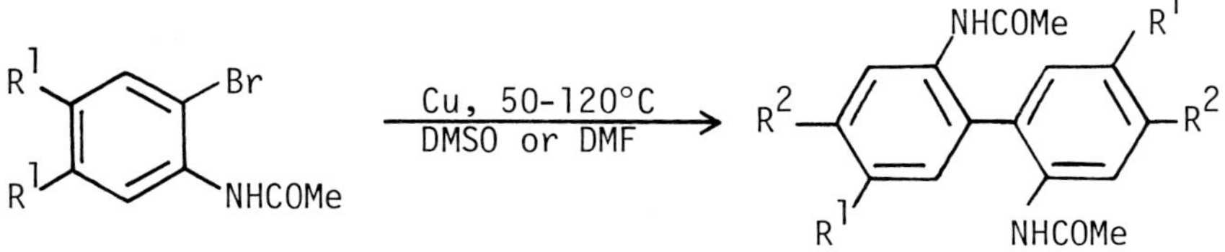

44-96%

I.F.2-3 B. Feringa and H. Wynberg, Tetrahedron Lett., 4447 (1977).

[Reaction: phenol with R^1, R^2, R^3, R^4 substituents, $Cu(NO_3)_2$, PhCHMeNH$_2$, MeOH or H$_2$O → biphenyl-diol product, 20-70%]

The reaction must be carried out in the absence of oxygen.

I.F.2-4 A. McKillop, A. G. Turrell, and E. C. Taylor, J. Org. Chem., 42, 764 (1977).

[Reaction: anisole derivative with R^1, R^2, R^3 substituents, $Tl(O_2CCF_3)_3$, CF_3CO_2H → biaryl product]

R^1	R^2	R^3	% Yield
MeO	H	Br	88
Me	Me	Me	74
MeO	H	I	89

[Reaction: naphthalene derivative with R^1, R^2 substituents, $Tl(O_2CCF_3)_3$, CF_3CO_2H → binaphthyl product]

R^1 = Me, R^2 = H, 93%
R^1 = MeO, R^2 = Br, 81%

I.F.2-5 H. Imaizumi, S. Sekiguchi, and K. Matsui, Bull. Chem. Soc. Japan, 50, 948 (1977).

$$PhNR^1R^2 \xrightarrow[150°C, O_2]{\text{molten salt} \atop (AlCl_3:NaCl:KCl = 3:1:1)} R^1R^2N-\underset{}{\bigcirc}-\underset{}{\bigcirc}-NR^1R^2$$

$R^1 = R^2 = H$, 78%

$R^1 = R^2 = Me$, 65%

$R^1 = H, R^2 = Me$, 67%

I.F.2-6 A. Sekiya and N. Ishikawa, J. Organomet. Chem., 125, 281 (1977).

$$ArI \xrightarrow[\text{2) Ar'MgBr}]{\text{1) PdCl}_2 \text{ (cat.), THF}} Ar-Ar'$$

Ar = Ph, Ar' = p-FC$_6$H$_4$, 80%

Ar = p-FC$_6$H$_4$, Ar' = Ph, 92%

The use of PdCl$_2$ as catalyst gives yields comparable to more complex Pd catalyst.

I.F.2-7 S. H. Korzeniowski, L. Blum, and G. W. Gokel, Tetrahedron Lett., 1871 (1977).

$$Y-C_6H_4N_2^+ \ BF_4^- \xrightarrow[\text{KOAc, ArH}]{\text{18-crown-6}} Y-C_6H_4-Ar$$

55-83%

I.F.2-8 T. Sheradsky and E. Nov, *J. Chem. Soc., Perkin Trans. I*, 1977, 1296.

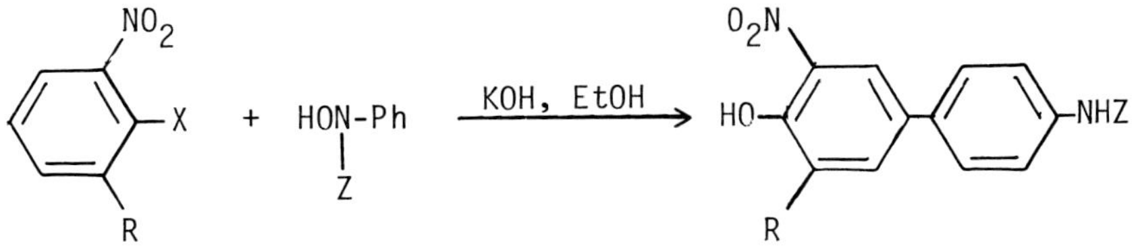

[Ar-X with NO$_2$ ortho and R] + HON-Ph(Z) →(KOH, EtOH)→ HO-Ar(O$_2$N, R)-Ar'-NHZ

Z = PhCH$_2$OCO

R = H, X = F, 75%
R = NO$_2$, X = Cl, 72%

I.F.2-9 E.-I. Negishi, A. O. King, and N. Okukado, *J. Org. Chem.*, **42**, 1821 (1977).

$$\text{RZnX + ArX'} \xrightarrow[\text{cat. Cl}_2\text{Pd(PPh}_3)_2 + \underline{i}\text{-Bu}_2\text{AlH}]{\text{cat. Ni(PPh}_3)_4 \text{ or}} \text{R-Ar}$$

RZnX	ArX'	% Yield
PhZnCl	p-MeOC$_6$H$_4$OMe	85
PhZnCl	p-NCC$_6$H$_4$Zr	90
PhZnCl	p-O$_2$NC$_6$H$_4$I	90
PhCH$_2$ZnBr	p-MeO$_2$CC$_6$H$_5$Br	85
PhCH$_2$ZnBr	p-O$_2$NC$_6$H$_4$I	88

I.F.2-10   J. F. Fauvarque and A. Jutland, *J. Organomet. Chem.*, **132**, C17 (1977).

$$BrZnCH_2CO_2Et \xrightarrow{\text{ArMX}(Ph_3P)_2 / \text{methylal/HMPA}} ArCH_2CO_2Et$$

Ar	MX	% Yield
Ph	PdX(I, Br, Cl)	43-53%
Ph	NiX(I, Br, Cl)	55-70%
p-HO$_2$CC$_6$H$_4$	PdI	85

I.F.2-11   M. F. Semmelhack and G. Clark, *J. Am. Chem. Soc.*, **99**, 1675 (1977).

$$\underset{Cr(CO)_3}{\underset{|}{\text{Ar}(R^1,R^2,R^3,R^4)}} \xrightarrow[2) I_2]{1) R^5Li} \text{Ar}(R^1,R^2,R^3,R^4,R^5)$$

R^1	R^2	R^3	R^4	R^5	% Yield
Me	H	H	H	NC(Me)$_2$C	95
MeO	H	H	H	NC(Me)$_2$C	93
MeO	H	H	MeO	NC(Me)$_2$C	92
H	H	MeO	MeO	NC(Me)$_2$C	83 (13% 1,3,4-isomer)
MeO	H	MeO	H	Bu(CN)(OR)C	>73

I.F.2-12 M. P. Doyle, B. Siegfried, R. C. Elliot, and J. F. Dellaria, Jr., J. Org. Chem., 42, 2431 (1977).

$$ArNH_2 + H_2C=CHZ \xrightarrow[CuCl_2, MeCN]{t-BuONO} ArCH_2CH(Cl)Z$$

Ar	Z	% Yield
p-O$_2$NC$_6$H$_4$	CN	93
p-MeOC$_6$H$_4$	CN	32
Ph	CN	71
Ph	Ph	53
p-MeC$_6$H$_4$	Ph	51
p-O$_2$NC$_6$H$_4$	CONH$_2$	49

I.F.2-13 M. Kosugi, K. Sasazawa, Y. Shimizu, and T. Migita, Chem. Lett., 1977, 301.

$$ArBr + CH_2=CHCH_2SnBu_3 \xrightarrow[PhH]{(Ph_3P)_4Pd} ArCH_2CH=CH_2$$

Ar	% Yield
Ph	96
p-ClC$_6$H$_4$	100
p-MeOC$_6$H$_4$	98
p-O$_2$NC$_6$H$_4$	72

I.F.2-14 T. C. Zebovitz and R. F. Heck, J. Org. Chem., 42, 3907 (1977).

$$\text{ArBr} + \text{CH}_2=\text{CHCH(OMe)}_2 \xrightarrow[\text{Et}_3\text{N, P-o-tol}_3]{\text{Pd(OAc)}_2,\ 100°\text{C}} \text{ArCH=CHCH(OMe)}_2$$

Ar = Ph, 92%

Ar = 4-Me$_2$NC$_6$H$_4$, 80%

I.F.2-15 B. A. Patel, C. B. Ziegler, N. A. Cortese, J. E. Plevyak, T. C. Zebovitz, M. Terpko, and R. F. Heck, J. Org. Chem., 42, 3903 (1977).

$$\text{ArBr} + \text{CH}_2=\text{CR}^1\text{CO}_2\text{R}^2 \xrightarrow[\text{P-o-tol}_3,\ \text{Et}_3\text{N}]{\text{Pd(OAc)}_2,\ 100°\text{C}} \text{ArCH=CR}^1\text{CO}_2\text{R}^2$$

Ar	R^1	R^2	% Yield
m-HO$_2$CC$_6$H$_4$	H	Me	67
o-MeO$_2$CC$_6$H$_4$	H	Me	69
Ph	H	H	98
Ph	Me	H	65
p-OHCC$_6$H$_4$	H	H	72

I.F.2-16 F. Akiyama, S. Teranishi, Y. Fujiwara, and H. Taniguchi, J. Organomet. Chem., 140, C7 (1977).

$$\text{PhCH=CH}_2 + \text{p-XC}_6\text{H}_4\text{NH}_2 \xrightarrow[\text{dioxane-HOAc}]{\text{Pd(OAc)}_2} \text{PhCH=CHC}_6\text{H}_4\text{X-p}$$

11-40%

I.F.2-17 A. Kasahara, T. Izumi, and N. Fukuda, Bull. Chem. Soc. Japan, 50, 551 (1977).

$$R^1CH=CHOAc + PhI \xrightarrow[\text{MeCN, 100°C}]{\text{Pd(OAc)}_2 \text{ (cat.)} \atop \text{Et}_3N, \text{Ph}_3P} PhCH=CHPh$$

R^1 = H, 52%
R^1 = Ph, 58%

I.F.2-18 G. Casiraghi, G. Casnati, G. Puglia, G. Sartori, and G. Terenghi, Synthesis, 1977, 122.

20-70%

I.F.2-19 E. Negishi and D. E. Van Horn, J. Am. Chem. Soc., 99, 3168 (1977).

$$R^1C\equiv CR^2 \xrightarrow[\text{2) ArX, Ni(PPh}_3)_4 \text{ cat.}]{\text{1) Cl(H)ZrCp}_2} \begin{array}{c} R^1 \\ \diagdown \\ H \end{array} C=C \begin{array}{c} R^2 \\ \diagup \\ Ar \end{array}$$

R^1	R^2	ArX	% Yield
$\underline{n}$-C_5H_{11}	H	PhI	96
$\underline{n}$-Bu	H	$\underline{p}$-MeOC$_6$H$_4$I	80
$\underline{n}$-Bu	H	$\underline{p}$-MeO$_2$CC$_6$H$_4$Br	92
$\underline{n}$-C_5H_{11}	Me	PhI	35

I.F.2-20 J. S. Kiely, P. Boudjouk, and L. L. Nelson, J. Org. Chem., 42, 2626 (1977).

$$HC\equiv CCH(OEt)_2 \xrightarrow[\substack{2) \text{ CuI} \\ 3) \text{ ArI}}]{1) \text{ n-BuLi, THF}} ArC\equiv CCH(OEt)_2$$

50-97%

The in situ generation of the copper acetylide increased the yield of coupled product substantially.

I.F.2-21 M. F. Semmelhack, Y. Thebtaranonth, and L. Keller, J. Am. Chem. Soc., 99, 959 (1977).

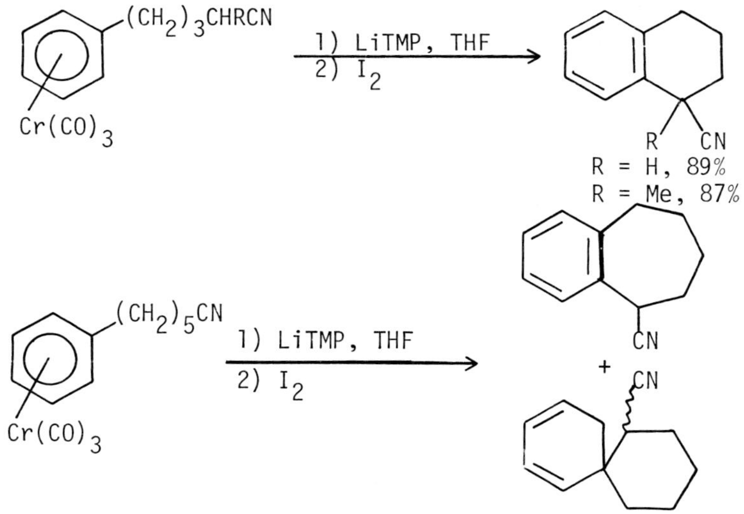

R = H, 89%
R = Me, 87%

Choice of reaction conditions allows one to favor either the fused or spiroannelated product.

I.F.2-22 C. Iwata, M. Yamada, Y. Shinoo, K. Kobayashi, and H. Okada, J. Chem. Soc., Chem. Commun., 1977, 888.

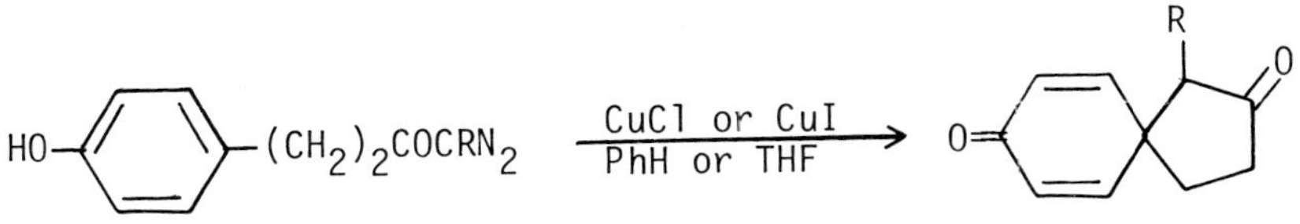

HO–⟨C₆H₄⟩–(CH$_2$)$_2$COCRN$_2$ $\xrightarrow{\text{CuCl or CuI}}{\text{PhH or THF}}$ [spiro dienone product]

R = H, 55-80%

R = Me, 60-90%

I.F.2-23 M. A. Schwartz, B. F. Rose, R. A. Holton, S. W. Scott, and B. Vishnuvajjala, J. Am. Chem. Soc., 99, 2571 (1977).

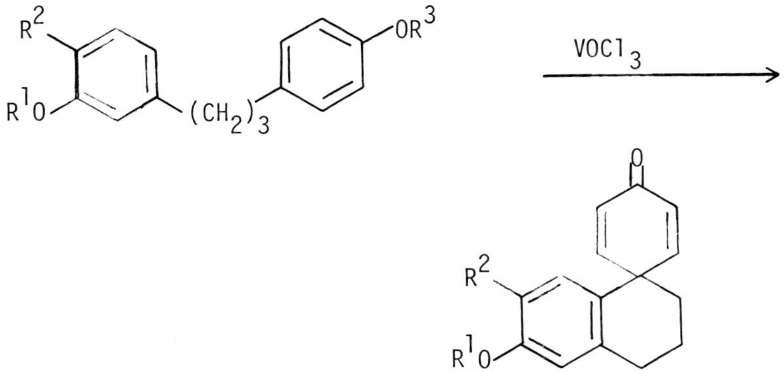

R^2–⟨Ar⟩–(CH$_2$)$_3$–⟨Ar⟩–OR^3, R^1O– $\xrightarrow{\text{VOCl}_3}$ [tricyclic spirodienone product with R^2, R^1O]

A full paper describing effect of metal on intramolecular oxidative doupling of diphenolic, monophenolic, and nonphenolic substrates is presented.

I.F.3-1 P. G. Gassman and R. L. Parton, Tetrahedron Lett., 2055 (1977); see also: P. G. Gassman and R. J. Balchunis, ibid, 2235 (1977).

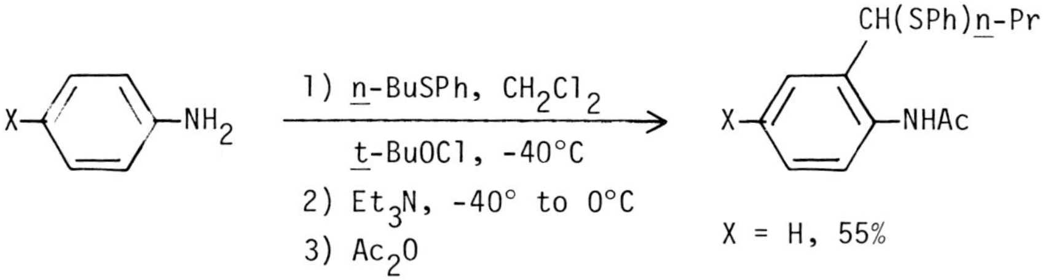

$X-C_6H_4-NH_2$

1) n-BuSPh, CH_2Cl_2, t-BuOCl, -40°C
2) Et_3N, -40° to 0°C
3) Ac_2O

→ $X-C_6H_3(CH(SPh)\underline{n}\text{-Pr})-NHAc$

X = H, 55%

X = CO_2Et, 54%

X = OMe, 40%

I.F.3-2 M. F. Semmelhack and T. M. Bargar, J. Org. Chem., **42**, 1481 (1977).

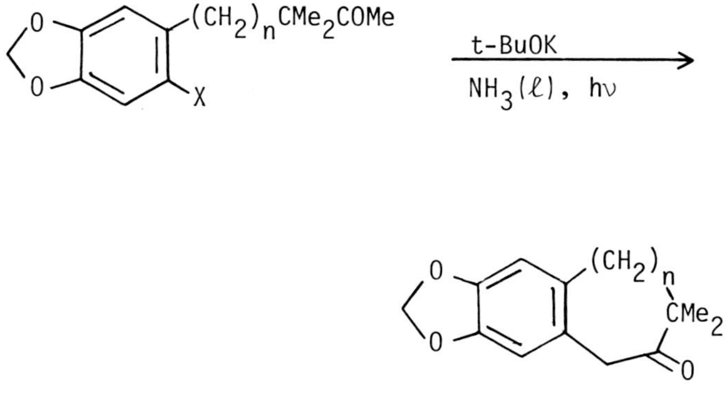

methylenedioxybenzene-$(CH_2)_n CMe_2 COMe$, X $\xrightarrow{\text{t-BuOK}}{NH_3(\ell),\ h\nu}$

methylenedioxybenzene fused ring with $(CH_2)_n$, CMe_2, C=O

n = 1, 3, or 5

99-35%

Systems which could give mixtures of ring sizes did.

I.F.3-3 M. Shamma, J. L. Moniot, L. A. Smeltz, W. A. Shores, and
L. Töke, Tetrahedron, 33, 2907 (1977).

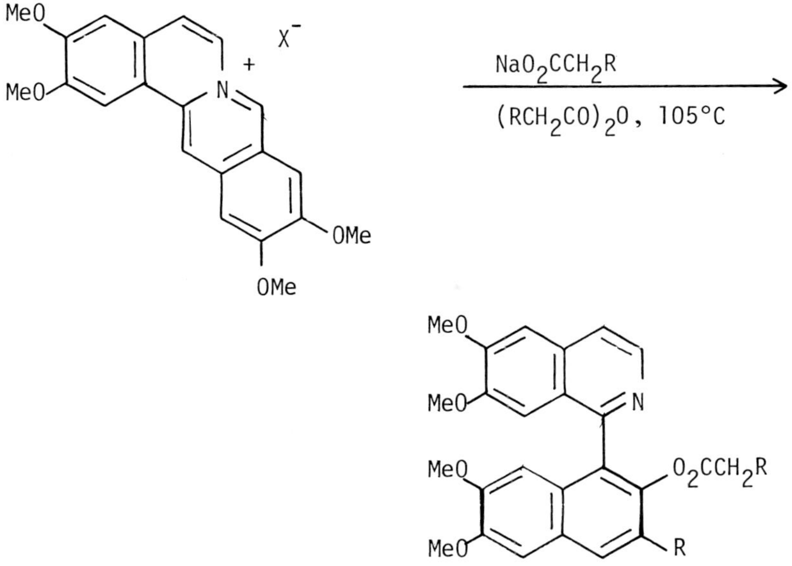

R = H, 70%
R = Me, 73%

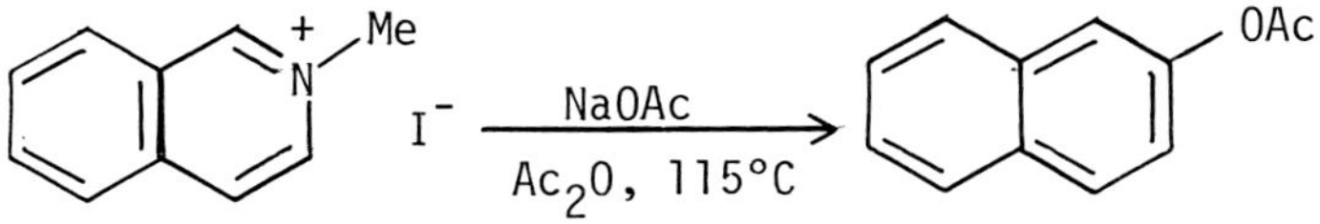

15%

I.G.1-1 P. Jacob, III and H. C. Brown, J. Org. Chem., 42, 579 (1977).

$$R^1C \equiv CR^2 \xrightarrow[\substack{2) \ R^3CHO, \ reflux \\ 3) \ H_2O_2, \ NaOH}]{1) \ 9\text{-}BBN, \ THF} \underset{H}{\overset{R^1}{>}}C=C\underset{CH(OH)R^3}{\overset{R^2}{<}}$$

R^1	R^2	R^3	% Yield
Et	H	H	48
n-Bu	H	Ph	86
t-Bu	Me	Me	69
$Cl(CH_2)_3$	H	Et	47

I.G.1-2 G. W. Kramer and H. C. Brown, J. Org. Chem., 42, 2292 (1977).

$$\text{\textcircled{C}}BCH_2CH=CHR^1 \xrightarrow[2) \ HOCH_2CH_2NH_2]{1) \ R^2COR^3} CH_2=CHCHR^1CR^2R^3OH$$

R^1	R^2	R^3	% Yield
H	H	t-Bu	97
H	Me	Ph	101
Me	Ph	H	87
H	MeCH=CH	Me	96

$$\text{\textcircled{C}}B\text{-}CH_2CH=CHR^1 \xrightarrow[\substack{2) \ i\text{-}PrOLi \\ 3) \ HOCH_2CH_2NH_2}]{1) \ R^2COX} HOCR^2(CHR^1CH=CH_2)_2$$

50-97%

I.G.1-3 K. Utimoto, K. Uchida, and H. Nozaki, Tetrahedron, 33, 1949 (1977).

$$R^1_3B \xrightarrow[\substack{2)\ R^2CHO \\ 3)\ H_2O_2,\ {}^-OH}]{1)\ CH_2=CHLi} R^1CH(OH)CH_2CH(OH)R^2$$

R^1	R^2	% Yield
Et	Ph	78
n-Bu	H	80
i-Pr	Me	52

I.G.1-4 J. A. Sinclair, G. A. Molander, and H. C. Brown, J. Am. Chem. Soc., 99, 954 (1977).

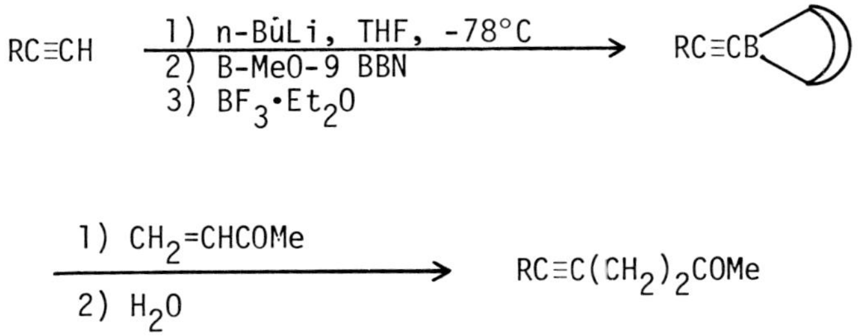

$$RC\equiv CH \xrightarrow[\substack{2)\ B-MeO-9\ BBN \\ 3)\ BF_3\cdot Et_2O}]{1)\ n-BuLi,\ THF,\ -78°C} RC\equiv CB\bigcirc$$

$$\xrightarrow[2)\ H_2O]{1)\ CH_2=CHCOMe} RC\equiv C(CH_2)_2COMe$$

R	% Yield
n-Bu	96
Ph	80
Cl(CH$_2$)$_3$	87
cyclohexyl	97

Other enones were also studied.

I.G.1-5 G. A. Molander and H. C. Brown, J. Org. Chem., 42, 3106 (1977).

$$R^1C\equiv C-B\bigcirc \cdot THF \xrightarrow[2)\ H_2O_2,\ NaOH]{1)\ R^2C(OMe)=CHCOR^3} R^1C\equiv CR^2=CHCOR^3$$

R^1	R^2	R^3	% Yield
n-Bu	H	Me	85
Ph	H	Me	88
t-Bu	H	Me	98
t-Bu	H	t-Bu	100
t-Bu	Me	Me	17

I.G.1-6 A. Pelter and L. Hughes, J. Chem. Soc., Chem. Commun., 1977, 913.

$$LiR^1_3BC\equiv CR^2 + R^3R^4C=C(NO_2)R^5 \xrightarrow{1)\ THF} \begin{array}{c} \xrightarrow[H_2O_2]{2)\ NaOAc} R^1COCHR^2CR^3R^4CHR^5NO_2 \quad A \\ \xrightarrow{2)\ \underline{i}-PrCO_2H} R^1CH=CR^2CR^3R^4CHR^5NO_2 \quad B \end{array}$$

					% Yield	
R^1	R^2	R^3	R^4	R^5	A	B
C_6H_{13}	C_6H_{13}	H	H	H	80	-
C_6H_{13}	C_6H_{13}	H	Me	H	93	82
C_6H_{13}	C_6H_{13}	H	H	Me	57	5

I.G.1-7 N. Miyaura, N. Sasaki, M. Itoh, and A. Suzuki, <u>Tetrahedron Lett.</u>, 173 (1977).

$$R_3B \xrightarrow[\substack{\text{2) CuX} \\ \text{3) ArCOCl}}]{\text{1) MeLi}} ArCOR$$

R	Ar	% Yield
n-Pr	Ph	83
i-Bu	Ph	92
n-Pr	4-ClC$_6$H$_4$	90

I.G.1-8 N. Miyaura, N. Sasaki, M. Itoh, and A. Suzuki, <u>Tetrahedron Lett.</u>, 3369 (1977).

$$R_3B \xrightarrow[\substack{\text{2) CuX} \\ \text{3) BrCH=CHCO}_2\text{Et}}]{\text{1) MeLi}} RCH=CHCO_2Et \quad 42\text{-}98\%$$

I.G.1-9 K. Utimoto, K. Uchida, M. Yamaya, and H. Nozaki, <u>Tetrahedron</u>, <u>33</u>, 1945 (1977).

$$R^1_3B \xrightarrow[\substack{\text{2) } R^3\triangle R^2 \\ \text{3) H}_2\text{O}_2\text{, }^-\text{OH}}]{\text{1) CH}_2=\text{CHLi}} R^1CH(OH)CH_2CHR^2CH(OH)R^3$$

R^1	R^2	R^3	% Yield
n-Bu	H	H	100
n-Bu	H	Me	93
i-Pr	H	Me	100
n-Bu	-(CH$_2$)$_4$-		77

Vinyl Grignard reagents work also.

I.G.L-10 N. Miyaura, M. Itoh, and A. Suzuki, Bull. Chem. Soc. Japan, 50, 2199 (1977).

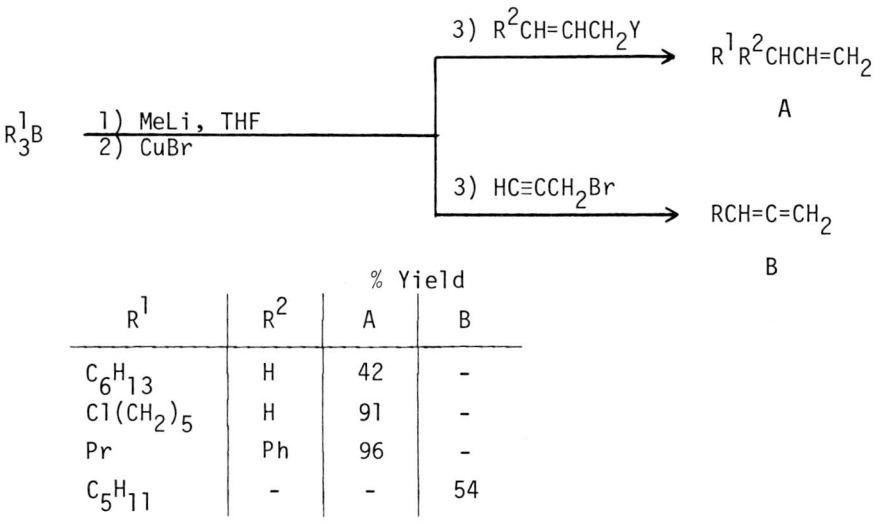

R^1	R^2	% Yield A	B
C_6H_{13}	H	42	-
$Cl(CH_2)_5$	H	91	-
Pr	Ph	96	-
C_5H_{11}	-	-	54

I.G.1-11 R. E. Merrill, J. L. Allen, A. Abramovitch, and E. Negishi, Tetrahedron Lett., 1019 (1977).

$$HC\equiv CCH_2CH_2OTs \xrightarrow[\text{2) } R_3B, -78°C]{\text{1) n-BuLi, }-78°C} \triangleright-COR$$
$$\text{3) NaOAc, } H_2O_2, -20°C \qquad 55-75\%$$

$$HC\equiv CCH_2CH_2OTs \xrightarrow[\text{2) } R_3B, -78 \text{ to } 25°C]{\text{1) n-BuLi, }-78°C} RC\equiv CCH_2CH_2OH$$
$$\text{3) NaOAc, } H_2O_2$$

I.G.1-12 K. Yamada, N. Miyaura, M. Itoh, and A. Suzuki, Synthesis, 1977, 679.

$$HC\equiv CCOR^1 \xrightarrow[\substack{2)\ R_3^2B \\ 3)\ I_2}]{1)\ LDA} R^2C\equiv CCOR^1$$

72-81%

R^1 = OEt or Ph, R^2 = alkyl

Use of LDA allows carbonyl-containing acetylenes to be utilized.

I.G.1-13 H. Yatagai, Y. Yamamoto, K. Maruyama, A. Sonoda, and S.-I. Murahashi, J. Chem. Soc., Chem. Commun., 1977, 852.

$$R^1C\equiv CH \xrightarrow[\substack{2)\ Pd(OAc)_2 \\ Et_3N,\ THF}]{1)\ R_2^2BH} \begin{array}{c} R^1 \\ \diagdown C=C \diagup H \\ H \diagup \quad \diagdown R^3 \end{array}$$

R^1	R^2	% Yield
Ph	$CHMeCHMe_2$	98
Ph	cyclohexyl	58
t-Bu	$CHMeCHMe_2$	74
$Cl(CH_2)_3$	$CHMeCHMe_2$	61

I.G.1-14 Y. Yamamoto, H. Yatagai, K. Maruyama, A. Sonoda, and S.-I. Murahashi, J. Am. Chem. Soc., 99, 5652 (1977).

$$\begin{array}{c} R^1 \diagdown \quad \diagup R^2 \\ C=C \\ H \diagup \quad \diagdown BCl \end{array} \xrightarrow[Et_2O,\ 0°C]{3\ eq.\ MeCu} \begin{array}{c} R^1 \diagdown \quad \diagup R^2 \\ C=C \diagdown \\ H \diagup \quad \quad C=C \diagup H \\ \quad R^2 \diagup \quad \diagdown R^1 \end{array}$$

64-99%

A full paper describing this reaction is presented.

I.G.1-15 Y. Takahashi, M. Tokuda, M. Itoh, and A. Suzuki, Chem. Lett., 1977, 999.

$$(R^1CH_2CHR^2)_3B + PhC{\equiv}CH \xrightarrow[\text{TBAI, THF}]{\text{Pt-Pt electrolysis}} PhC{\equiv}CCHR^2CH_2R^1$$

70-94%

All three alkyl groups of the organoborane are utilized.

I.G.1-16 M. M. Midland, J. Org. Chem., 42, 2650 (1977).

$$R^1R^2C(OAc)C{\equiv}CH \xrightarrow[\text{2) n-BuLi, -120°C to r.t.}]{\text{1) } R^3_3B, \text{ THF, Et}_2O}$$

$$R^3_2BCR^3{=}C{=}CR^1R^2 \begin{array}{c} \xrightarrow{\text{HOAc}} R^3CH{=}C{=}CR^1R^2 \\ \text{A} \\ \\ \xrightarrow{H_2O} R^3C{\equiv}CCHR^1R^2 \\ \text{B} \end{array}$$

R^1	R^2	R^3	% Yield A	B
Me	Me	i-Bu	90	-
Ph	H	sec-Bu	82	91
-(CH$_2$)$_5$-		n-Bu	82	84
-(CH$_2$)$_5$-		MeO$_2$C(CH$_2$)$_{10}$	73	-

I.G.1-17 G. Zweifel, S. J. Backlund, and T. Leung, J. Am. Chem. Soc., 99, 5192 (1977).

$$R^1CH=CR^2CH=CXBR^3R^4 \xrightarrow[\substack{2)\ CH_3CO_2H \\ 3)\ H_2O_2,\ ^-OH}]{1)\ h\nu\ or\ (\underline{sec}\text{-Bu})_3BHK} R^1CH=CR^2CH_2CHR^3OH$$

R^1	R^2	R^3	R^4	X	Conditions	% Yield
H	Me	cyclohexyl	cyclohexyl	H	$h\nu$	69
H	Me	cyclohexyl	cyclohexyl	I	$(\underline{sec}\text{-Bu})_3BHK$	77
H	Me	cyclopentyl	thexyl	H	$h\nu$	70
$-(CH_2)_4-$		cyclopentyl	thexyl	I	$(\underline{sec}\text{-Bu})_3BHK$	73

I.G.1-18 R. J. Hughes, S. Ncube, A. Pelter, K. Smith, E. Negishi, and T. Yoshida, J. Chem. Soc., Perkin Trans. I, 1977, 1172.

$$R^2_3B + LiCR^1(SPh)_2 \xrightarrow[\substack{2)\ HgCl_2 \\ 3)\ H_2O_2,\ ^-OH}]{1)\ THF} R^1R^2_2COH$$

R^1	R^2	% Yield
H	$\underline{n}\text{-}C_6H_{13}$	85
H	cyclohexyl	88
$\underline{n}\text{-Pr}$	$\underline{n}\text{-}C_8H_{17}$	90
$\underline{n}\text{-Pr}$	cyclopentyl	85

I.G.1-19 M. W. Rathke, E. Chao, and G. Wu, J. Organomet. Chem., 122, 145 (1977).

$$Cl_2CHB(O\underline{i}\text{-}Pr)_2 \xrightarrow[2)\ H_2O_2,\ pH\ 8.8]{1)\ RM} RCHO$$

RM = $\underline{n}$-BuLi, 64%

RM = PhMgBr, 70%

$$Cl_2CHB(O\underline{i}\text{-}Pr)_2 \xrightarrow[2)\ H_2O_2,\ pH\ 8.8]{1)\ 2\ eq.\ \underline{n}\text{-}BuLi} (\underline{n}\text{-}Bu)_2CHOH$$

40%

I.G.1-20 D. S. Matteson and R. J. Moody, J. Am. Chem. Soc., 99, 3196 (1977).

$$CH_2\left[B{<}^{O-}_{O-}\right]_2 \xrightarrow[2)\ R^1X]{1)\ LiTMP,\ THF} RCH\left[B{<}^{O-}_{O-}\right]_2 \xrightarrow{[OX]} R^1CHO$$

71-86%

$$\xrightarrow[\substack{2)\ R^2X \\ 3)\ [OX]}]{1)\ LiTMP,\ THF} R^1COR^2$$

66-99%

$$R^1CH\left[B{<}^{O-}_{O-}\right]_2 \xrightarrow[\substack{2)\ R^2CO_2Me \\ 3)\ H_2O}]{1)\ LiTMP,\ THF} R^1CH_2COR^2$$

CARBON—CARBON BOND FORMING REACTIONS

I.G.1-21   H. C. Brown and N. Ravindran, Synthesis, 1977, 695.

H_2BBr, SMe_2 (MBBS)

The preparation and utility of this hydroborating agent is described.

I.G.1-22   R. Köster, Pure and Appl. Chem., 49, 765 (1977).

Lecture: "Organoboranes in Synthesis and Analysis".

I.G.1-23   H. C. Brown and E. Negishi, Tetrahedron, 33, 2331 (1977).

Review: "Boraheterocycles via Cyclic Hydroboration".

I.G.1-24   G.M.L. Cragg and K. R. Koch, Chem. Soc. Rev., 6, 393 (1977).

Review: "Organoborates in Organic Synthesis: The Use of Alkenyl-, Alkynyl-, and Cyanoborates as Synthetic Intermediates".

I.G.2-1 A. Guinot, P. Cadiot, and J. L. Roustan, *J. Organomet. Chem.*, 128, C35 (1977).

$$RX \xrightarrow[\substack{2)\ CH_2=C=CH_2 \\ 3)\ H^+ \\ 4)\ Me_3NO}]{1)\ Na_2Fe(CO)_4} CH_2=CMeCOR$$

RX = EtBr, 30%

RX = $\underline{n}$-C_5H_{11}Br, 55%

I.G.2-2 M. P. Cooke, Jr. and R. M. Parlman, *J. Am. Chem. Soc.*, 99, 5222 (1977).

$$R^1X \xrightarrow[2)\ R^2CH=CR^3Z]{1)\ Na_2Fe(CO)_4} R^1COCHR^2CHR^3Z$$

R^1X	R^2	R^3	Z	% Yield
$\underline{n}$-C_6H_{13}Br	H	H	CO_2Et	92
$\underline{n}$-BuI	H	Me	CO_2Et	82
EtI	H	H	COMe	58
$\underline{n}$-BuI	Me	H	CN	87
$\underline{n}$-BuI	\multicolumn{2}{c}{-$CH_2CH_2CH_2CO$-}		94

I.G.2-3 M. Yamashita, K. Mizushima, Y. Watanabe, T. Mitsudo, and Y. Takegami, *Chem. Lett.*, 1977, 1355.

$$R^1ONa + Fe(CO)_5 \xrightarrow[2)\ R^2X]{1)\ THF-NMP} R^2CO_2R^1$$

20-80%

I.G.2-4 M. Yamashita and R. Suemitsu, J. Chem. Soc., Chem. Commun., 1977, 691.

$$R^1 \overset{X}{\underset{X}{CH}} \quad \xrightarrow[\substack{2) \text{ Fe(CO)}_5, \text{ r.t.} \\ 3) R^2I, \text{ N-methylpyrrolidone}}]{1) \text{ n-BuLi, THF, 0°C}} \quad R^1COCOR^2$$

R^1	R^2	X	% Yield
Ph	Me	S	68
Ph	Me	O	51
Ph	Et	S	65
Et	Et	S	70

I.G.2-5 S. Sarel and M. Langbeheim, J. Chem. Soc., Chem. Commun., 1977, 827.

$$\underset{n=2}{\text{(structure with (CH}_2)_n\text{ ring, cyclopropane, CH}_2\text{ and Me groups)}} \xrightarrow[h\nu]{\text{Fe(CO)}_5} \text{(bicyclic cyclooctenone with (CH}_2)_n\text{ and Me)}$$

n = 2

I.G.2-6 H. Alper and H. Des Abbayes, J. Organomet. Chem., 134, C11 (1977); see also: L. Cassar and M. Foa, ibid, C15 (1977).

$$ArCH_2Br \xrightarrow[\substack{\text{CO (atm. pressure)} \\ PhCH_2NEt_3Cl, 5 \underline{N} \text{ NaOH}}]{Co_2(CO)_8 \text{ (cat.)}} ArCH_2CO_2H$$

Ar = Ph, 85%
Ar = p-NCC$_6$H$_4$, 50%

I.G.2-7 P. S. Braterman, B. S. Walker, and T. H. Robertson, J. Chem. Soc., Chem. Commun., 1977, 651.

$$Co_2(CO)_8 \text{ or } Co_4(CO)_{12} \xrightarrow[PhH]{[K(crown)]_2[CoX_4]} [Co(CO)_4]^- \xrightarrow{\begin{array}{c}\text{o-}C_6H_4(CH_2X)_2\\ \hline PhH \text{ or } THF\end{array}}$$

80%

X = Cl or Br

(2-indanone product shown, 80%)

I.G.2-8 E. Colomer, R.J.P. Corriu, and J. C. Young, J. Chem. Soc., Chem. Commun., 1977, 73.

$$R^1R^2R^3SiH \xrightarrow{\begin{array}{c}1)\ Co_2(CO)_8\\ 2)\ PhLi,\ -78°C\\ 3)\ heat\end{array}} R^1R^2R^3SiCOPh$$

39-47%

I.G.2-9 Y. Seki, S. Murai, I. Yamamoto, and N. Sonoda, Angew. Chem. Int. Ed. Engl., 16, 789 (1977).

$$\underset{O}{\overset{(CH_2)_n}{\bigcap}} \xrightarrow{\begin{array}{c}HSiEt_2Me,\ CO\\ \hline PhH,\ Co_2(CO)_8\end{array}} Et_2MeSiO(CH_2)_{n+2}CHO$$

n = 1, 40%
n = 2, 53%

(cyclohexene oxide) $\xrightarrow{\begin{array}{c}HSiMe_2Me,\ CO\\ \hline PhH,\ Co_2(CO)_8\end{array}}$ (trans-2-(OSiEt_2Me)cyclohexyl-CHO)

51%

I.G.2-10 Y. Seki, A. Hidaka, S. Murai, and N. Sonoda, Angew. Chem. Int. Ed. Engl., 16, 174 (1977); see also: ibid, 881 (1977); Y. Seki, A. Hidaka, S. Makino, S. Murai, and N. Sonoda, J. Organometal. Chem., 140, 361 (1977).

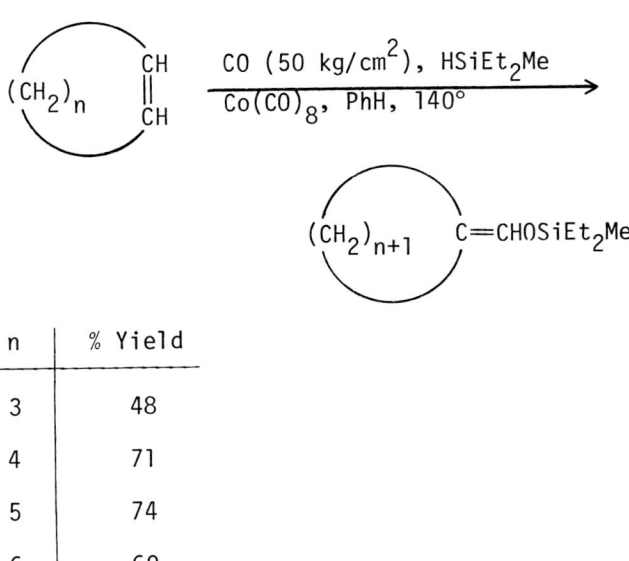

n	% Yield
3	48
4	71
5	74
6	69

I.G.2-11 L. V. Morozova, G.K.-I. Magomedov, and V. D. Sheludyakov, J. Gen. Chem. USSR, 47, 1345 (1977).

$$Me_3SiCH=CH_2 \xrightarrow[\substack{CO\ (170\ atm.),\ EtOH \\ 180°C}]{Co_2(CO)_8\ or\ Rh(CO)_4}$$

$$Me_3SiCHMeCO_2Et$$

93%

I.G.2-12   A. Spencer, J. Organomet. Chem., 124, 85 (1977).

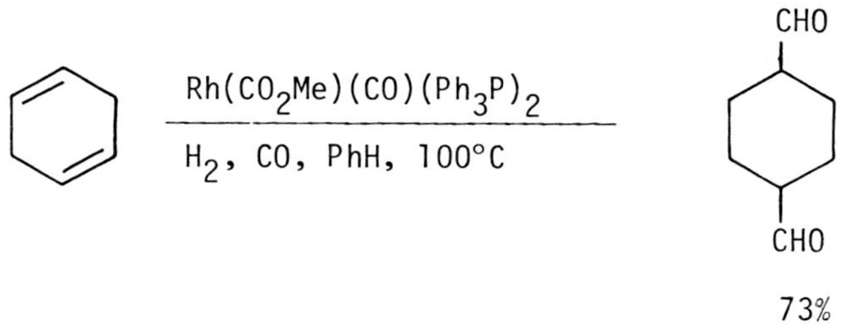

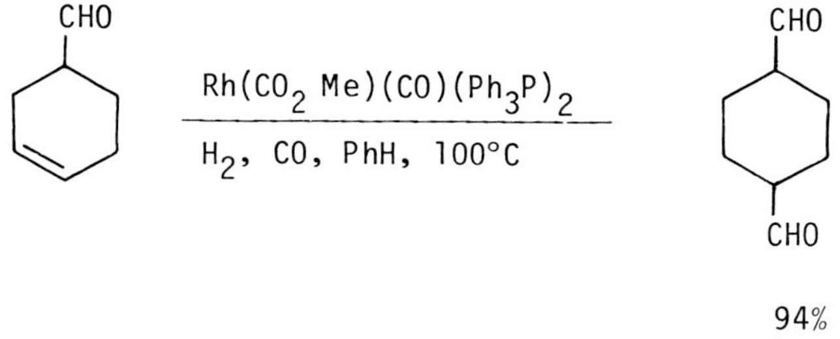

[1,4-cyclohexadiene] $\xrightarrow{\text{Rh}(CO_2Me)(CO)(Ph_3P)_2}{H_2, CO, PhH, 100°C}$ [cyclohexane-1,4-dicarbaldehyde]

73%

[3-cyclohexenecarbaldehyde] $\xrightarrow{\text{Rh}(CO_2\,Me)(CO)(Ph_3P)_2}{H_2, CO, PhH, 100°C}$ [cyclohexane-1,4-dicarbaldehyde]

94%

I.G.2-13   I. Rhee, M. Ryang, T. Watanabe, H. Omura, S. Murai, and N. Sonoda, Synthesis, 1977, 776.

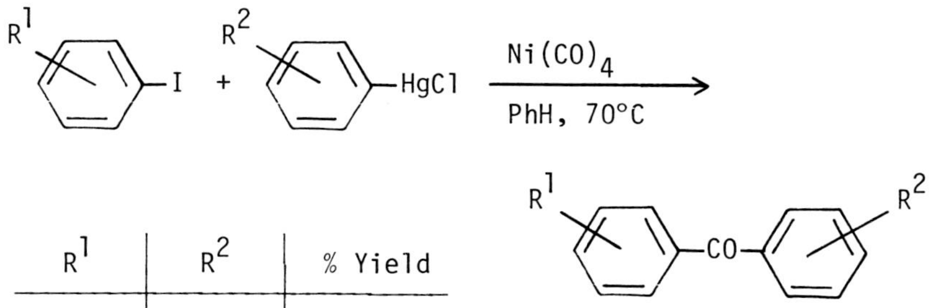

R^1–[C$_6$H$_4$]–I + R^2–[C$_6$H$_4$]–HgCl $\xrightarrow{\text{Ni(CO)}_4}{\text{PhH, 70°C}}$ R^1–[C$_6$H$_4$]–CO–[C$_6$H$_4$]–R^2

R^1	R^2	% Yield
H	4-Me	88
H	4-Cl	90
4-H_2N	4-Me	60

164

I.G.2-14 A. Kasahara, T. Izumi, and A. Suzuki, Bull. Chem. Soc. Japan, 50, 1639 (1977).

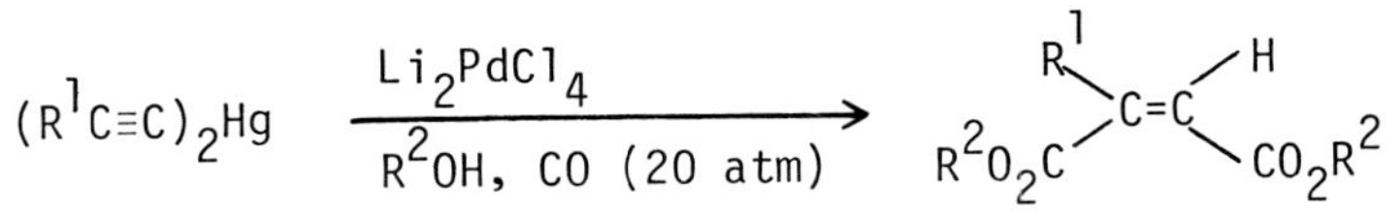

$$(R^1C{\equiv}C)_2Hg \xrightarrow{Li_2PdCl_4, \; R^2OH, \; CO \; (20 \; atm)} \underset{R^2O_2C}{\overset{R^1}{\diagdown}} C{=}C \underset{CO_2R^2}{\overset{H}{\diagup}}$$

R^1 = Ph, 43-62%
R^1 = Et, 40-46%

I.G.2-15 F. Calderazzo, Angew. Chem. Int. Ed. Engl., 16, 299 (1977).

Review: "Synthetic and Mechanistic Aspects of Inorganic Insertion Reactions. Insertion of Carbon Monoxide".

I.G.3-1 G. A. Olah, T.-L. Ho, G. K. Surya Prakash, and B.G.B. Gupta, Synthesis, 1977, 677.

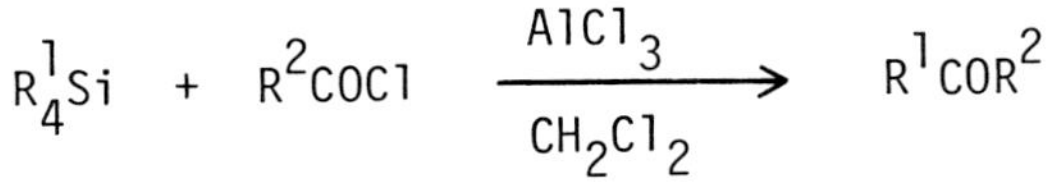

$$R^1_4Si \; + \; R^2COCl \xrightarrow[CH_2Cl_2]{AlCl_3} R^1COR^2$$

R^1	R^2	% Yield
Et	Et	79
Et	Ph	41
Et	i-Pr	82
n-Bu	cyclo-C_6H_{11}	50

I.G.3-2 D. B. Carr and J. Schwartz, J. Am. Chem. Soc., 99, 638 (1977).

$$Cp_2Zr\begin{smallmatrix}Cl\\R\end{smallmatrix} \xrightarrow[\text{2) MeCOCl, CH}_2\text{Cl}_2,\ -30°C]{\text{1) AlCl}_3,\ \text{CH}_2\text{Cl}_2,\ 0°C} RCOMe$$

R	% Yield
t-BuCH$_2$CH$_2$	98
t-BuCH=CH	97
i-PrCH=CMe	98

Organozirconium(IV) complexes can be transmetalated to give organoaluminum compounds which react better than the initial zirconium compounds.

I.G.3-3 M. L. Saïhi and M. Pereyre, Bull. Soc. Chim. Fr., 1977, 1251.

$$Bu_3SnCH=CR^1R^2 + R^3COX \xrightarrow[CH_2Cl_2,\ -10°C]{AlCl_3} R^3COCH=CR^1R^2$$

R^1	R^2	R^3	X	% Yield
H	H	Me	Cl	23
H	Me	Me	Cl	65
H	Ph	Ph	Cl	45
Me	Me	Ph	COPh	51
H	Ph	Me	Ac	46

I.G.3-4 M. Kosugi, Y. Shimizu, and T. Migita, J. Organomet. Chem., 129, C36 (1977).

$$R^1COCl \xrightarrow[(Ph_3P)_3RhCl\ (cat.)]{R^2CH=CHCH_2SnBu_3} R^1COCH_2CH=CHR^2$$

R^1	R^2	% Yield
Me	H	51
Ph	H	86
Et	Me	64

I.G.3-5 M. Kosugi, Y. Shimizu, and T. Migita, Chem. Lett., 1977, 1423.

$$R^1COCl\ +\ R^2_4Sn \xrightarrow[PhH,\ \Delta]{(Ph_3P)_4Pd} R^1COR^2$$

42-87%

R = alkyl, aryl, or vinyl

I.G.3-6 Y. Tohda, K. Sonogashira, and N. Hagihara, Synthesis 1977, 777.

$$R^1C\equiv CH\ +\ R^2COCl \xrightarrow[CuI\ (cat.),\ Et_3N]{(Ph_3P)_2PdCl_2} R^1C\equiv CCOR^2$$

R^1	R^2	% Yield
Ph	Ph	96
Ph	t-Bu	79
n-Bu	Ph	81
n-Bu	i-Pr	61
Ph	Me$_2$N	92

I.G.3-7 C. U. Pittman, Jr. and R. M. Hanes, J. Org. Chem., 42, 1194 (1977).

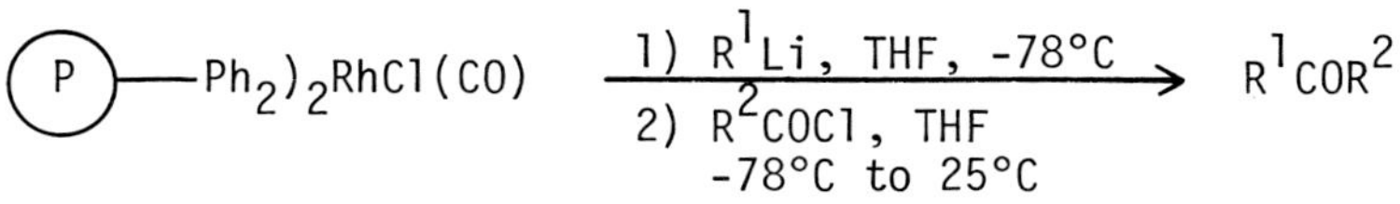

$$\text{(P)}-Ph_2)_2RhCl(CO) \xrightarrow[\substack{2)\ R^2COCl,\ THF \\ -78°C\ to\ 25°C}]{1)\ R^1Li,\ THF,\ -78°C} R^1COR^2$$

R^1	R^2	% Yield
Bu	Ph	61
Bu	$C_{11}H_{23}$	58
Ph	Me	60
Ph	$MeO_2C(CH_2)_4$	56

I.G.3-8 J. Blum, M. Weitzberg, and R. J. Mureinik, J. Organomet. Chem., 122, 261 (1977).

$$PhX \xrightarrow[2)\ PhCH_2COCl,\ 220°C]{1)\ IrCl(CO)(PPh_3)_2} PhCH_2COPh$$

"reasonable yield"

I.G.3-9 M. Kubota, A. Miyashita, and A. Yamamoto, J. Organomet. Chem., 139, 111 (1977).

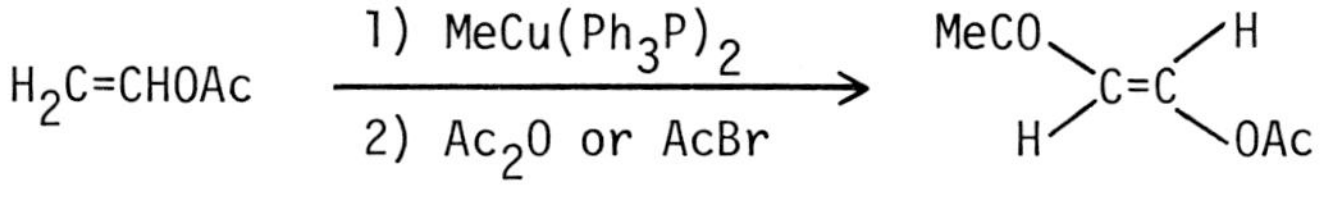

$$H_2C=CHOAc \xrightarrow[2)\ Ac_2O\ or\ AcBr]{1)\ MeCu(Ph_3P)_2} \underset{H}{\overset{MeCO}{>}}C=C\underset{OAc}{\overset{H}{<}}$$

58-61%

I.G.3-10   T. C. Flood and A. Sarhangi, Tetrahedron Lett., 3861 (1977).

$$RCOCl \xrightarrow[Et_2O, \text{ reflux}]{Fe_2(CO)_9} RCOR$$

R	% Yield
$\underline{n}\text{-}C_7H_{15}$	71
$PhCH_2$	33
$CH_2=CH(CH_2)_8$	40

I.G.3-11   J. E. McMurry and K. L. Kees, J. Org. Chem., **42** 2655 (1977).

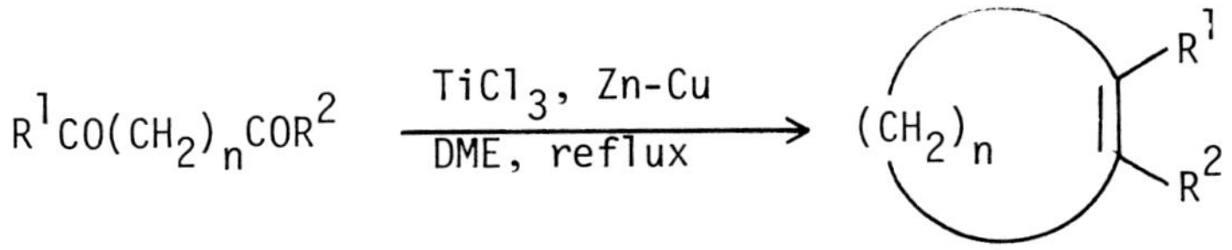

$$R^1CO(CH_2)_nCOR^2 \xrightarrow[\text{DME, reflux}]{TiCl_3, \text{ Zn-Cu}} (CH_2)_n\!\!\begin{array}{c}R^1\\R^2\end{array}$$

n	R^1	R^2	% Yield
2	Ph	Ph	87
3	Me	Ph	70
4	Me	$\underline{n}$-Bu	79
6-9	Bu	Bu	67-76
10-12	H	H	52-76
13	H	Ph	80

I.G.3-12   D. Lenoir, Synthesis, 1977, 553.

$$R^1COR^2 \xrightarrow[\text{py, THF}]{\text{TiCl}_4/\text{Zn}} R^1R^2C=CR^1R^2$$

R^1	R^2	% Yield
Me	Me_3CCH_2	48
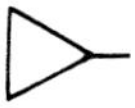 ▷	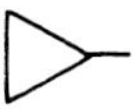 ▷	25
		51
$CH_2CMe(CH_2)_2CH(\underline{i}\text{-Pr})$		

I.G.3-13   H. Alper and H.-N. Paik, J. Org. Chem., 42, 3522 (1977).

$$R_2CS \xrightarrow[\text{PhH, reflux}]{Co_2(CO)_8} R_2C=CR_2$$

$R = Ph, 71\%$

$R = p\text{-MeC}_6H_4, 83\%$

I.G.3-14   S. Inaba and I. Ojima, Tetrahedron Lett., 2009 (1977).

$$R^1R^2C=C(OMe)OSiMe_3 \xrightarrow{TiCl_4} R^1R^2C(CO_2Me)C(CO_2Me)R^1R^2$$

73-80%

I.G.3-15 A. Ishida and T. Mukaiyama, Bull. Chem. Soc. Japan, 50, 1161 (1977).

$$R^1CH(OR^2)_2 + R^3CH=CR^4CH=CHOSiMe_3 \xrightarrow{TiCl_4 \text{ or } Ti(\underline{i}\text{-PrO})_4}$$

$$R^1CH(OR^2)CHR^3CR^4=CHCHO$$

A full paper describing the scope of this procedure is presented.

I.G.3-16 E. Nakamura and I. Kuwajima, J. Am. Chem. Soc., 99, 7360 (1977).

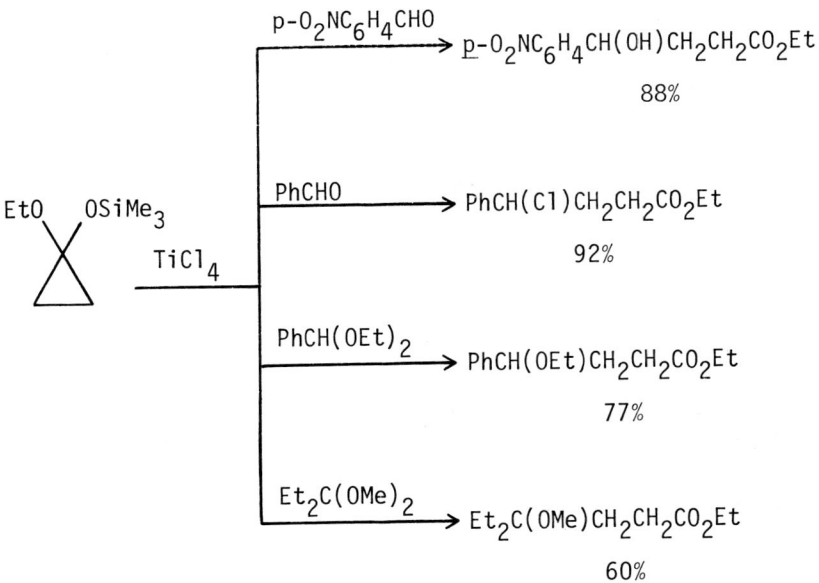

I.G.3-17   K. Yamamoto, O. Nunokawa, and J. Tsuji, Synthesis, 1977, 720.

$$\underset{H}{\overset{R^1}{>}}C=C\underset{SiMe_3}{\overset{R^2}{<}} \quad \xrightarrow[\text{2) } H_2O]{\text{1) } Cl_2CHOMe,\ TiCl_4,\ CH_2Cl_2} \quad \underset{H}{\overset{R^1}{>}}C=C\underset{CHO}{\overset{R^2}{<}}$$

R^1	R^2	% Yield
n-Bu	H	76
Ph	H	72
n-Pr	n-Pr	79

I.G.3-18   H. Alper and H.-N. Paik, J. Chem. Soc., Chem. Commun., 1977, 126.

$$(p\text{-}RC_6H_4)_2C=S \; + \; [Cp\text{-}Fe(CO)_2]^- \; \xrightarrow{THF} \; \underset{58\text{-}82\%}{\bigcirc=C(C_6H_4R\text{-}p)_2}$$

I.G.3-19   R. A. Holton, Tetrahedron Lett., 355 (1977).

$$\underset{NMe_2}{\overset{Cl\diagdown Pd/2}{\bigodot}} \; \xrightarrow{R^1COCR^2=CH_2} \; \underset{NMe_2}{\overset{CH=CR^2COR^1}{\bigodot}}$$

R^1	R^2	% Yield
Me	H	95
Ph	H	94
Me	Ph	80

I.G.3-20 R. A. Holton and R. A. Kjonaas, J. Organomet. Chem., 133, C5 (1977); see also: ibid, 142, C15 (1977); ibid, J. Am. Chem. Soc., 99, 4177 (1977).

$$\underset{\text{NMe}_2}{\overset{\text{MeO}}{\underset{R^1}{\diagup}}\hspace{-0.5em}\overset{\text{Pd}}{\diagdown}\hspace{-0.5em}\overset{\text{Cl}}{\diagup}_2} \quad \xrightarrow[\text{PhH, Et}_3\text{N, reflux}]{R^2\text{COCH=CH}_2}$$

$$Me_2NCH_2CR^1(OMe)CH_2CH=CHCOR^2$$

R^1	R^2	% Yield
Me	Me	86
Me	Et	60
H	Me	70

I.G.3-21 M. J. Loots and J. Schwartz, J. Am. Chem. Soc., 99, 8045 (1977).

$$R^1C\equiv CR^2 \quad \xrightarrow[\substack{2)\ R^3CH=CHCOR^4 \\ Ni(AcAc)_2,\ THF}]{1)\ Cp_2Zr(H)Cl} \quad R^1CH=CR^2CHR^3CH_2COR^4$$

R^1	R^2	R^3	R^4	% Yield
t-Bu	H	H	Me	> 95
t-Bu	H		$-(CH_2)_3-$	77
t-Bu	H		$-(CH_2)_2-$	78

With R^2 = alkyl yields are $\sim 10\%$

I.G.3-22 M. Yoshifuji, M. J. Loots, and J. Schwartz, Tetrahedron Lett., 1303 (1977).

$$\text{t-BuC}\equiv\text{CH} \xrightarrow[\text{2) CuCl}_2]{\text{1) Cp}_2\text{Zr(H)Cl}} \text{t-BuCH=CHCH=CHBu-t}$$
$$90\%$$

$$\text{t-BuC}\equiv\text{CH} \xrightarrow[\substack{\text{2) LiI, MeCOCH=CH}_2 \\ \text{3) CuOT}_f}]{\text{1) Cp}_2\text{Zr(H)Cl}} \text{t-BuCH=CH(CH}_2)_2\text{COMe}$$
$$73\%$$

Transmetalation from Zr to Cu can be accomplished readily.

I.G.3-23 M. Obayashi, K. Utimoto, and H. Nozaki, Tetrahedron Lett., 1805 (1977).

$$\text{HC}\equiv\text{CSiMe}_3 \xrightarrow[\text{2) R}^2\text{X}]{\text{1) R}^1\text{Cu·MgBr}_2} \text{R}^1\text{CH=CR}^2\text{SiMe}_3$$
$$39-76\%$$

I.G.3-24 H. Westmijze, J. Meijer, and P. Vermeer, Tetrahedron Lett., 1823 (1977).

$$\text{Ph}_3\text{SiC}\equiv\text{CH} \xrightarrow[\text{2) Electrophile}]{\text{1) RCu·Mg(Cl or Br)}_2} \text{Ph}_3\text{SiCX=CHR}$$

X = H, Cl, Br, I, or Me

80-95%

I.G.3-25 A. Alexakis, G. Cahiez, J. F. Normant, and J. Villieras, Bull. Soc. Chim. Fr., 1977, 693.

$$R^1C\equiv C-Z \xrightarrow[2)\ H_2O]{1)\ R^2Cu} \underset{A}{\overset{R^1}{\underset{H}{>}}C=C\overset{R^2}{\underset{Z}{<}}} + \underset{B}{\overset{R^1}{\underset{R^2}{>}}C=C\overset{Z}{\underset{H}{<}}}$$

R^1	R^2	Z	% Yield A	B
H	Et	SEt	-	91
Me	Me	SEt	-	79
H	$\underline{n}$-C_7H_{15}	OEt	80	-
H	Et	NPh_2	80	-
Me	$\underline{n}$-Bu	NEt_2	54	-

The intermediate vinylcuprates could also react with electrophiles.

I.G.3-26 G. Courtois and L. Miginiac, Comptes rendus, Ser. C, 285, 207 (1977).

$$RC\equiv CH \xrightarrow[THF]{\underline{t}-Bu_2Zn} \underline{t}-BuCH=CHR$$

R	% Yield	Z:E
$\underline{n}$-C_6H_{13}	40	55:45
CH_2OH	65	12:88
CH_2CH_2OH	40	60:40
CH_2OTHP	30	35:65
CH_2NEt_2	43	35:65

I.G.3-27 J. Auger, G. Courtois, and L. Miginiac, J. Organomet. Chem., 133, 285 (1977); see also: B. Mauze', ibid, 321 (1977).

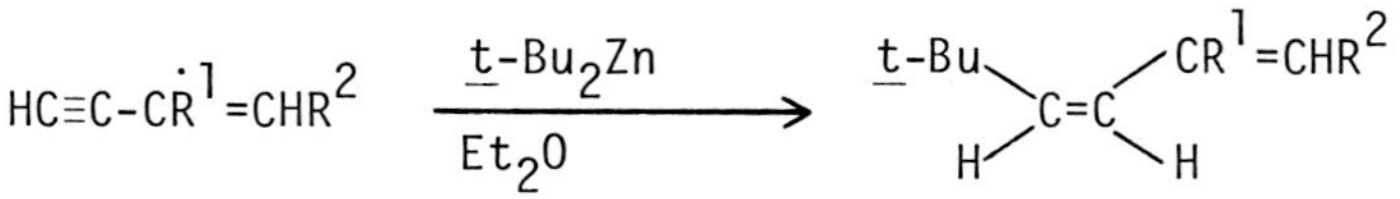

$$HC \equiv C-CR^1=CHR^2 \xrightarrow[Et_2O]{t-Bu_2Zn} \underset{H}{\overset{t-Bu}{>}}C=C\underset{H}{\overset{CR^1=CHR^2}{<}}$$

R^1	R^2	% Yield
H	n-Bu	30
H	CH_2OH	45
H	CH_2NEt_2	47
$-(CH_2)_4-$		55

I.G.3-28 H. E. Tweedy, R. A. Coleman, and D. W. Thompson, J. Organomet. Chem., 129, 69 (1977).

$$TiCl(OCHR^1(CH_2)_nC \equiv CR^2)(C_5H_7O_2)_2 \xrightarrow[2) \; 6 \; \underline{N} \; NaOH, \; O_2]{1) \; Et_2AlCl, \; CH_2Cl_2}$$

$$HOCHR^1(CH_2)_nCH=CR^2Et$$

n	R^1	R^2	% Yield
1	H	H	56
1	H	Ph	50
1	Me	H	36
2	H	H	21

I.G.3-29 R. C. Larock and J. C. Bernhardt, J. Org. Chem., **42**, 1680 (1977).

$$R^1CH=CR^2HgCl \xrightarrow[\text{HMPA}]{[ClRh(CO)_2]_2} R^1CH=CR^2CR^2=CHR^1$$

R^1	R^2	% Yield
Ph	H	100
n-Bu	H	100
Et	Et	35

$$ArHgCl \xrightarrow[\substack{\text{LiCl, HMPA} \\ 80°C}]{[ClRh(CO)_2]_2} ArAr \quad 40\text{-}96\%$$

I.G.3-30 K. Takagi, N. Hayama, T. Okamoto, Y. Sakakibara, and S. Oka, Bull. Chem. Soc. Japan, **50**, 2741 (1977).

$$R_2Hg \xrightarrow[\text{HMPA or MeCN, 80°C}]{(Ph_3P)_3RhCl} R\text{-}R$$

R	% Yield
Ph	100
p-MeC$_6$H$_4$	86
p-ClC$_6$H$_4$	91
PhC≡C	95
PhCH$_2$	44

I.G.3-31 T.-L. Ho and G. Olah, Synthesis, 1977, 170.

$$R^1R^2CHBr \xrightarrow[\text{[V(II)]}]{VCl_3/LiAlH_4} R^1R^2CHCHR^1R^2$$

$R^1 = R^2 = Ph, 95\%$

$R^1 = H, R^2 = Ph, 82\%$

$R^1, R^2 = -CH=CH(CH_2)_3, 80\%$

I.G.3-32 Y. Ito, T. Konoike, T. Harada, and T. Saegusa, J. Am. Chem. Soc., 99, 1487 (1977).

$$R^1CH_2COR^2 \xrightarrow[\text{2) CuCl}_2\text{, DMF}]{\text{1) LDA, THF, -78°C}} R^2COCHR^1CHR^1COR^2$$

A full paper describing the scope of this reaction in intermolecular, intramolecular, crossed coupling, and vinylog cases is presented.

I.G.3-33 Y. Kobayashi, T. Taguchi, and E. Tokuno, Tetrahedron Lett., 3741 (1977).

$$R^1\overset{OLi}{\underset{|}{C}}=CR^2R^3 \xrightarrow[\text{THF, i-BuCn}]{Cu(OTf)_2} R^1COCR^2R^3CR^2R^3COR^1$$

R^1	R^2	R^3	% Yield
Ph	H	H	83
Ph	H	Me	80
Et	H	Me	63
$-(CH_2)_4-$		H	73

The reaction was also successful using the trimethylsilyl enol ether instead of the lithium enolate.

I.G.3-34 T. Sakakibara, J. Kotobuki, and Y. Dogomori, Chem. Lett., 1977, 25.

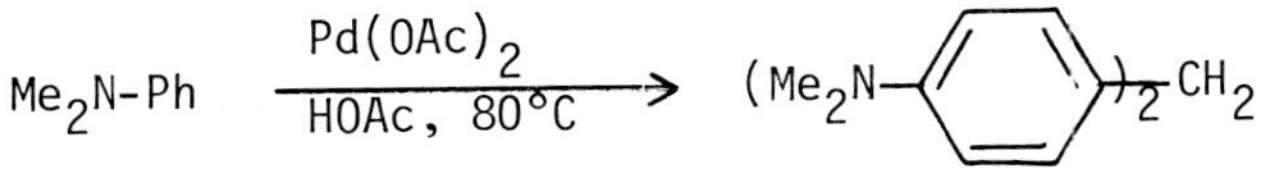

$$Me_2N-Ph \xrightarrow[\text{HOAc, 80°C}]{Pd(OAc)_2} (Me_2N-\underset{}{\underset{}{\bigcirc}})_2 CH_2$$

72%

I.G.3-35 Y. Tanigawa, H. Kanamaru, A. Sonoda, and S.-I. Murahashi, J. Am. Chem. Soc., 99, 2361 (1977).

$$R^1OH \xrightarrow[\substack{2) CuI \\ 3) R^2Li \\ 4) [Ph_3PN(Me)Ph]^+I^-}]{1) MeLi} R^1-R^2$$

R^1	R^2	% Yield
MeCH=CHCH=CHCH$_2$	n-Bu	81
PhCHMe	Et	65
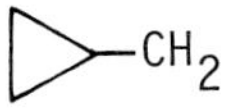 ▷—CH$_2$	Ph	73

I.G.3-36 E. C. Ashby and J. J. Lin, J. Org. Chem., 42, 2805 (1977).

$$R-X \xrightarrow{Li_wCu_yMe_z} RMe$$

R = alkyl, cycloalkyl, and aryl; $Li_wCu_yMe_z$ = $LiCu_2Me_3$, Li_2CuMe, and $Li_2Cu_3Me_5$

I.G.3-37 F. Derguini-Boumechal, R. Lorne, and G. Linstrumelle, Tetrahedron Lett., 1181 (1977).

$$R^1R^2C=CHCH_2MgX \xrightarrow{R^3X / CuI} R^1R^2C=CHCH_2R^3$$

$$R^1R^2C=CHCH_2MgX \xrightarrow{R^3OTs} R^1R^2R^3CCH=CH_2$$

I.G.3-38 C. Chuit, H. Felkin, C. Frajerman, G. Roussi, and G. Swierczewski, J. Organomet. Chem., 127, 371 (1977).

$$R^1CH=CHC(OH)R^2R^3 \xrightarrow{R^4MgBr / (Ph_3P)_2NiCl_2} \begin{array}{c} R^1CH=CHCR^2R^3R^4 \\ A \\ + \\ R^1R^4CHCH=CR^2R^3 \\ B \end{array}$$

R^1	R^2	R^3	R^4	% Yield	A:B
Me	H	H	Ph	> 73	66:34
H	H	Me	Me	82	69:31
H	$-(CH_2)_5-$		Me	64	75:25

I.G.3-39 H. Kleijn, J. Meijer, H. Westmijze, and P. Vermeer, Recl. Trav. Chim. Pays-Bas, 96, 251 (1977).

$$R^1R^2C=C=C=CHOMe \quad \begin{array}{c} \xrightarrow{R^3MgX \ (2 \ eq.) \ / \ CuBr \ (0.2 \ eq.)} R^1R^2C=CR^3C\equiv CH \quad 80\text{-}90\% \\ \\ \xrightarrow{[RCuBr]M \ / \ THF} R^1R^2C=CR^3CH=CHOMe \quad 80\text{-}90\% \end{array}$$

M = Li or MgX

I.G.3-40 J. F. Normant, J. Villieras, and F. Scott, Tetrahedron Lett., 3263 (1977).

$$R^1MgCl \xrightarrow[5\% \ Cu(I), \ THF]{BrCH_2CH_2OR^2} R^1CH_2CH_2OR^2$$

R^2 = alkyl, 48-83%

R^2 = H, 56-88%

R^2 = Ac, 53-80%

I.G.3-41 R.-D. Acker, Tetrahedron Lett., 3407 (1977).

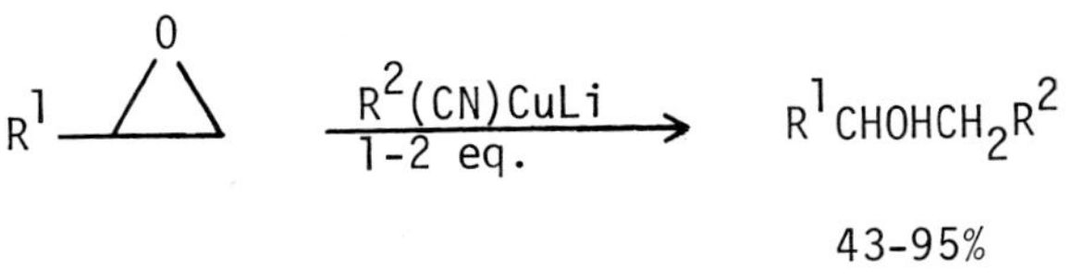

$$R^1{-}\triangle\!{}^O \xrightarrow[1\text{-}2 \ eq.]{R^2(CN)CuLi} R^1CHOHCH_2R^2$$

43-95%

I.G.3-42 D. B. Malpass, S. C. Watson, and G. S. Yeargin, J. Org. Chem., 42, 2712 (1977).

$$R^1_2Al\text{-}C(H)=C(H)\text{-}R^2 \xrightarrow[2)\ H_3O^+]{1)\ \triangle O} R^2\text{-}C(H)=C(H)\text{-}(CH_2)_2OH$$

48-70%

I.G.3-43 A. O. King, N. Okukado, and E. Negishi, J. Chem. Soc., Chem. Commun., 1977, 683.

$$\underset{R^2}{\overset{R^1}{>}}C=C\underset{X}{\overset{R^3}{<}} + ClZnC{\equiv}CR^4 \xrightarrow[THF]{Pd\ catalyst} \underset{R^2}{\overset{R^1}{>}}C=C\underset{C{\equiv}CR^4}{\overset{R^3}{<}}$$

R^1	R^2	R^3	R^4	X	% Yield
H	n-Bu	H	H	I	83
n-Bu	H	H	n-C_5H_{11}	I	87
CO_2Me	Me	H	n-Bu	Br	87

I.G.3-44 A. Commercon, J. F. Normant, and J. Villieras, J. Organomet. Chem., 128, 1 (1977).

$$\underset{R^2}{\overset{R^1}{>}}C=C\underset{I}{\overset{H}{<}} \xrightarrow[Cu^+\ (cat.),\ THF]{R^3MgX} \underset{R^2}{\overset{R^1}{>}}C=C\underset{R^3}{\overset{H}{<}}$$

R^1	R^2	R^3	% Yield
n-Bu	Et	n-Bu	70
n-C_7H_{15}	H	n-Bu	73
n-Bu	Et	$PhCH_2$	53

I.G.3-45 A. A. Millard and M. W. Rathke, J. Am. Chem. Soc., 99, 4833 (1977).

$$LiCR^1R^2CO_2R^3 + R^4X \xrightarrow[\underline{n}\text{-BuLi}]{NiBr_2} R^4CR^1R^2CO_2R^3$$

R^1	R^2	R^3	R^4X	% Yield
H	H	$\underline{t}$-Bu	PhI	73
H	H	$\underline{t}$-Bu	PhCH=CHBr	83
H	Me	$\underline{t}$-Bu	MeCH=CHBr	53
Me	Me	Et	MeCH=CHBr	72

I.G.3-46 T. Hayashi and L. S. Hegedus, J. Am. Chem. Soc., 99, 7093 (1977).

$$NaCR^2(COR^3)R^4 \xrightarrow[\substack{PdCl_2(MeCN)_2 \\ THF, -50°C}]{CH_2=CHR^1, Et_3N} \begin{array}{c} \xrightarrow{H_2} MeCHR^1CR^2(COR^3)R^4 \quad A \\ \\ \xrightarrow{25°C} CH_2=CR^1CR^2(COR^3)R^4 \quad B \end{array}$$

R^1	R^2	R^3	R^4	% Yield A	B
H	H	OMe	CO_2Me	65	53
Me	Me	OEt	CO_2Et	-	90
NHAc	H	OMe	CO_2Me	88	-
H	H	Me	$CO_2\text{-}\underline{t}\text{-Bu}$	60	-
H	H	Me	Ph	72	-

I.G.3-47 S. Cacchi, M. Felici, and G. Rosini, J. Chem. Soc., Perkin Trans. I., 1977, 1260.

$$R^1\overset{NNHTs}{\overset{\|}{C}}CH_2R^2 \xrightarrow[2)\ R^3_2CuLi]{1)\ TAB,\ THF} R^1\overset{NNHTs}{\overset{\|}{C}}CHR^2R^3$$

$R^1 = Ph, R^2 = Ph, R^3 = Me, 64-68\%$

$R^1 = Ph, R^2 = H, R^3 = Ph, 35-40\%$

$R^1, R^2 = -CHMe(CH_2)_3-, R^3 = Me, 55-60\%$

$$Pr\overset{NNHTs}{\overset{\|}{C}}Pr \xrightarrow[2)\ Ph_2CuLi]{1)\ TAB\ (2\ eq.),\ THF} EtCHPh\overset{NNHTs}{\overset{\|}{C}}CHPhEt$$

50-55%

I.G.3-48 R. F. Lockwood and K. M. Nicholas, Tetrahedron Lett., 4163 (1977).

$$[HC\equiv CC(OH)R^1R^2]Co_2(CO)_6 \xrightarrow[HBF_4 \cdot Me_2O]{PhOMe} HC\overset{}{\underset{Co(CO)_6}{=}}CCR^1R^2Ar$$

R^1	R^2	Ar	% Yield
H	H	$MeOC_6H_4$	35 (2-MeO)
			+ 49 (4-MeO)
Me	Me	$4-MeOC_6H_4$	78

I.G.3-49   A. J. Pearson, Aust. J. Chem., 30, 345 (1977).

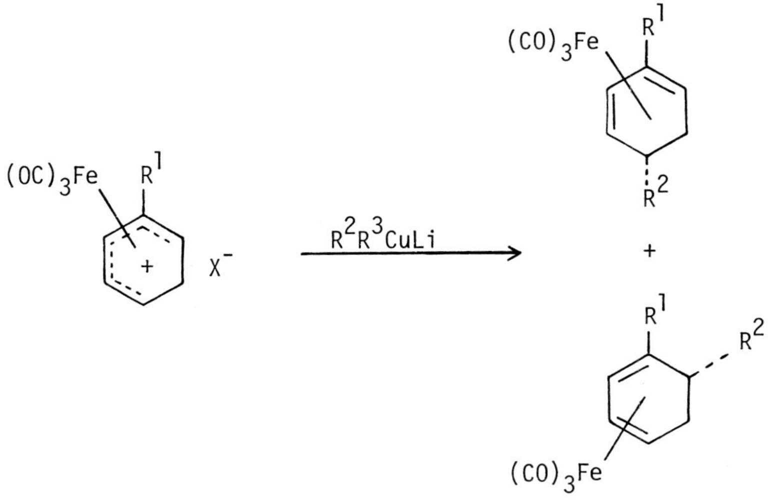

$$(OC)_3Fe\underset{+}{\bigcirc}R^1 \quad X^- \quad \xrightarrow{R^2R^3CuLi} \quad (CO)_3Fe\underset{R^2}{\bigcirc}R^1 \quad + \quad \underset{(CO)_3Fe}{\bigcirc}R^1\text{---}R^2$$

A complete study on product dependence on mixed alkylcuprate reagents was carried out.

I.G.3-50   A. J. Pearson, J. Chem. Soc., Chem. Commun., 1977, 339.

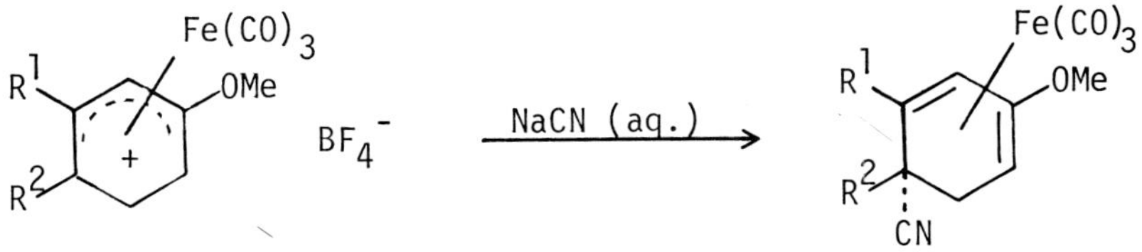

$$R^1\underset{R^2}{\underset{+}{\bigcirc}}\overset{Fe(CO)_3}{\text{OMe}} \quad BF_4^- \quad \xrightarrow{NaCN\ (aq.)} \quad R^1\underset{R^2}{\bigcirc}\overset{Fe(CO)_3}{\underset{CN}{\text{OMe}}}$$

$R^1 = H, R^2 = Me, 50\%$

$R^1, R^2 = -(CH_2)_4-, 80\%$

I.G.3-51 A. J. Pearson, J. Chem. Soc., Perkin Trans. I, 1977, 2069; see also: ibid, J. Chem. Soc., Chem. Commun., 1977, 339.

$$\underset{(CH_2)_n}{\text{[Fe(CO)}_3\text{-OMe cation]}} \xrightarrow[\text{THF}]{\text{NaCH(CO}_2\text{Me)}_2} \underset{(CH_2)_n}{\text{[Fe(CO)}_3\text{-OMe, CH(CO}_2\text{Me)}_2\text{]}}$$

$\underline{n}$ = 1, 77%

$\underline{n}$ = 2, 57%

I.G.3-52 B. W. Roberts and J. Wong, J. Chem. Soc., Chem. Commun., 1977, 20.

$$H_2C\!\!=\!\!CHR^1 \underset{Fe(CO)_4}{} + [R^2C(CO_2R^3)_2]Na \xrightarrow[\substack{1)\ THF,\ 0°C \\ 2)\ CF_3CO_2H \\ 3)\ H_3O^+,\ H_2O_2 \\ 4)\ Ce^{4+}}]{}$$

$$(R^3O_2C)_2CR^2CH_2CH_2R^1$$

R^1	R^2	R^3	% Yield
H	H	Me	68
CO$_2$Me	H	Me	91
CO$_2$Me	Me	Et	88

I.G.3-53 P. Lennon, A. M. Rosan, and M. Rosenblum, <u>J. Am. Chem. Soc.</u>, <u>99</u>, 8426 (1977).

$$R^1CH{\stackrel{\shortmid}{=}}CHR^2 + {}^-CHXY \longrightarrow FpCHR^1CHR^2CHXY$$
$$\quad\ \ Fp^+$$

X and Y = CN, CO_2Me, COMe or NO_2

60-99%

$$R^1CH{\stackrel{\shortmid}{=}}CHR^2 + \underset{\ }{\boxed{}}N{-}CR^3{=}CHR^4 \xrightarrow[\text{2) NaOH}]{\text{1) MeCN, 0°C}}$$
$$\quad\ \ Fp^+$$

$$RCOCHR^4CHR^2CHR^1Fp$$

31-98%

$$R^1CH{\stackrel{\shortmid}{=}}CHR^2 + R^3M \longrightarrow R^3CHR^2CHR^1Fp$$
$$\quad\ \ Fp^+$$

M = MgX or R^3CuLi

10-72%

I.G.3-54 W. R. Jackson and J. U. Strauss, <u>Aust. J. Chem.</u>, <u>30</u>, 553 (1977); see also: <u>ibid</u>, 2167 (1977).

$$\underset{PdCl)_2}{\underset{\ }{CH_2{\stackrel{R^1}{\diagdown\!\!\diagup}}CHCOR^2}} \xrightarrow[\text{DMSO}]{NaCH(CO_2Et)R^3} R^3(EtO_2C)CHCH_2CR^1{=}CHCOR^2$$

R^1	R^2	R^3	% Yield
H	OEt	CO_2Et	100
H	OEt	CN	70
H	OEt	SO_2Me	52
Me	Me	CO_2Et	90
Me	Me	CN	90

I.G.3-55 B. M. Trost and T. R. Verhoeven, J. Am. Chem. Soc., 99, 3867 (1977).

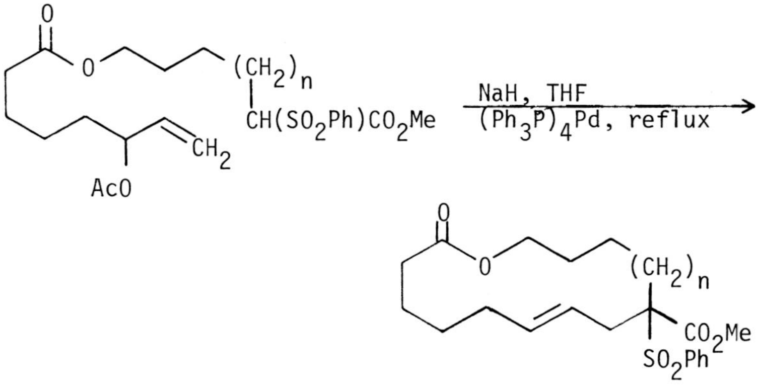

$$\text{macrolactone with CH(SO}_2\text{Ph)CO}_2\text{Me side chain, AcO and =CH}_2 \xrightarrow[\text{(Ph}_3\text{P)}_4\text{Pd, reflux}]{\text{NaH, THF}} \text{cyclized product}$$

n = 1 or 3, 49-69%

I.G.3-56 R. L. Hillard, III and K.P.C. Vollhardt, J. Am. Chem. Soc., 99, 4058 (1977).

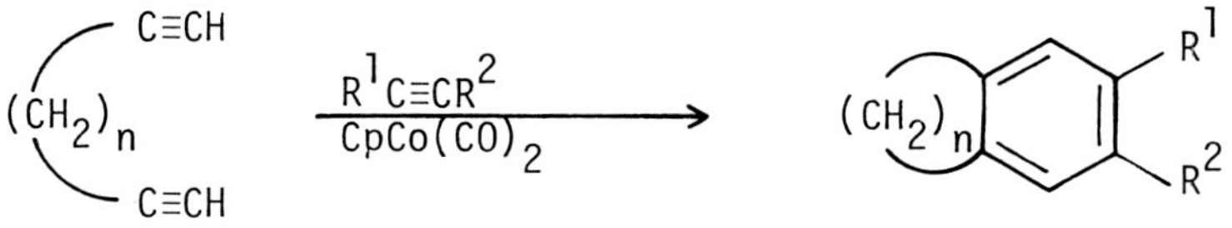

$$(\text{CH}_2)_n\text{(C≡CH)}_2 \xrightarrow[\text{CpCo(CO)}_2]{R^1\text{C≡C}R^2} (\text{CH}_2)_n\text{-benzo-}R^1,R^2$$

n = 2, 3, 4, or 5

A full paper describing the synthetic utility of this reaction is presented.

I.G.3-57 R. S. Dickson, C. Mok, and G. Connor, Aust. J. Chem., 30, 2143 (1977).

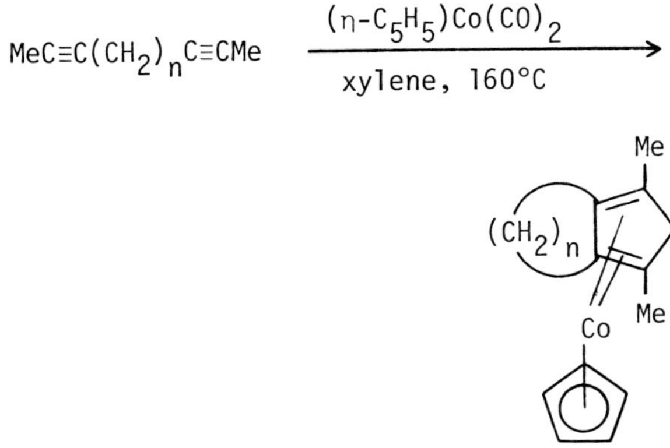

$$MeC\equiv C(CH_2)_nC\equiv CMe \xrightarrow{(\eta-C_5H_5)Co(CO)_2}_{xylene,\ 160°C}$$

n = 3, 38%
n = 4, 16%

Lower yields of a similar iron complex can be formed using $Fe(CO)_5$.

I.G.3-58 K. H. Dötz and R. Dietz, Chem. Ber., 110, 1555 (1977).

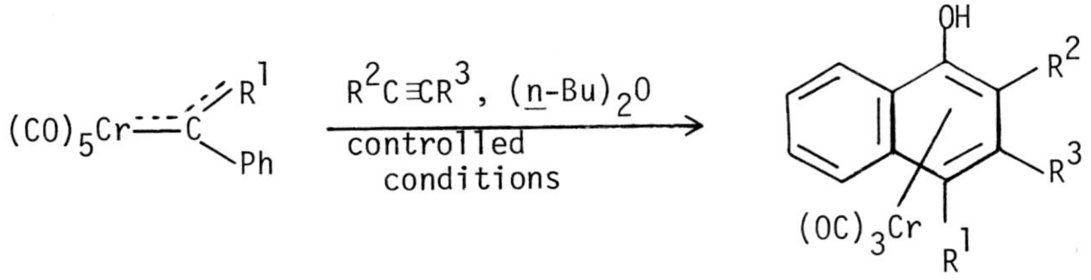

$$(CO)_5Cr\!=\!C\!\begin{smallmatrix}R^1\\Ph\end{smallmatrix} \xrightarrow[\text{conditions}]{R^2C\equiv CR^3,\ (\underline{n}\text{-Bu})_2O}$$

R^1	R^2	R^3	% Yield
MeO	H	Et	35
MeO	Me	Me	68
Ph	Me	Me	12
Ph	Ph	H	27

I.G.3-59 W. Münzenmaier and H. Straub, Justus Liebigs Ann. Chem., 1977, 313.

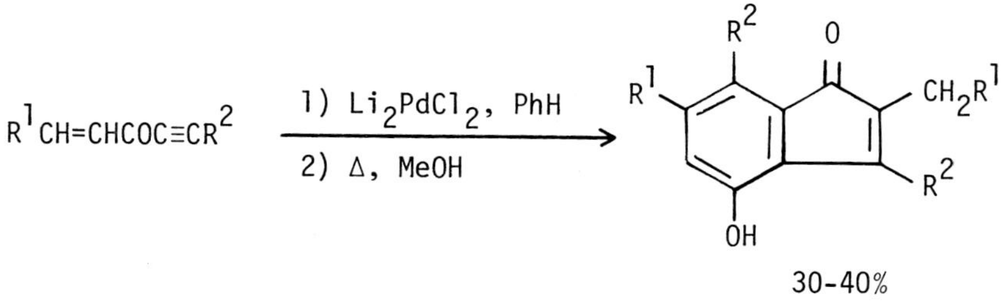

R1CH=CHCOC≡CR2
1) Li2PdCl2, PhH
2) Δ, MeOH
R2
O
R1
CH2R1
R2
OH
30-40%

I.G.3-60 M. E. Kuehne and W. H. Parsons, J. Org. Chem., 42, 3408 (1977).

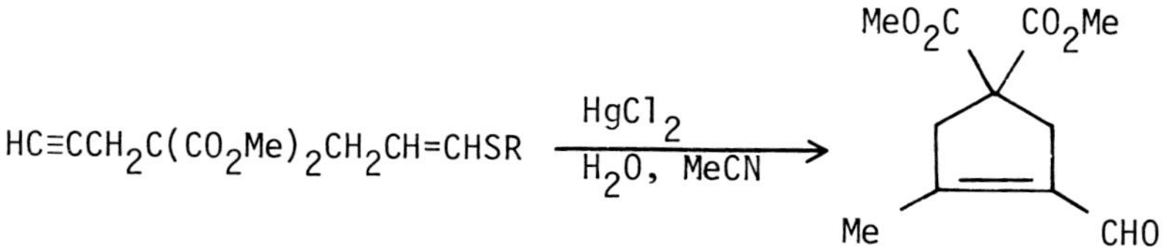

HC≡CCH2C(CO2Me)2CH2CH=CHSR
HgCl2
H2O, MeCN
MeO2C
CO2Me
Me
CHO

I.G.4-1 L. S. Hegedus, J. Organomet. Chem., 126, 151 (1977).

Review: "Transition Metal Derivatives in Organic Synthesis; Annual Survey Covering the Year 1975."

I.G.4-2 L. S. Hegedus, J. Organomet. Chem., 143, 309 (1977).

Review: "Transition Metal Derivatives in Organic Synthesis; Annual Survey Covering the Year 1976."

CARBON−CARBON BOND FORMING REACTIONS

I.G.4-3   B. M. Trost, Tetrahedron, 33, 2615 (1977).

Review: "Organopalladium Intermediates in Organic Synthesis."

I.G.4-4   K.P.C. Vollhardt, Acc. Chem. Res., 10, 1 (1977).

Review: "Transition-Metal-Catalyzed Acetylene Cyclizations in Organic Synthesis."

I.G.4-5   T. Mukaiyama, Angew. Chem. Int. Ed. Engl., 16, 817 (1977).

Review: "Titanium Tetrachloride in Organic Synthesis."

I.G.4-6   C. Blomberg and F. A. Hartog, Synthesis, 1977, 18.

Review: "The Barbier Reaction — A One-Step Alternative for Syntheses via Organomagnesium Compounds."

I.G.4-7   S. S. Washburne, J. Organomet. Chem., 123, 1 (1977).

Review: "Silicon-Application to Organic Synthesis; Annual Survey Covering the Year 1974."

II.A.1-1   J. Jallabert and H. Riviere, Tetrahedron Lett., 1215 (1977).

$$R\text{-}CH_2OH \xrightarrow[K_2CO_3, \text{ benzene}]{CuCl, \text{ phenanthroline, } O_2} R\text{-}CHO$$

R = Ph, Bz, -CH=CHPh, alkyl   50-83%

$$\underset{\underset{OH}{|}}{PhCH\text{-}CH_3} \xrightarrow{\text{same conditions}} PhCOCH_3$$

93%

II.A.1-2   V. I. Stenberg et al., J. Org. Chem., 42, 171 (1977).

$$R\text{-}CH_2OH \xrightarrow[h\nu]{FeCl_3} R\text{-}CHO$$

R = 1°, 2° alkyl   52-100%

II.A.1-3   R. O. Hutchins, N. R. Natale, and W. J. Cook, Tetrahedron Lett., 4167 (1977).

$$\underset{R \quad R'}{\overset{H \quad OH}{\diagdown \; \diagup}{C}} \xrightarrow[\text{phase-transfer catalyst}]{K_2Cr_2O_7, \text{ benzene}} \underset{R \quad R'}{\overset{O}{\overset{\|}{C}}}$$

R = Ph, cinnamyl, cyclic   Widely varying yields.
R'= H, Me, cyclic

II.A.1-4   T.-L. Ho and G. A. Olah, J. Org. Chem., 42, 3097 (1977).

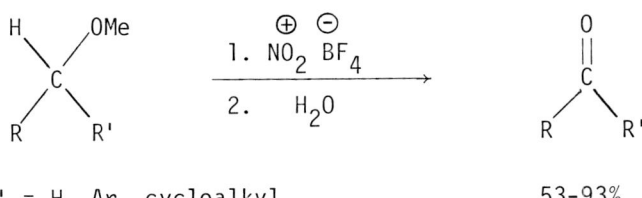

R,R' = H, Ar, cycloalkyl   53-93%

II.A.1-5   F. Yoneda and Y. Sakuma, J. Chem. Soc., Chem. Commun., 825 (1977).

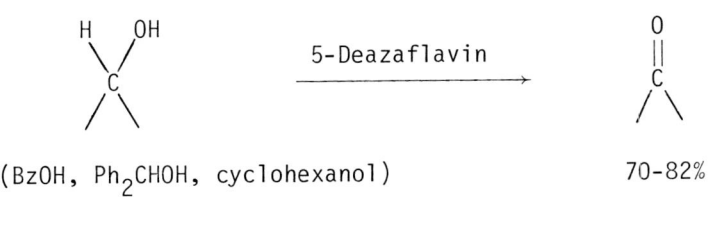

(BzOH, Ph$_2$CHOH, cyclohexanol)   70-82%

II.A.1-6   T. F. Blackburn and J. Schwartz, J. Chem. Soc., Chem. Commun., 157 (1977).

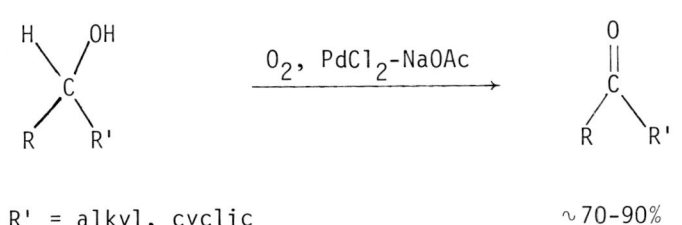

R, R' = alkyl, cyclic   ~70-90%

II.A.1-7 G. H. Posner and M. J. Chapdelaine, Synthesis, 555 (1977).

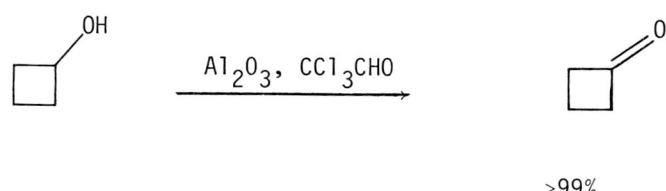

>99%

II.A.1-8 R. A. El-Zaru and A. A. Jarrar, Chem. and Ind., 741 (1977).

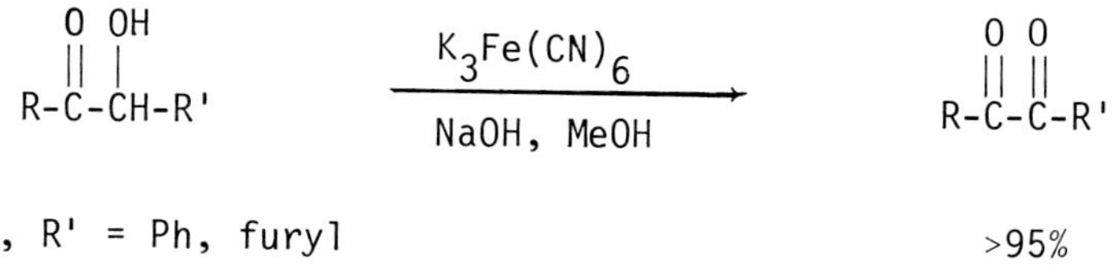

$$\underset{R-C-CH-R'}{\overset{O\ OH}{\|\ |}} \xrightarrow[\text{NaOH, MeOH}]{K_3Fe(CN)_6} \underset{R-C-C-R'}{\overset{O\ O}{\|\ \|}}$$

R, R' = Ph, furyl   >95%

II.A.1-9 Z. Yoshida, Y. Yamada, and Y. Tamaru, Chem. Lett., 423 (1977).

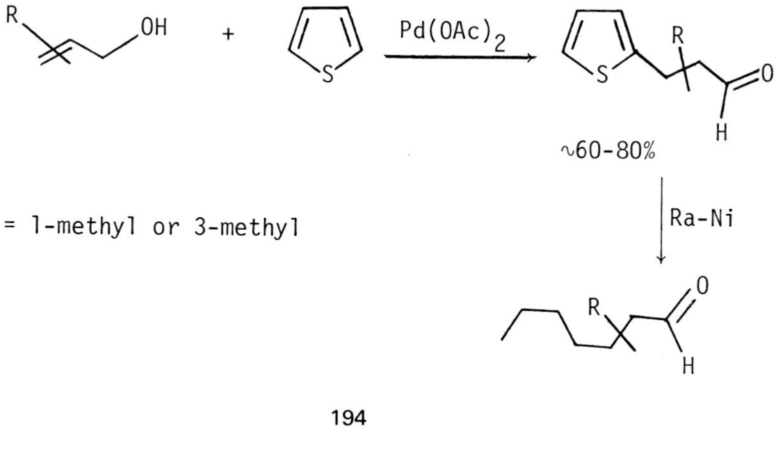

R = 1-methyl or 3-methyl

∼60-80%

Ra-Ni

OXIDATIONS

II.B.1-1 N. C. Deno et al., Tetrahedron, 33, 2503 (1977).

$$R-H \xrightarrow{H_2O_2, \text{ TFA}} R-OH$$

R = 2° alkyl, ~50%
generally n-alkanes (mixture of 2° alcohols)

II.B.1-2 J. Wellmann and E. Steckhan, Chem. Ber., 110, 3561 (1977).

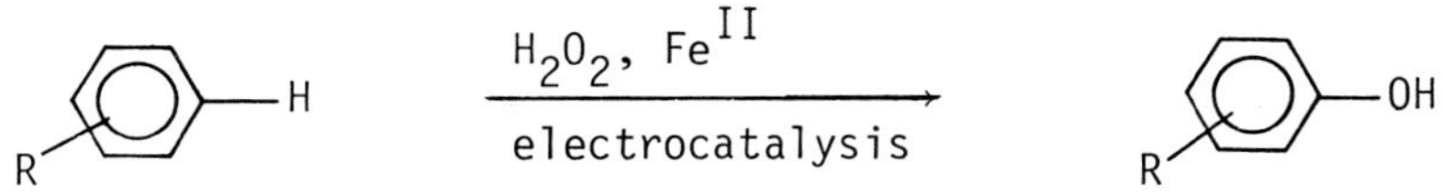

R–C₆H₄–H $\xrightarrow[\text{electrocatalysis}]{H_2O_2, \text{ Fe}^{II}}$ R–C₆H₄–OH

R = Cl, F, CN 20-80%

II.B.1-3 E. H. Appelman, R. Bonnett, and B. Mateen, Tetrahedron, 33, 2119 (1977).

R–C₆H₄–H $\xrightarrow{\text{HOF}}$ R–C₆H₄–OH

R = H, alkyl, F, Cl, Widely varying yields.
    OMe, NO_2

II.B.1-4   J. V. Crivello, Synth. Commun., 6, 543 (1976).

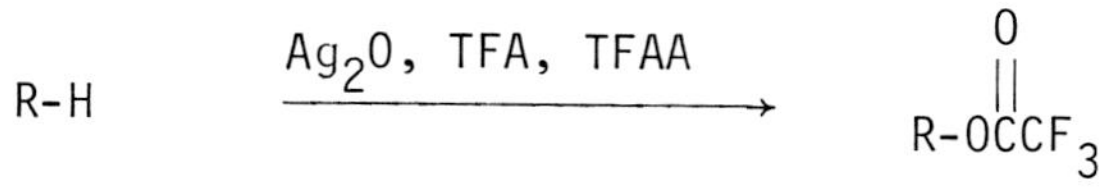

$$R-H \xrightarrow{Ag_2O,\ TFA,\ TFAA} R-O\overset{O}{\underset{\|}{C}}CF_3$$

R = 1-adamantyl, exo-norbornyl, cyclooctyl   57-98%

II.B.1-5   W. Adam and O. Cueto, J. Org. Chem., 42, 38 (1977).

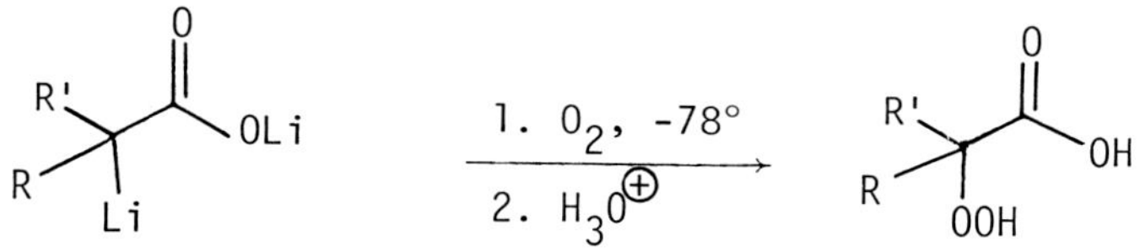

R = aryl
R' = H, aryl

66-82%

II.B.1-6   K. Oka and S. Hara, Tetrahedron Lett., 695 (1977).

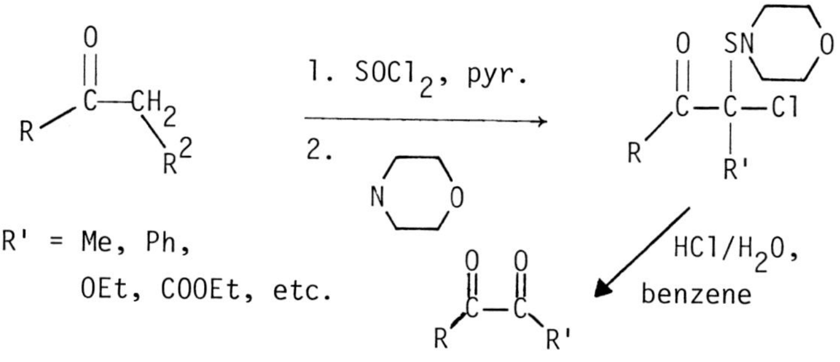

R,R' = Me, Ph, OEt, COOEt, etc.

~90%

OXIDATIONS

II.B.1-7   M. A. Umbreit and K. B. Sharpless, *J. Am. Chem. Soc.*, **99**, 5526 (1977).

$$\text{C=C-CH}_2 \xrightarrow[\underline{t}\text{-BuOOH}]{\text{SeO}_2} \text{C=C-C=O}$$

Widely varying yields.

A comparison of methods using catalytic and stoichiometric amounts of SeO_2.

II.B.1-8   B. M. Trost and G. S. Massiot, *J. Am. Chem. Soc.*, **99**, 4405 (1977).

$$\text{O=}\bigcirc \xrightarrow[\text{2. Pb(OAc)}_4]{\text{1. PhSSPh}} \underset{99\%}{\text{PhS, AcO intermediate}} \longrightarrow \text{O=}\bigcirc\text{=O}$$

II.B.2-1   A. S. Erickson and N. Kornblum, *J. Org. Chem.*, **42**, 3764 (1977).

$$R\text{-CH}_2\text{-NO}_2 \xrightarrow[\text{2. X}_2,\ \text{CH}_2\text{Cl}_2(-78°)]{\text{1. KOH, H}_2\text{O}} R\text{-CH(X)-NO}_2$$

R = $\underline{n}$-alkyl   82-94%

X = Cl, Br, I

II.B.2-2   E. C. Taylor et al., J. Org. Chem., 42, 362 (1977).

$$ArH \xrightarrow{TTFA} ArTl(OCOCF_3)_2 \xrightarrow{KF} ArTlF_2 \xrightarrow{BF_3} ArF$$

Ar = alkyl substituted benzenes   45-71%

II.B.2-3   R. Louw and P. W. Franken, Chem. and Ind., 127 (1977).

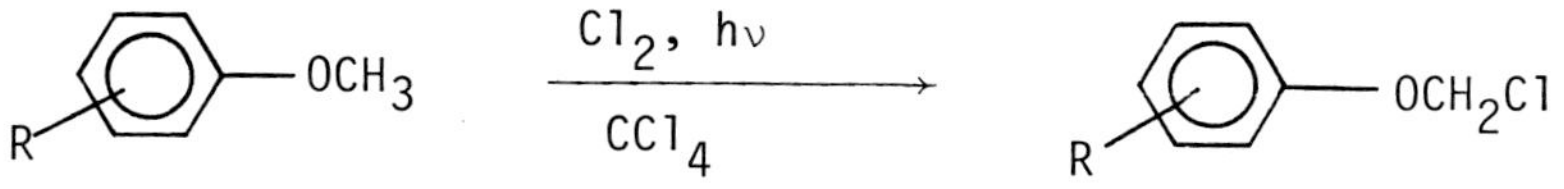

R–C$_6$H$_4$–OCH$_3$ $\xrightarrow[CCl_4]{Cl_2,\ h\nu}$ R–C$_6$H$_4$–OCH$_2$Cl

Widely varying yields.

II.B.2-4   T. Yajima and K. Munakata, Chem. Lett., 891 (1977).

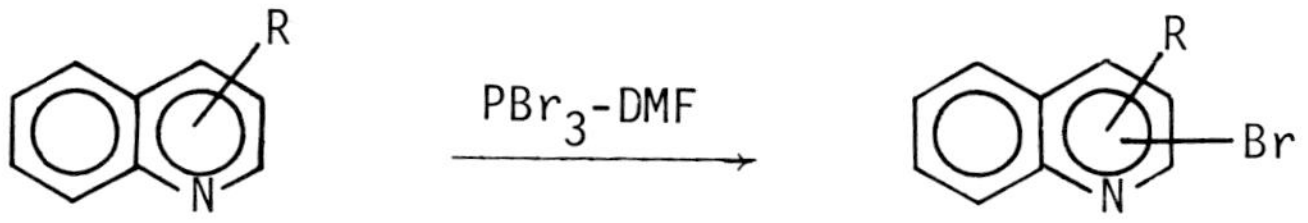

$$\text{(quinoline-R)} \xrightarrow{PBr_3\text{-DMF}} \text{(quinoline-R-Br)}$$

R = Cl, OMe   68-87%

II.B.2-5   R. Breslow et al., J. Am. Chem. Soc., 99, 905 (1977).

"Selective Halogenation of Steroids Using Attached Aryl Iodide Templates"

Full paper with many examples, experimental details.

198

II.B.2-6 F. M. Laskovics and E. M. Schulman, J. Am. Chem. Soc., 99, 6672 (1977).

$$\underset{R'}{\underset{|}{R-\overset{O}{\overset{\|}{C}}-CH}} \longrightarrow \underset{R'}{\underset{|}{R-C=CH}}-N(CH_3)_2 \xrightarrow[\text{3. NaHCO}_3]{\text{1. CCl}_3\text{COCCl}_3 \quad \text{2. H}_3O^{\oplus}} \underset{R'}{\underset{|}{R-\overset{O}{\overset{\|}{C}}-CH}}-Cl$$

~90%

II.B.2-7 B. Modarai and E. Khoshdel, J. Org. Chem., 42, 3527 (1977).

$$\underset{X}{\underset{|}{R-\overset{O}{\overset{\|}{C}}-\overset{H}{\overset{|}{C}}-R'}} \xrightarrow{\text{NBS}} \underset{X}{\underset{|}{R-\overset{O}{\overset{\|}{C}}-\overset{Br}{\overset{|}{C}}-R'}}$$

>90%

$$R-\overset{O}{\overset{\|}{C}}-CH_2-R' \xrightarrow{\text{NBS}} R-\overset{O}{\overset{\|}{C}}-CBr_2-R'$$

>90%

R,R' = Ph, Me

X = F, Cl

II.B.2-8   V. Calo, L. Lopez, and G. Pesce, J. Chem. Soc., Perkin I, 501 (1977).

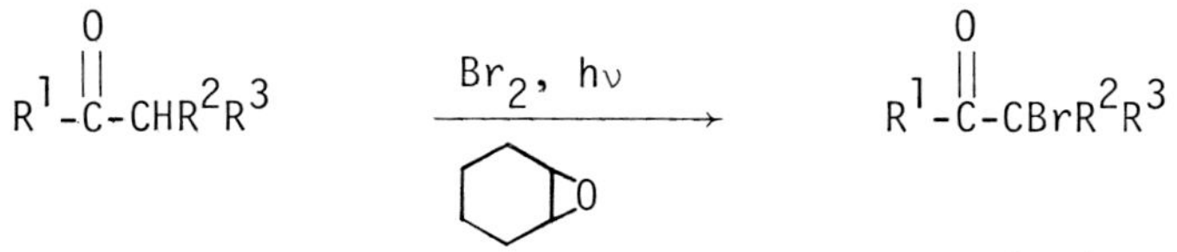

$$R^1\text{-}\underset{\underset{O}{\|}}{C}\text{-}CHR^2R^3 \xrightarrow{Br_2, h\nu, \text{ cyclohexene oxide}} R^1\text{-}\underset{\underset{O}{\|}}{C}\text{-}CBrR^2R^3$$

R^1 = Me, Bz, n-pentyl

R^2, R^3 = H, Me, Ar

~60-100%

II.B.3-1   W. Schroth et al., Z. Chem., 17, 411 (1977).

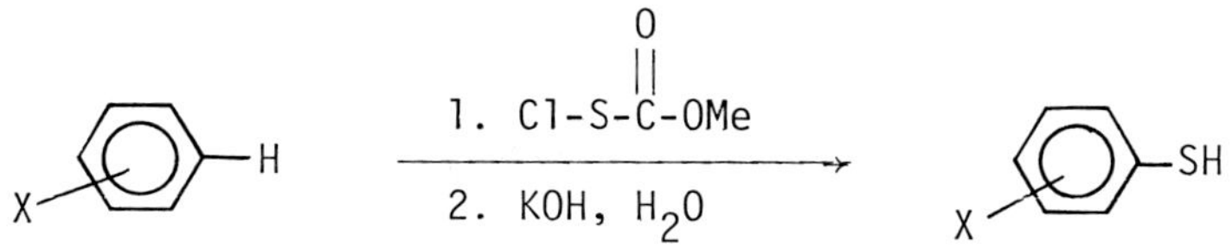

$$X\text{-}C_6H_4\text{-}H \xrightarrow[\text{2. KOH, } H_2O]{\text{1. Cl-S-C(=O)-OMe}} X\text{-}C_6H_4\text{-}SH$$

X = alkyl, alkoxyl, Br, fused aromatic

80-98%

II.B.3-2   K. Oka and S. Hara, Tetrahedron Lett., 2939 (1977).

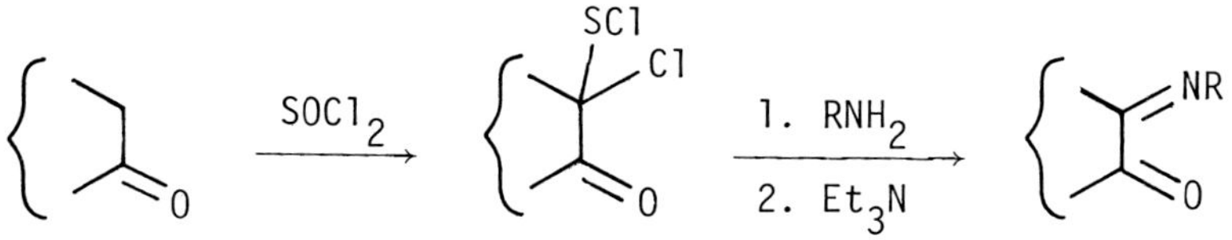

$$\text{cyclic ketone} \xrightarrow{SOCl_2} \text{α-Cl, α-SCl ketone} \xrightarrow[\text{2. } Et_3N]{\text{1. } RNH_2} \text{α-NR imine}$$

R = Ph, Bz, cyclohexyl   46-94%

OXIDATIONS

II.C-1 J. Bensoam and F. Mathey, *Tetrahedron Lett.*, 2797 (1977).

[cyclohexene-NR$_2$] $\xrightarrow{\text{1. } N_2F_2, CH_2Cl_2, \text{pyr} \quad \text{2. } H_2O}$ [cyclohexanone-F]   ~40-60%

II.C-2 E. Keinan and Y. Mazur, *J. Am. Chem. Soc.*, **99**, 3861 (1977).

$$\underset{RR'}{H\diagdown\!\!\diagup NO_2} \xrightarrow[\text{ether}]{\text{basic silica gel}} \underset{RR'}{\overset{O}{\|}C}$$

R,R' = H, n-alkyl, cyclic   ~80-90%

II.C-3 P. A. Bartlett et al., *Tetrahedron Lett.*, 331 (1977).

$$\underset{RR'}{H\diagdown\!\underset{C}{}\!\diagup NO_2} \xrightarrow[\substack{\text{2. t-BuOOH,}\\ \text{VO(acac)}_2,\\ \text{benzene}}]{\text{1. t-BuOK}} \underset{RR'}{\overset{O}{\|}C}$$

R,R' = H, alkyl   ~50-80%

II.D-1   E. Keinan and Y. Mazur, J. Org. Chem., **42**, 844 (1977).

$$R-NH_2 \xrightarrow[\text{2. } O_3]{\text{1. adsorbed on silica gel}} R-NO_2$$

R = 1°,2°,3° alkyl, benzyl

∼60-70%

II.D-2   E. E. Gilbert, Synthesis, 315 (1977).

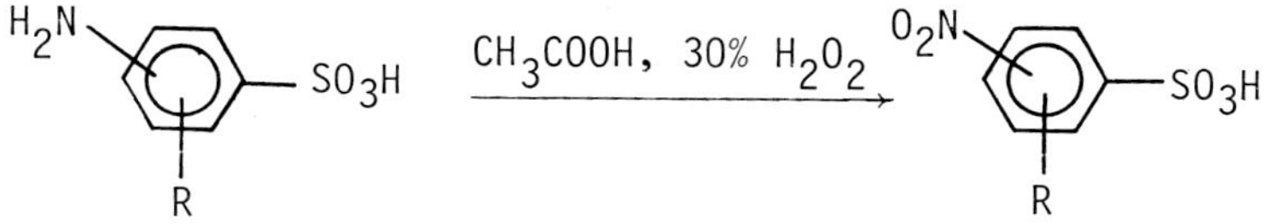

$$H_2N\text{-}C_6H_3(R)\text{-}SO_3H \xrightarrow{CH_3COOH,\ 30\%\ H_2O_2} O_2N\text{-}C_6H_3(R)\text{-}SO_3H$$

R = H, CH_3

33-74%

II.D-3   L. W. Deady, Synth. Commun., **7**, 509 (1977).

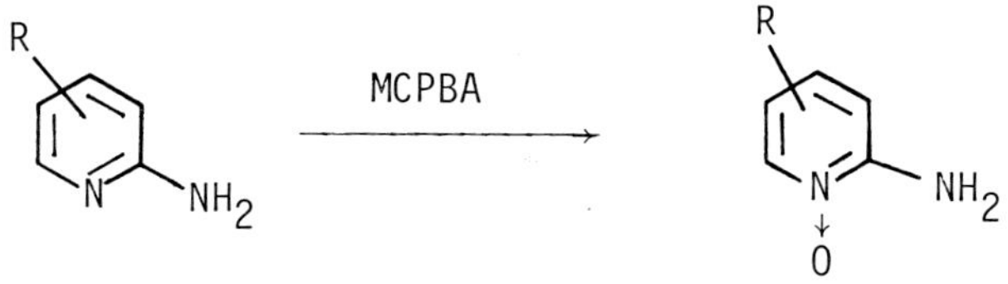

$$\underset{\text{2-aminopyridine (R)}}{\text{pyridine}} \xrightarrow{MCPBA} \text{N-oxide}$$

R = H, NO_2, Br, etc.

∼70-75%

Also works with other N-heterocycles.

OXIDATIONS

II.E-1 E. Fujita et al., Tetrahedron Lett., 1345 (1977).

$$R-S-R \xrightarrow[CHCl_3/HOAc]{Tl(NO_3)_3} R-\underset{\underset{O}{\|}}{S}-R$$

R = n-alkyl, Ph   82-94%

II.E-2 Y. Ueno, T. Inoue, and M. Okawara, Tetrahedron Lett., 2413 (1977).

$$R-S-R' \xrightarrow{(Bu_3Sn)_2O, Br_2} R-\underset{\underset{O}{\|}}{S}-R'$$

R,R' = Ph, Me, Bz, n-alkyl   78-92%
                              18% for R=R'=Ph

II.E-3 N. K. Gusarova, et al., Bull. Akad. USSR Chem., 25, 1536 (1976).

$$R-S-R \xrightarrow[170°]{Me_2SO} R-\underset{\underset{O}{\|}}{S}-R$$

R = Bu, Pr   55-59%

II.E-4   P. Geneste et al., Bull Soc. Chim. France, 271 (1977).

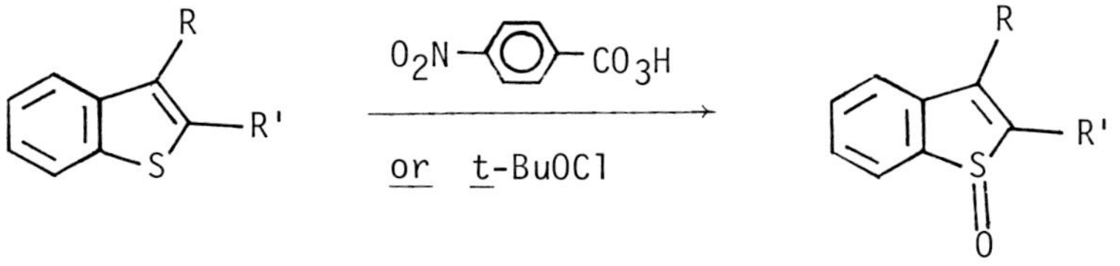

$$\text{benzothiophene-R,R'} \xrightarrow{O_2N-C_6H_4-CO_3H \text{ or } t\text{-BuOCl}} \text{benzothiophene-S-oxide-R,R'}$$

~50-70%

II.E-5   T.-L. Ho, Synth. Commun., 7, 363 (1977).

$$2\text{RSH} \xrightarrow{ClSO_2CCl_3, \text{ pyridine}} \text{R-S-S-R}$$

R = n-Bu, t-Bu, Ph, Bz   >90%

II.E-6   S. Oae et al., Chem. Lett., 893 (1977).

$$\text{RSH} \xrightarrow{N_2O_4} [\text{R-S-N=O}] \xrightarrow{\text{R'SH}} \text{R-S-S-R'}$$

R,R' = alkyl, aryl $\searrow$ R"SO$_2$H   ~50-90%

R' = aryl

                                       R-SSO$_2$-R"

                                       61-82%

OXIDATIONS

II.E-7   S. Oae, D. Fukushima, and Y. H. Kim, J. Chem. Soc., Chem. Commun., 407 (1977).

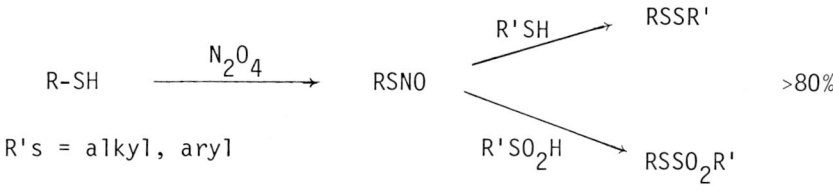

R-SH $\xrightarrow{N_2O_4}$ RSNO

R's = alkyl, aryl

R'SH → RSSR'

R'SO$_2$H → RSSO$_2$R'

>80%

II.E-8   S. Uemura, S. Tanaka, and M. Okano, Bull. Chem. Soc. Japan, 50, 220 (1977).

2 R-SH $\xrightarrow{Tl(OAc)_3}$ R-S-S-R

R = alkyl, aryl, heterocyclic, containing OH, COOR, NH$_2$ groups

68-100%

II.F.1-1   H. Kropf and M. R. Yazdanbakhch, Synthesis, 711 (1977).

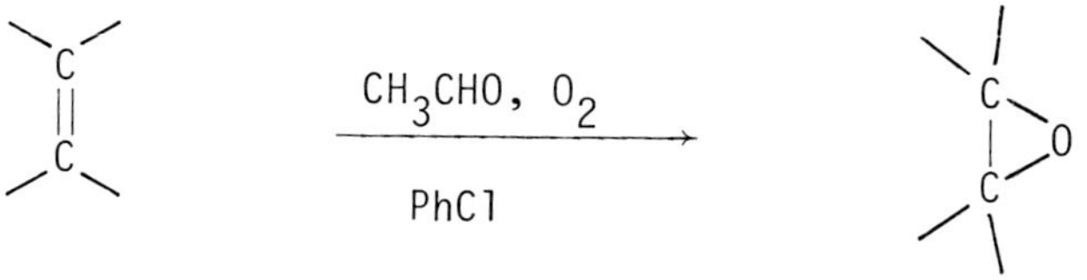

$$\underset{C}{\overset{C}{\|}} \xrightarrow[PhCl]{CH_3CHO,\ O_2} \underset{C}{\overset{C}{\diagdown}}\!O$$

(alkyl, Ph substituted)

∿70-90%

II.F.1-2 T. Yamamoto and M. Kimura, J. Chem. Soc., Chem. Commun., 948 (1977).

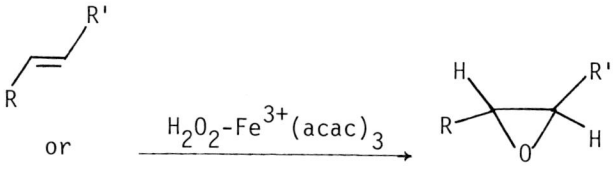

R,R' = alkyl, Ph

30-78%

>80% trans

II.F.1-3 P. A. Grieco et al., J. Org. Chem., 42, 2034 (1977).

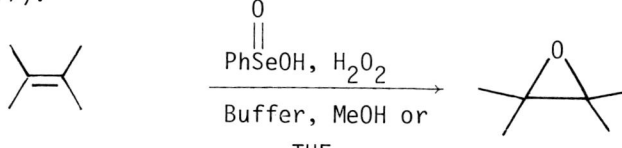

Mono-, di-, and trisubstituted olefins.

47-85%

II.F.1-4 J. A. Cella et al., J. Org. Chem., 42, 2077 (1977).

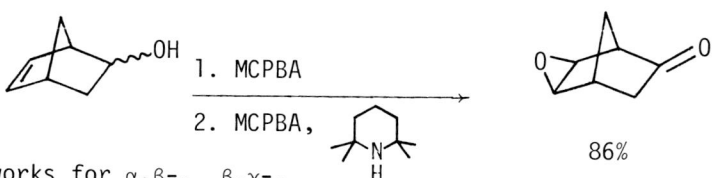

Also works for α,β-, β,γ-, and other epoxy ketones.

86%

OXIDATIONS

II.F.1-5   P. A. Bartlett and K. K. Jernstedt, *J. Am. Chem. Soc.*, **99**, 4829 (1977).

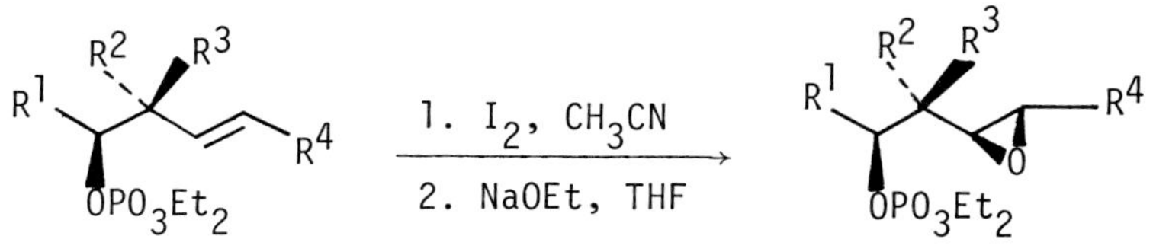

$$R^1\text{-}C(R^2)(R^3)\text{-}CH\text{=}CH\text{-}R^4, \text{ OPO}_3\text{Et}_2 \xrightarrow[\text{2. NaOEt, THF}]{\text{1. I}_2, \text{ CH}_3\text{CN}} \text{epoxide product, OPO}_3\text{Et}_2$$

∼70%

II.F.1-6   R. Breslow and L. M. Maresca, *Tetrahedron Lett.*, 623 (1977).

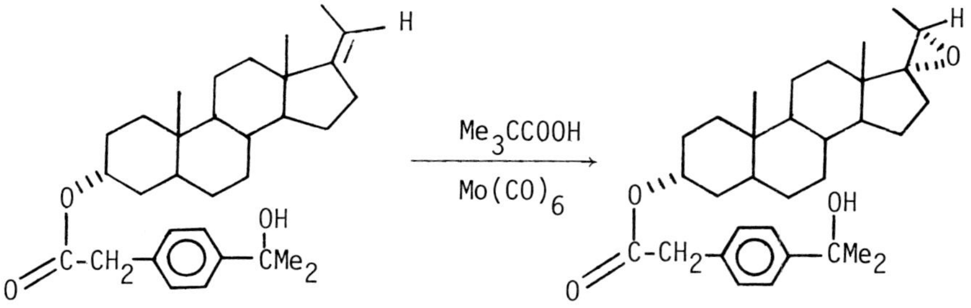

Reagents: Me$_3$CCOOH, Mo(CO)$_6$

60% (remainder recovered starting material)

II.F.1-7   K. Ishikawa, H. C. Charles, and G. W. Griffin, *Tetrahedron Lett.*, 427 (1977).

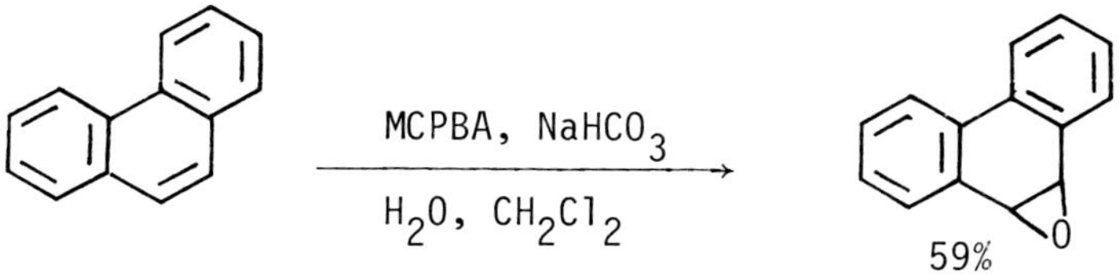

Phenanthrene $\xrightarrow[\text{H}_2\text{O, CH}_2\text{Cl}_2]{\text{MCPBA, NaHCO}_3}$ K-region arene oxide

59%

Also provides other K-region arene oxides, yields ∼50%

II.F.1-8 S. Krishnan, D. G. Kuhn, and G. A. Hamilton, Tetrahedron Lett., 1369 (1977).

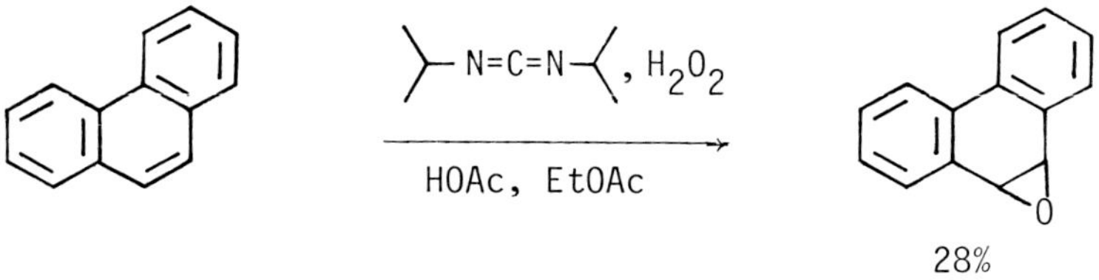

phenanthrene + (iPr)N=C=N(iPr), H_2O_2 / HOAc, EtOAc → phenanthrene 1,2-epoxide

28%

II.F.1-9 S. Krishnan, D. G. Kuhn, and G. A. Hamilton, J. Am. Chem. Soc., 99, 8121 (1977).

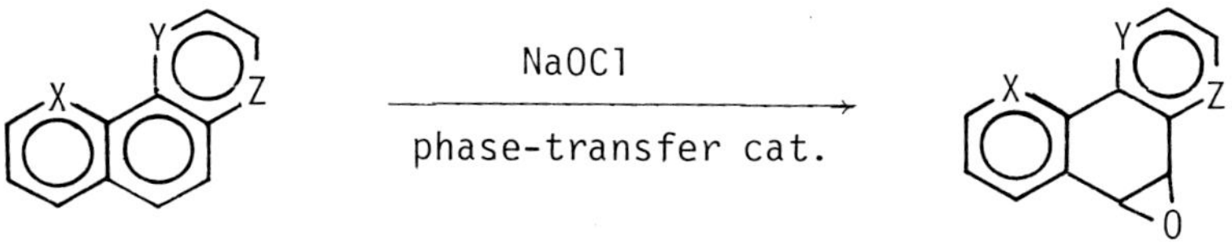

$\underset{\text{phase-transfer cat.}}{\xrightarrow{\text{NaOCl}}}$

X,Y,Z = C,N

II.F.2-1 A. A. Frimer, Synthesis, 578 (1977).

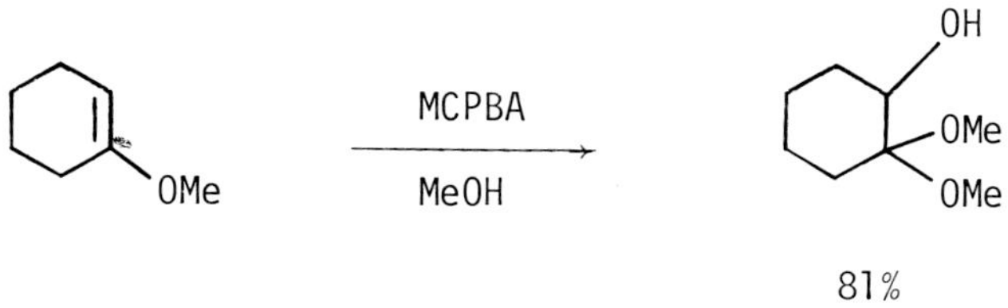

1-methoxycyclohexene + MCPBA / MeOH → 1,2-dimethoxycyclohexan-1-ol

81%

OXIDATIONS

II.F.3-1   G. M. Rubottom and J. M. Gruber, <u>J. Org. Chem.</u>, <u>42</u>, 1051 (1977).

[structure: 1-(trimethylsilyloxy)cyclohexadiene with R substituent]   1. Pb(OCOPh)$_4$
2. Et$_4$N$^{\oplus}$ F$^{\ominus}$
→ [cyclohexanone with OCOPh groups and R]   ~60%

II.F.3-2   G. Piancatelli, A. Scettri, and M. D'Auria, <u>Tetrahedron Lett.</u>, 3483 (1977).

$$\underset{\text{R-C=C-OR'}}{\overset{\text{H H}}{|\,|}} \xrightarrow{C_5H_5NHCrO_3Cl} R-CH_2-\overset{O}{\underset{||}{C}}-OR'$$

R = H   75-95%
R'= Et, steroidal,
-(CH$_2$)$_2$-, -(CH$_2$)$_3$-

II.F.3-3   P. D. Woodgate <u>et al.</u>, <u>J. Chem. Soc., Perkin I</u>, 2231 (1977).

[cycloalkene] $\xrightarrow{I(OAc)_3}$ [cycloalkane with OAc and I substituents]

II.F.3-4   A. J. Duggan and S. S. Hall, J. Org. Chem., 42, 1057 (1977).

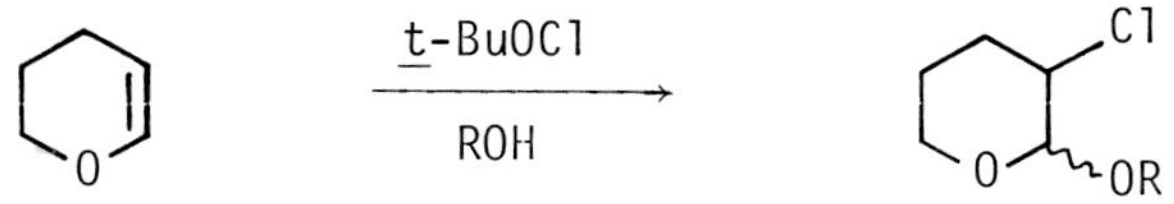

[dihydropyran] + t-BuOCl / ROH → [2-chloro-3-alkoxy tetrahydropyran]

R = 1°, 2° alkyl, Bz   50-85%

II.F.3-5   J. E. Bäckvall, M. W. Young, and K. B. Sharpless, Tetrahedron Lett., 3523 (1977).

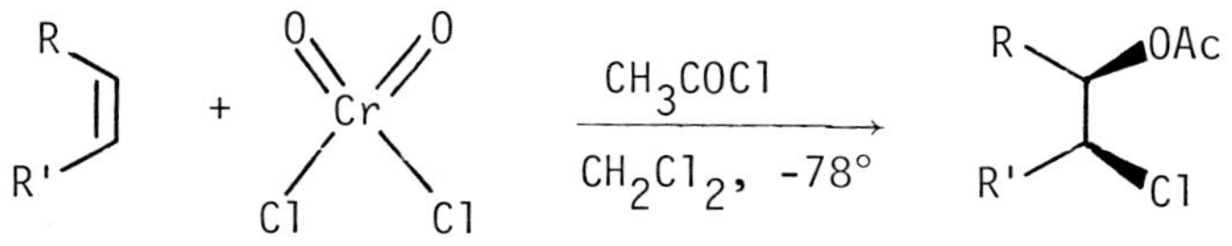

R−CH=CH−R' + CrO₂Cl₂ —(CH₃COCl, CH₂Cl₂, −78°)→ R−CH(OAc)−CH(Cl)−R'

55-88%

II.F.3-6   G. Cardillo and M. Shimizu, J. Org. Chem., 42, 4268 (1977).

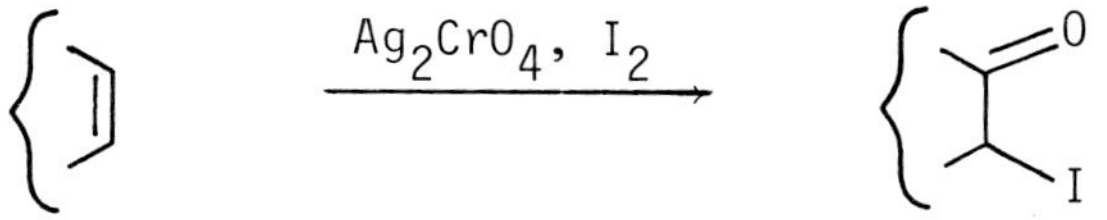

[cyclopentene] —(Ag₂CrO₄, I₂)→ [2-iodocyclopentanone]

60-65%

R−CH=CH₂ —(Ag₂CrO₄, I₂)→ R−C(=O)−CH₂I

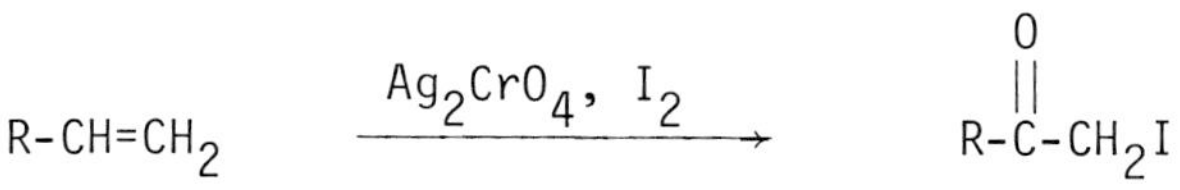

R = alkyl, Ph, cinnamyl, CH₂OBz   ~70-80%

OXIDATIONS

II.F.3-7 H. Driguez, J. M. Paton, and J. Lessard, Can. J. Chem., 55, 700 (1977).

[steroid with AcO and double bond] $\xrightarrow[\text{chromous chloride}]{\text{RO-C(=O)-NHCl}}$ [steroid with AcO, Cl, NHCOOR]

79%

II.F.3-8 V. Jager and J. Günther, Angew. Chem. Int. Ed., 16, 246 (1977).

$CH_2=CH-(CH_2)_n-CH=CH_2$ $\xrightarrow[\text{Ether, 0°C}]{N_2O_4/I_2}$ $O_2N-CH_2-CHI-(CH_2)_n-CH=CH_2$

n = 2,3,4   >80%

II.F.3-9 T.-L. Ho, B.G.B. Gupta, and G. A. Olah, Synthesis, 676 (1977).

[cycloalkene $(CH_2)_n$ with CH=CH] $\xrightarrow[\text{phase-transfer cat.}]{HX, H_2O_2}$ [cycloalkane $(CH_2)_n$ with CHX-CHX]

n = 4,5,6

∼75% (X=Cl)
∼95% (X=Br)

II.F.3-10 L. Rasteikiene et al., Russ. Chem. Rev., 46, 548 (1977).

Review: "The Addition of Sulphenyl Chlorides to Unsaturated Compounds"

II.G-1 A. McKillop and D. W. Young, Synth. Commun., 7, 467 (1977).

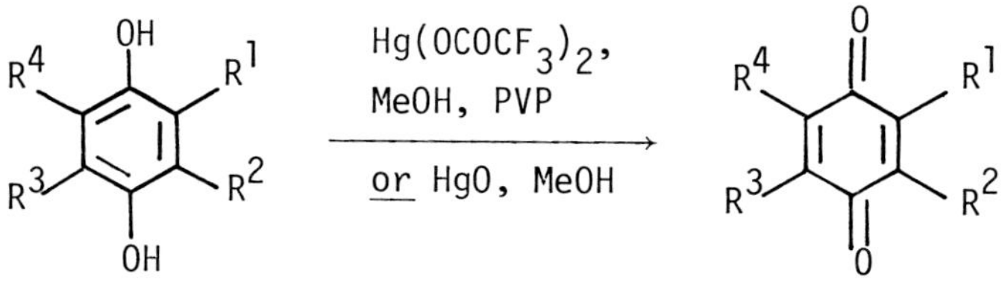

[hydroquinone with R¹, R², R³, R⁴ substituents] → Hg(OCOCF₃)₂, MeOH, PVP or HgO, MeOH → [benzoquinone with R¹, R², R³, R⁴ substituents]

R's = H, Me, t-Bu, Cl   ~70-90%

II.G-2 A. McKillop and S. J. Ray, Synthesis, 847 (1977).

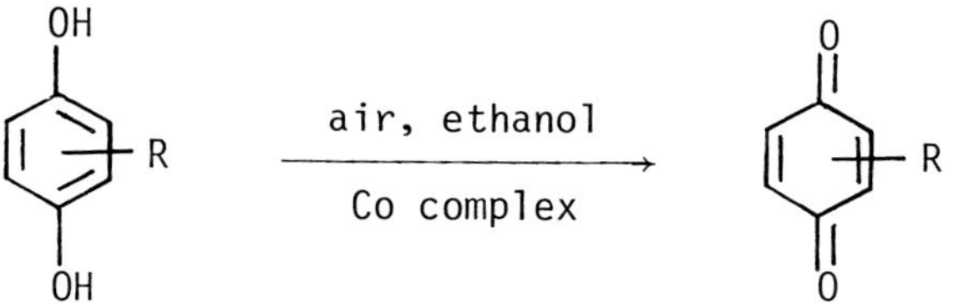

[hydroquinone with R] → air, ethanol, Co complex → [benzoquinone with R]

R = H, Me, t-Bu, Cl   42-96%

II.G-3 Y. Yamada and K. Hosaka, Synthesis, 53 (1977).

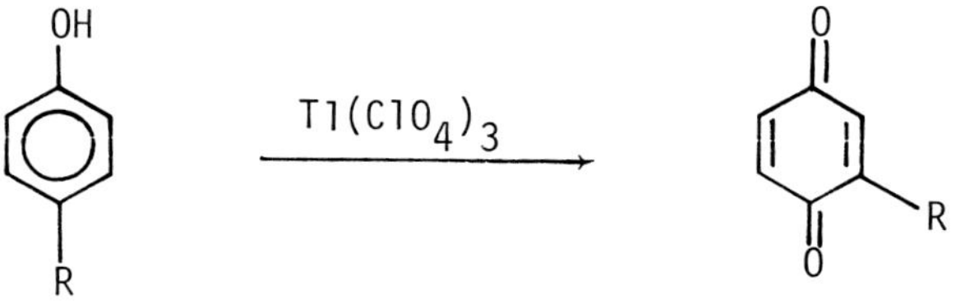

[4-substituted phenol] → Tl(ClO₄)₃ → [substituted benzoquinone with R]

R = CH₃, CH₂CH₂OCOCH₂Cl   65-70%

OXIDATIONS

II.G-4   A. Nishinaga et al., Synthesis, 270 (1977).

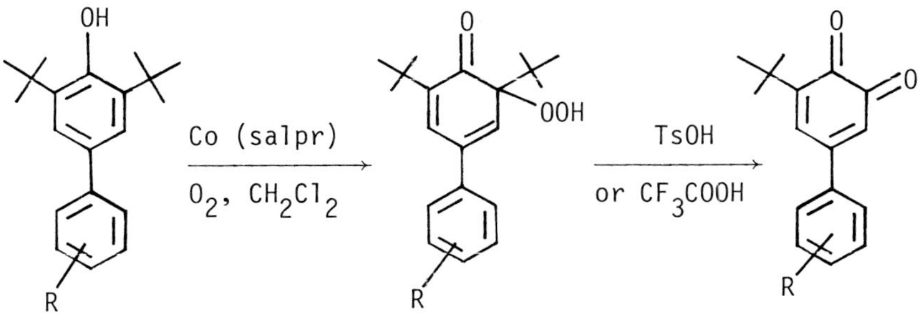

R = H, Cl, Me, OMe   62-84%   100%

II.G-5   M. J. Manning, D. R. Henton, and J. S. Swenton, Tetrahedron Lett., 1679 (1977).

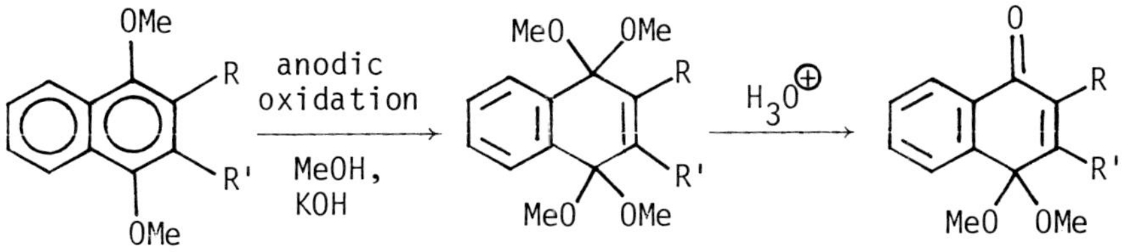

R = H, Me   74-95%   56-94%
R'= H, Me, Br

II.G-6   I. Tabushi, K. Fujita, and H. Kawakubo, J. Am. Chem. Soc., 99, 6456 (1977).

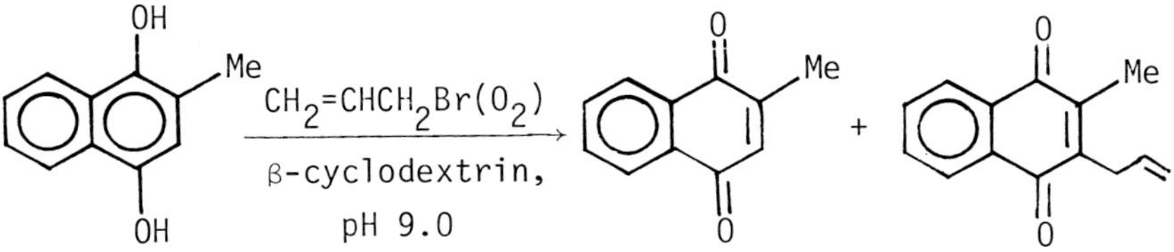

[Reaction: 1,4-dihydroxy-2-methylnaphthalene + $CH_2=CHCH_2Br(O_2)$, β-cyclodextrin, pH 9.0 → 2-methyl-1,4-naphthoquinone (22%) + 2-methyl-3-allyl-1,4-naphthoquinone (60%)]

II.H-1   V. N. Odinokov, L. P. Zhemaiduk, and G. A. Tolstikov, Bull. Akad. USSR Chem., 25, 1790 (1977).

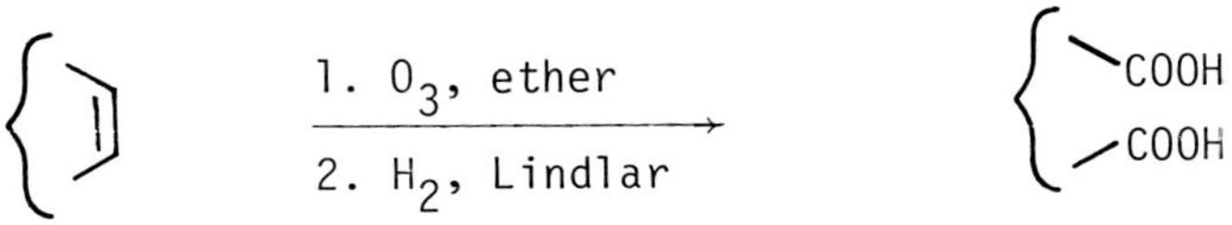

[Reaction: cyclopentene → 1. O_3, ether; 2. H_2, Lindlar → dicarboxylic acid (COOH, COOH), 80-99%]

II.H-2   B. M. Trost and G. S. Massiot, J. Am. Chem. Soc., 99, 4405 (1977).

[Reaction: cyclic ketone → 1. PhSSPh; 2. $Pb(OAc)_4$ → α-PhS, OAc substituted ketone (55-99%) → H_2O_2, NaOH → dicarboxylic acid (HOOC, HOOC) (65-99%)]

OXIDATIONS

II.H-3   A. P. Krapcho, J. R. Larson, and J. M. Eldridge, J. Org. Chem., 42, 3749 (1977).

$$R\text{-}CH=CH_2 \xrightarrow[H_2O]{KMnO_4, HOAc} R\text{-}COOH$$

R = alkyl, Ph   ~90%

II.H-4   N. C. Deno et al., Tetrahedron Lett., 1703 (1977).

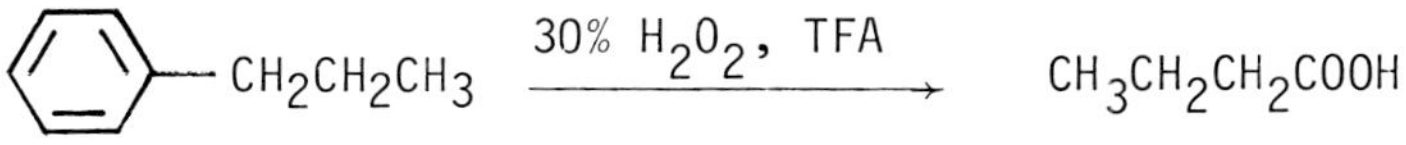

$$Ph\text{-}CH_2CH_2CH_3 \xrightarrow{30\% H_2O_2, TFA} CH_3CH_2CH_2COOH$$

~70%

II.H-5   B. M. Trost and Y. Tamaru, J. Am. Chem. Soc., 99, 3101 (1977).

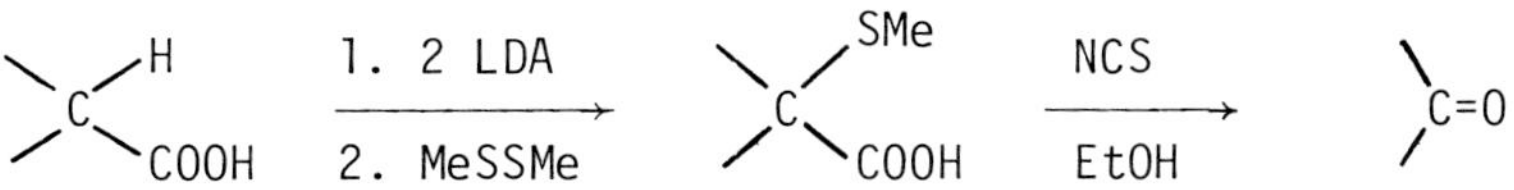

$$\underset{COOH}{\overset{H}{>}C<} \xrightarrow[\text{2. MeSSMe}]{\text{1. 2 LDA}} \underset{COOH}{\overset{SMe}{>}C<} \xrightarrow[EtOH]{NCS} >C=O$$

~50-60% overall

II.I-1   V. I. Stenberg et al., J. Org. Chem., 42, 171 (1977).

$$R\text{-}CH_2OH \xrightarrow[h\nu]{FeCl_3} R\text{-}CHO$$

R = 1°,2° alkyl

II.J-1   E. Mincione, G. Ortaggi, and A. Sirna, Synthesis, 773 (1977).

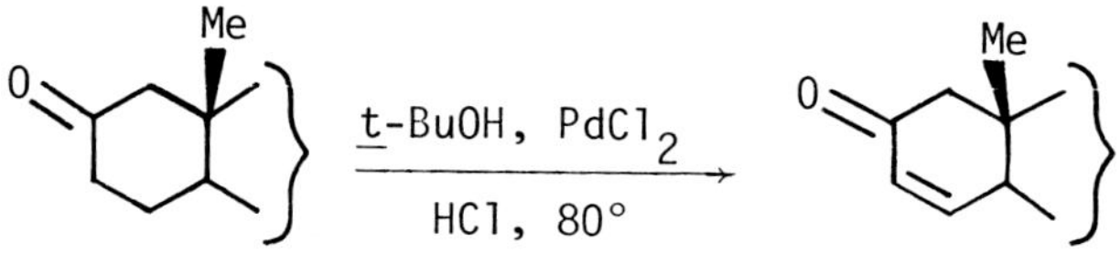

$$\text{ketone} \xrightarrow[\text{HCl, 80°}]{\underline{t}\text{-BuOH, PdCl}_2} \text{enone}$$

∼40-80%

II.J-2   D. Bondon, Y. Pietrasanta, and B. Pucci, Tetrahedron Lett., 821 (1977).

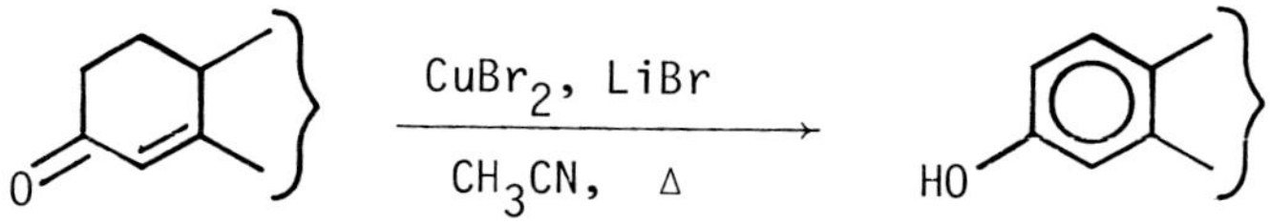

$$\text{enone} \xrightarrow[\text{CH}_3\text{CN, }\Delta]{\text{CuBr}_2,\text{ LiBr}} \text{phenol}$$

∼80%

II.J-3   K. Dimroth and W. Tuncher, Synthesis, 339 (1977).

$$\text{R-NH-NH-R'} \xrightarrow[\text{K}_3\text{Fe(CN)}_6,\text{ NaOH}]{(4-\underline{t}\text{-Bu-Ph})_3\text{C}_6\text{H}_2\text{OH}} \text{R-N=N-R'}$$

R,R' = Ph, t-Bu, alkyl   63-98%
containing CN groups

II.K-1   T. Matsuura, Tetrahedron, 33, 2869 (1977).

Review: "Bio-mimetic Oxygenation"

OXIDATIONS

II.K-2 K. J. Divakar and A. S. Rao, Indian J. Chem., 14B, 704 (1976).

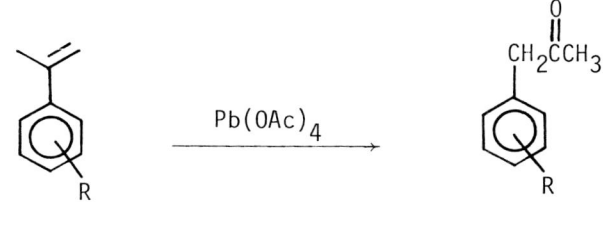

35-80%

II.K-3 E. Lee-Ruff, Chem. Soc. Rev., 6, 195 (1977).

Review: "The Organic Chemistry of Superoxide"

II.K-4 V. Karnojitzky, Russ. Chem. Rev., 46, 121 (1977).

Review: "The Autoxidation of Nitrogenous Compounds and Certain Aspects of Associated Biochemical Processes"

III.A-1   P. Caubere et al., Tetrahedron Lett., 1069 (1977).

$$\underset{R}{\overset{O}{\underset{\|}{C}}}\diagdown_{R'} \quad \xrightarrow[\text{THF}]{\text{NaH-}\underline{t}\text{-AmylONa-Ni(OAc)}_2} \quad \underset{R}{\overset{H\diagup\diagdown OH}{\underset{C}{}}}\diagdown_{R'}$$

R,R' = 1°,3° alkyl,   48-90% isolated
       Ph, cyclic

III.A-2   J. G. de Vries, T. J. van Bergen, and R. M. Kellogg, Synthesis, 246 (1977).

$$R-\overset{O}{\underset{\|}{C}}-R' \quad \xrightarrow[\text{H}_2\text{O/dioxane}]{\text{Na}_2\text{S}_2\text{O}_4} \quad R-\overset{OH}{\underset{|}{C}H}-R'$$

R,R' = cyclic, alkyl,   ~80-90%
       aryl, H

III.A-3   G. Mestroni, G. Zassinovich, and A. Camins. J. Organomet. Chem., 140, 63 (1977).

The $Rh(BiPy)_2^+$ system will hydrogenate ketones to alcohols in the presence of olefins.

III.A-4   O. Kriz, J. Stuchlik, and B. Casensky, Z. Chem., 17, 18 (1977).

$$\underset{RR'}{R-\underset{\underset{\displaystyle O}{\|}}{C}-R'} \xrightarrow{(HAlN-\underline{i}-Pr)_6} \underset{RR'}{\overset{HOH}{\underset{}{\diagdown C \diagup}}}$$

R = alkyl, aryl
R'= H, alkyl, Ph

50-98%

III.A-5   V. Z. Sharf et al., Bull. Akad. USSR Chem., 26, 995 (1977).

cyclohexenone + (PPh$_3$)$_3$RhCl

1. H$_2$SiR$_2$
2. H$^{\oplus}$ → cyclohexanone

1. H$_3$SiR
2. H$^{\oplus}$ → cyclohexenol (OH)

Widely varying yields.

III.A-6   R. S. Brinkmeyer and V. M. Kapoor, J. Am. Chem. Soc., 99, 8339 (1977).

$$R'C\equiv C-\underset{\underset{\displaystyle O}{\|}}{C}-R \xrightarrow{LiAlH_2 \cdot 2 \text{ chiral aminoalcohol}} R'-C\equiv C-\underset{R}{\overset{H}{\underset{|}{C}}}-OH$$

R = n-pentyl, i-Bu
R'= H, Me, n-pentyl, Me$_3$Si

∼80-90% chemical yields
∼30-80% ee

III.A-7   T. Saegusa et al., J. Org. Chem., 42, 2797 (1977).

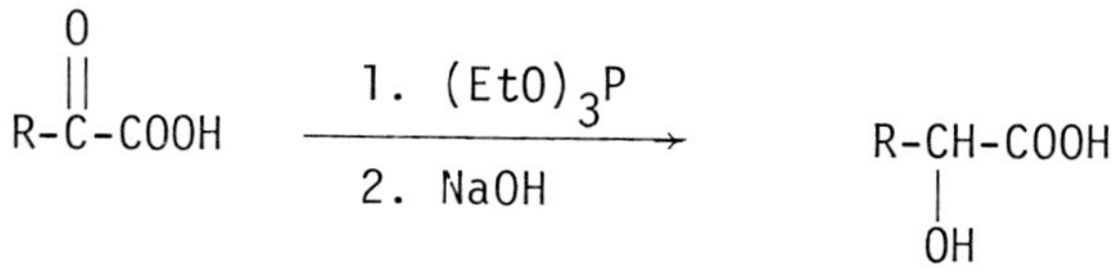

$$R-\overset{O}{\underset{\|}{C}}-COOH \xrightarrow[\text{2. NaOH}]{\text{1. (EtO)}_3P} R-\underset{OH}{\underset{|}{CH}}-COOH$$

R = Me, Ph, aliphatic acid   67-94%

III.A-8   A. Romeo, et al., Tetrahedron Lett., 2369 (1977).

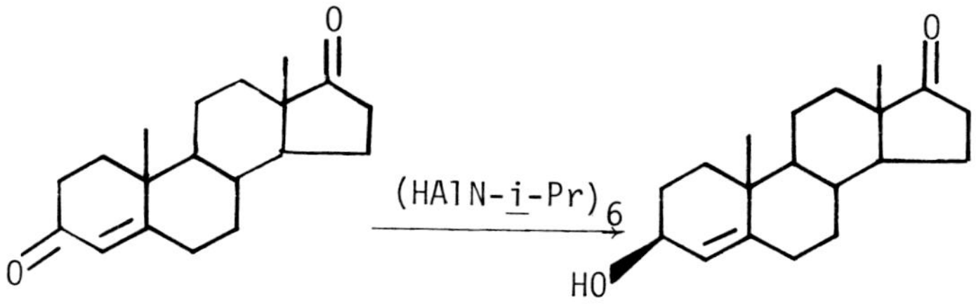

(HAlN-i-Pr)$_6$

~50%

III.A-9   S. Nishimura, M. Ishige, and M. Shiota, Chem. Lett., 963 (1977).

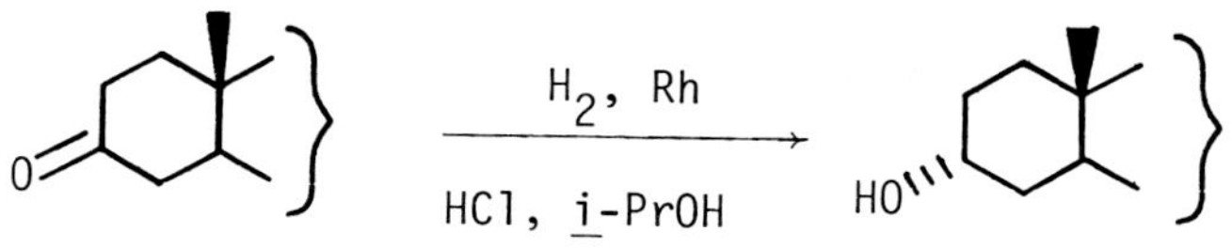

$$\xrightarrow[\text{HCl, i-PrOH}]{H_2, Rh}$$

>95%

III.A-10   S. Krishnamurthy, F. Vogel, and H. C. Brown, J. Org. Chem., 42, 2534 (1977).

$$\underset{R \quad R'}{\overset{O}{\underset{\|}{C}}} \xrightarrow[\text{THF, } -78°]{\text{Li(HB-IPC-9-BBN)}} \underset{R \quad R'}{\overset{H \quad OH}{\underset{*}{C}}}$$

70-80%

3-37% ee, predominantly (R)

III.A-11   N. Baggett and P. Stribblehill, J. Chem. Soc., Perkin I, 1123 (1977).

$$\underset{R \quad R'}{\overset{O}{\underset{\|}{C}}} \xrightarrow{\text{LiAlH}_4 \cdot \text{mannitol derivatives}} \underset{R \quad R'}{\overset{H \quad OH}{\underset{*}{C}}}$$

37-77%, up to 5% ee

III.A-12   T. Mukaiyama et al., Chem. Lett., 783 (1977).

$$\text{Ph-}\underset{\|}{\overset{O}{C}}\text{-CH}_3 \xrightarrow{\text{LiAlH}_4} \text{Ph-}\underset{*}{\text{CH}}(\text{OH})\text{-CH}_3 \quad 93\%$$

(with pyrrolidine derivative: N-H, CH₂NHPh, H)

(S) 92% ee

III.A-13   I. Ojima, T. Kogure, and M. Kumagai, J. Org. Chem., 42, 1671 (1977).

$$\underset{\text{R-C-C-OR}^1}{\overset{O\;\;O}{\|\;\;\|}} \xrightarrow[\text{DIOP or BMPP}]{R^2R^3SiH_2} \underset{*}{\overset{OH}{\underset{|}{\text{R-CH-COOR}^1}}}$$

R^1 = n-alkyl R^2, R^3 = Me, Ph, Et,   ~80-90%
R = Me, Ph   α-Np   10-80% ee

III.A-14   K. Achiwa, Tetrahedron Lett., 3735 (1977).

$$\underset{\text{CH}_3\text{CCOOR}}{\overset{O}{\|}} \xrightarrow{H_2,\; \text{CPPM-Rh}} \underset{*}{\overset{OH}{\underset{|}{\text{CH}_3\text{-CHCOOR}}}}$$

CPPM = [pyrrolidine with PPh$_2$, CH$_2$PPh$_2$, COOC$_{27}$H$_{45}$]

~99% yield,
~65% ee (R)

III.A-15   T. Harada et al., Chem. Lett., 1131 (1977).

$$\underset{\text{CH}_3\text{-C-CH}_2\text{C-OCH}_3}{\overset{O\;\;\;\;\;O}{\|\;\;\;\;\;\|}} \xrightarrow[\text{tartaric acid -}\atop\text{modified NiO}]{H_2,\; \text{EtCOOMe}} \underset{*}{\overset{H\;\;OH}{\diagdown\diagup}}\text{CH}_3\text{-C-CH}_2\text{-}\overset{O}{\overset{\|}{\text{C}}}\text{-OCH}_3$$

85% ee

REDUCTIONS

III.A-16   H. C. Brown and S. U. Kulkarni, J. Org. Chem., 42, 4169 (1977).

$$R\text{-}CHO \xrightarrow{\text{9-BBN·pyr}} R\text{-}CH_2OH$$

R = alkyl, aryl,   85-100%
cyclic, cinnamyl

Can reduce aldehydes selectively in the presence of ketones.

III.A-17   J. Tsuji and H. Suzuki, Chem. Lett., 1085 (1977).

$$R\text{-}CHO \xrightarrow{H_2, \; RuCl_2(PPh_3)_3} R\text{-}CH_2OH$$

R = aryl, n-alkyl   100%

Ketones and nitro compounds do not react.

III.A-18   W. Strohmeier and H. Steigerwald, J. Organomet. Chem., 129, C43 (1977).

$$IrH_3(PPh_3)_3 + HOAc \longrightarrow \text{catalyst}$$

$$R\text{-}CHO \xrightarrow[\text{catalyst}]{H_2, \; 10 \text{ atm (neat)}} RCH_2OH$$

Ketones are not reduced.

III.A-19   G. H. Posner, A. W. Runquist, and M. J. Chapdelaine, J. Org. Chem., 42, 1202 (1977).

$$\text{R-CHO} \xrightarrow[\text{or } \underline{i}\text{-Pr}_2\text{CHOH on alumina}]{\underline{i}\text{-PrOH on alumina}} \text{R-CH}_2\text{OH}$$

$$\underset{R \quad R'}{\overset{O}{\underset{\|}{C}}} \xrightarrow{\text{same conditions}} \underset{R \quad R'}{\overset{OH}{\underset{|}{CH}}}$$

Very selective; can be done in the presence of sensitive functional groups.

III.A-20   H. W. Gibson and F. C. Bailey, J. Chem. Soc., Chem. Commun., 815 (1977).

R-CHO $\xrightarrow{\text{[polymer-CH}_2\text{NMe}_3^+ \text{ BH}_4^-\text{]}}$ R-CH$_2$OH

(reagent: $-(\text{CHCH}_2)_n-$ polystyrene with $-\text{CH}_2\text{NMe}_3^{\oplus}\ \text{BH}_4^{\ominus}$)

III.A-21   R. C. Northrop, Jr. and P. L. Russ, J. Org. Chem., 42, 4148 (1977).

$$\underset{H \quad NH}{\overset{O}{\underset{\|}{C}}}\text{-peptide} \xrightarrow[\text{THF}]{\text{BH}_3} \text{CH}_3\text{-NH-peptide}$$

∼50%

III.A-22 A. M. Maione and M. G. Quaglia, Chem. and Ind., 230 (1977).

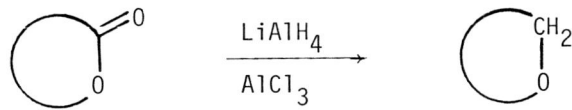

Widely varying yields.

III.A-23 M. E. Kuehne and P. J. Shannon, J. Org. Chem., 42, 2082 (1977).

$$\underset{\text{R-C-NR'}_2}{\overset{O}{\|}} \xrightarrow[\text{2. NaBH}_4]{\text{1. POCl}_3} \text{R-CH}_2\text{-NR'}_2$$

R = Ph, 1°,2°,3° alkyl, cyclic   generally ~60-80%
R'= H, Me, alkyl, cyclic

III.A-24 Y. Tsuda, T. Sano, and H. Watanabe, Synthesis, 652 (1977).

$$\underset{\text{NH}}{\overset{O}{\overset{\|}{C}}} \xrightarrow[\text{2. NaBH}_4,\ \text{SnCl}_4]{\text{1. Et}_3\text{O}^{\oplus}\ \text{BF}_4^{\ominus}} \underset{\text{NH}}{\text{CH}_2}$$

~70-80% overall

III.A-25   A. Basha and A. Rahman, Experientia, 33, 101 (1977).

$$\underset{R'}{\text{Ar}}-\overset{O}{\underset{\|}{C}}-NR_2 \xrightarrow[\text{2. Zn, EtOH}]{\text{1. POCl}_3} \underset{R'}{\text{Ar}}-CH_2-NR_2$$

$R_2 = H_2, -(CH_2)_5-$

$R' = H, Cl, Br$

~80-90%

III.A-26   T. Mizoroki et al., Bull. Chem. Soc. Japan, 50, 2148 (1977).

$$\text{PhCH=C-CHO} \xrightarrow[\text{Rh}_2\text{Cl}_2(\text{CO})_4, \text{Me-N}\bigcirc]{H_2} \text{PhCH=C-CH}_2\text{OH}$$
$$||$$
$$RR$$

R = H, Me

~85%

III.A-27   H. C. Brown and S. C. Kim, Synthesis, 635 (1977).

$$R-\overset{O}{\underset{\|}{C}}-NMe_2 \xrightarrow[\text{THF}]{\text{LiEt}_3\text{BH}} R-CH_2OH$$

R = Ph, 1°,2° alkyl, cycloalkyl

62-90%

REDUCTIONS

III.A-28 J. T. Traxler, Synth. Commun., 7, 161 (1977).

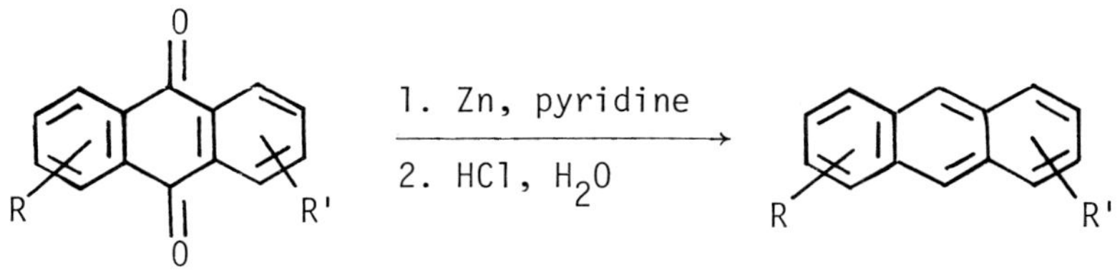

[anthraquinone with R, R' substituents] →(1. Zn, pyridine; 2. HCl, H₂O)→ [anthracene with R, R' substituents]

R,R' = H, Cl, Me, COOH   Widely varying yields.

III.A-29/III.B-1   J. W. Reed and W. L. Jolly, J. Org. Chem., 42, 3963 (1977).

$$Ph-C\equiv N \xrightarrow{BH_3OH^{\ominus}} Ph-CH_2NH_2$$

89%

$$Ph-COOEt \xrightarrow{BH_3OH^{\ominus}} PhCH_2OH$$

89%

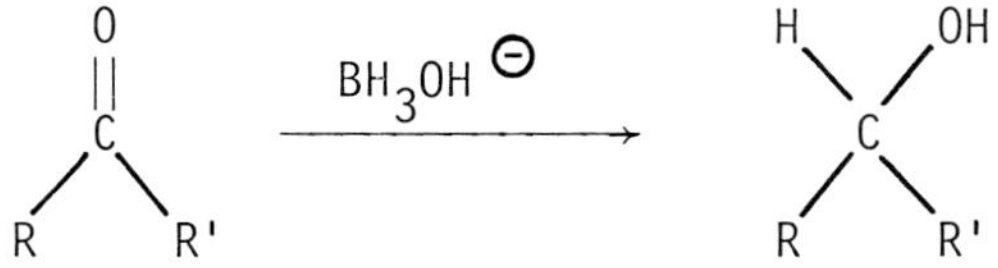

[ketone R-CO-R'] →(BH₃OH⁻)→ [secondary alcohol R-CH(OH)-R']

∼70-80%

R, R' = alkyl

III.C-1 T.-L. Ho, Synth. Commun., 7, 321 (1977).

$$\text{R-S(=O)-R'} \xrightarrow{\text{Mo(CO)}_6, \text{HOAc}} \text{R-S-R'}$$

R, R' = n-Bu, Ph, Bz,   75-90%
Me, -(CH$_2$)$_4$-

III.C-2 J. Drabowicz and S. Oae, Chem. Lett., 767 (1977).

$$\text{R-S(}\uparrow\text{O)-R'} + (CF_3CO)_2O + H_2S \xrightarrow[-60°]{CH_2Cl_2} \text{R-S-R'}$$

R = Ph, p-Tol   ∼90%
R' = Me, neopentyl

Can be done selectively in the presence of sulfinic esters and thiosulfinates.

III.C-3 R. Tanikaga et al., Chem. Lett., 395 (1977).

$$\text{R-S(=O)-R'} \xrightarrow[CH_3SCH_3 \ (1.2\text{-}2 \ eq)]{(CF_3CO)_2O} \text{R-S-R'}$$

R = CH$_3$, Ph   >95%
R' = alkyl, vinyl, aryl,
    containing ethers, esters,
    nitriles, etc.

III.C-4   D.L.J. Clive et al., J. Chem. Soc., Chem. Commun., 657 (1977).

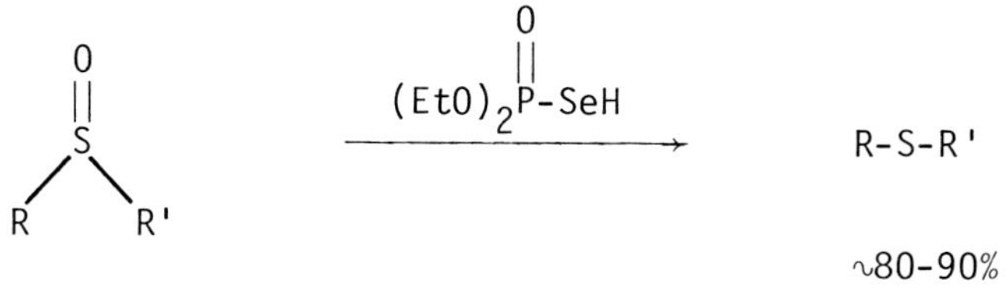

$$\underset{RR'}{\overset{O}{\underset{\|}{S}}} \quad \xrightarrow{(EtO)_2\overset{O}{\underset{\|}{P}}-SeH} \quad R-S-R'$$

$\sim$80-90%

R,R' = Me, Bu, Ph, Bz, etc.

III.C-5   J. Drabowicz and S. Oae, Synthesis, 404 (1977).

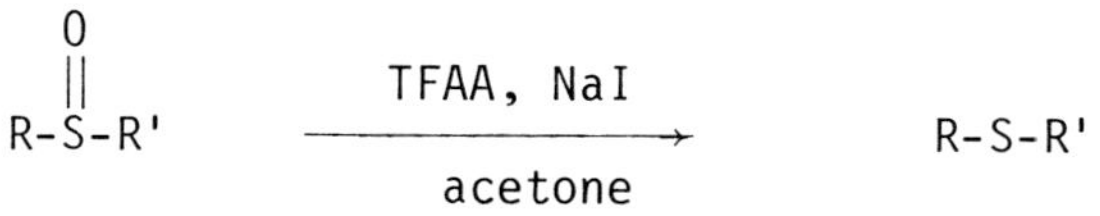

$$R-\overset{O}{\underset{\|}{S}}-R' \quad \xrightarrow[\text{acetone}]{\text{TFAA, NaI}} \quad R-S-R'$$

R,R' = Ph, Bz, 1°,3° alkyl   >90%

III.C-6   I.W.J. Still, S. K. Hansan, and K. Turnbull, Synthesis, 468 (1977).

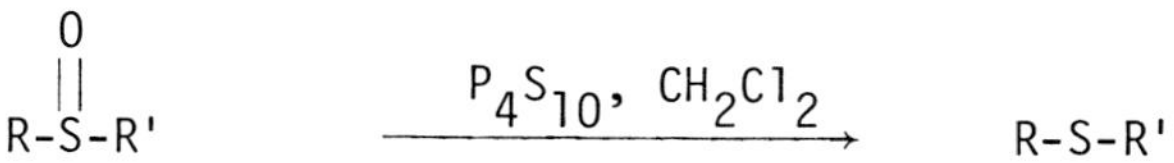

$$R-\overset{O}{\underset{\|}{S}}-R' \quad \xrightarrow{P_4S_{10},\ CH_2Cl_2} \quad R-S-R'$$

R,R' = alkyl, aryl, cyclic   45-99%

III.C-7 G. A. Olah, B.G.B. Gupta, and S. C. Narang, Synthesis, 583 (1977).

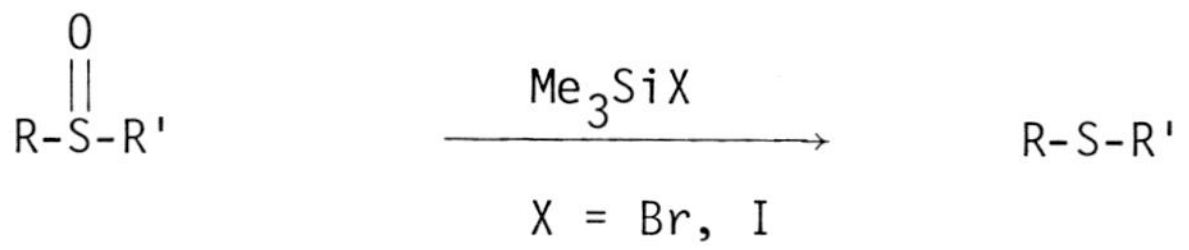

$$\underset{R,R' = 1° \text{ alkyl, Ph, Bz}}{\overset{O}{\underset{\|}{R-S-R'}}} \xrightarrow[X = Br, I]{Me_3SiX} \underset{\sim 70\text{-}90\%}{R-S-R'}$$

III.C-8 R. G. Nuzzo, H. J. Simon, and J. San Filippo, Jr., J. Org. Chem., 42, 568 (1977).

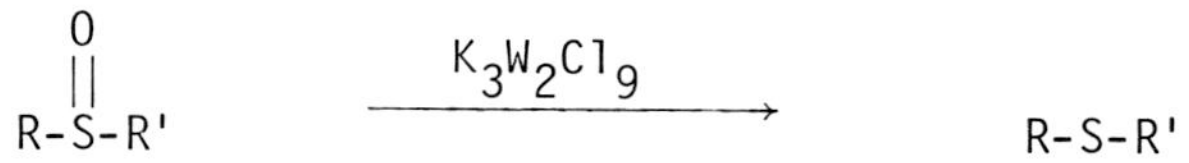

$$\overset{O}{\underset{\|}{R-S-R'}} \xrightarrow{K_3W_2Cl_9} R-S-R'$$

R,R' = alkyl, phenyl   >90%

III.C-9 T. Yamamoto, M. Kakimoto, and M. Okawara, Tetrahedron Lett., 1659 (1977).

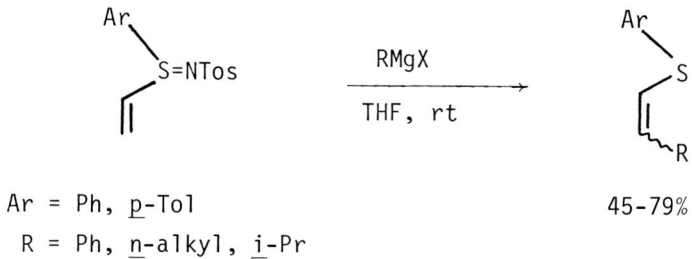

Ar = Ph, p-Tol   45-79%
R = Ph, n-alkyl, i-Pr

REDUCTIONS

III.D-1 H. des Abbayes and H. Alper, <u>J. Am. Chem. Soc.</u>, <u>99</u>, 98 (1977).

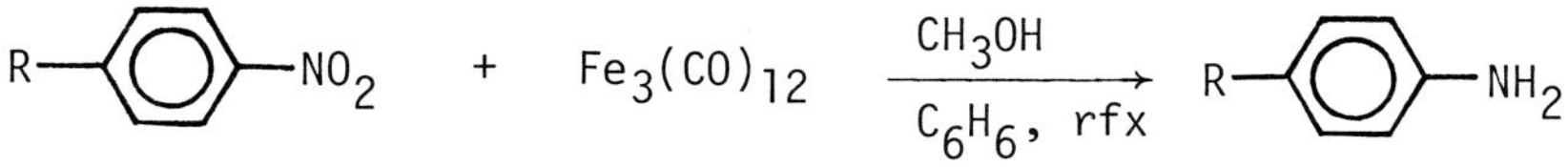

$$R-\text{C}_6\text{H}_4-NO_2 + Fe_3(CO)_{12} \xrightarrow[C_6H_6, \text{ rfx}]{CH_3OH} R-\text{C}_6\text{H}_4-NH_2$$

R = Me, OMe, Cl, COCH$_3$ 60-85%

III.D-2 N. A. Cortese and R. F. Heck, <u>J. Org. Chem.</u>, <u>42</u>, 3491 (1977).

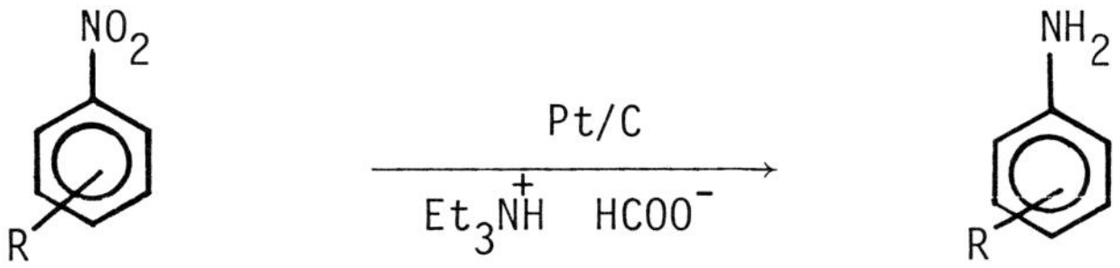

$$R-\text{C}_6H_4-NO_2 \xrightarrow[Et_3\overset{+}{N}H \ HCOO^-]{Pt/C} R-\text{C}_6H_4-NH_2$$

~80%

Compatible with -COOR, -CH=CHCOOR, etc.

III.D-3 A. Ohta <u>et al.</u>, <u>Synthesis</u>, 792 (1977).

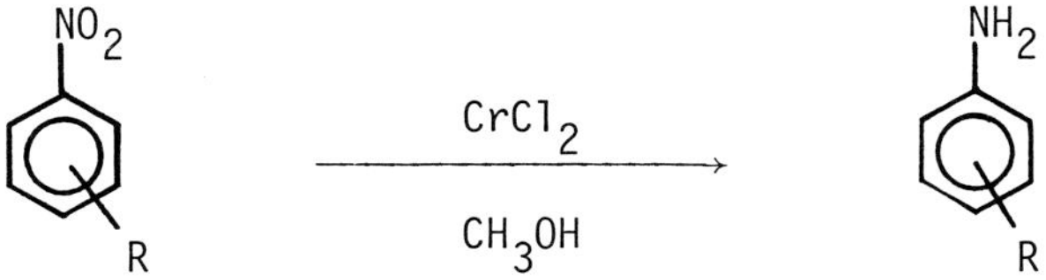

$$R-\text{C}_6H_4-NO_2 \xrightarrow[CH_3OH]{CrCl_2} R-\text{C}_6H_4-NH_2$$

R = H, OH, Cl 58-97%

III.D-4   H. Alper, D. Des Roches, and H. des Abbayes, Angew. Chem. Int. Ed., 16, 41 (1977).

$$ArNO_2 \xrightarrow[\text{18-crown-6, benzene}]{Fe_3(CO)_{12},\ KOH} Ar-NH_2$$

Ar = Me, OMe,   60-84%
Cl-subst. benzene

III.D-5   D. C. Owsley and J. J. Bloomfield, Synthesis, 118 (1977).

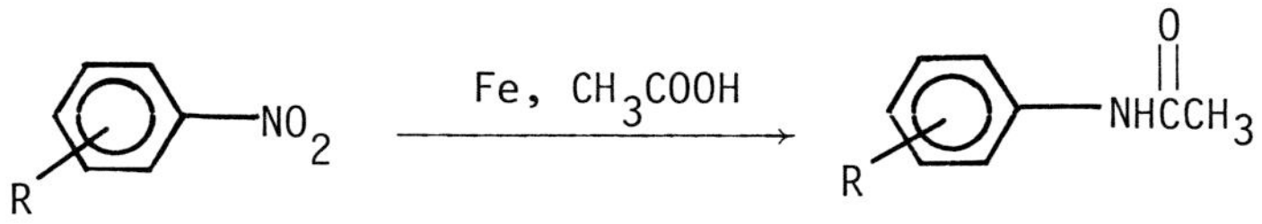

$$R\text{-}C_6H_4\text{-}NO_2 \xrightarrow{Fe,\ CH_3COOH} R\text{-}C_6H_4\text{-}NHCOCH_3$$

R = Me, OEt, Cl   ~70-90%

III.D-6   C. S. Rondestvedt, Jr. and T. A. Johnson, Synthesis, 850 (1977).

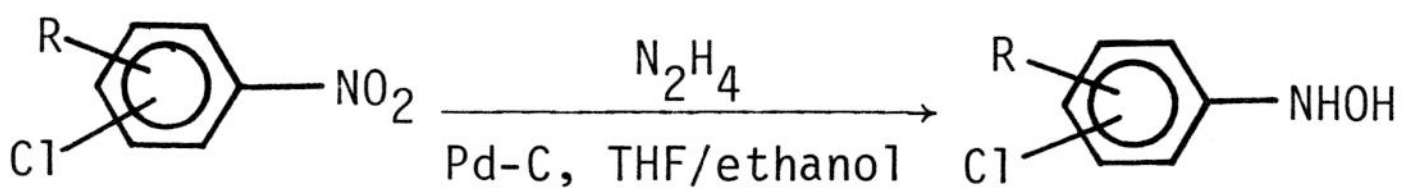

$$\underset{Cl}{\overset{R}{C_6H_3}}\text{-}NO_2 \xrightarrow[\text{Pd-C, THF/ethanol}]{N_2H_4} \underset{Cl}{\overset{R}{C_6H_3}}\text{-}NHOH$$

R = H, Cl, Me   58-100%

REDUCTIONS

III.E.1-1   K. Kochloefl and W. Liebelt, J. Chem. Soc., Chem. Commun., 510 (1977).

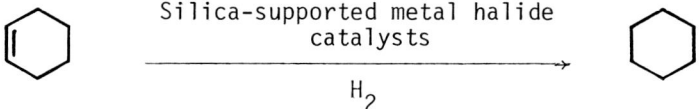

$$\text{cyclohexene} \xrightarrow[H_2]{\text{Silica-supported metal halide catalysts}} \text{cyclohexane}$$

(Rates are 2-4 orders of magnitude greater than those for the homogeneous counterparts.)

III.E.1-2   P. Caubere et al., Tetrahedron Lett., 1069 (1977).

$$\text{Me}_2\text{C=CMe}_2 \xrightarrow[\text{PhOCH}_3]{\text{NaH-}\underline{t}\text{-BuONa-Ni(OAc)}_2} \text{Me}_2\text{CH-CHMe}_2$$

>90% for monosub. and disub. alkenes

25-30% for methylcyclohexene

III.E.1-3   K. Kondo, S. Murai, and N. Sonoda, Tetrahedron Lett., 3727 (1977).

$$\text{>C=C<} + O_2 \xrightarrow[\text{Se}]{\text{NH}_2\text{NH}_2} -\overset{|}{\underset{H}{C}}-\overset{|}{\underset{H}{C}}-$$

III.E.1-4   W. Strohmeier and M. LuKacs, J. Organomet. Chem., 129, 331 (1977).

$Ir(CO)Cl(PPh_3)_2$ catalyzed hydrogenation of cyclopentene, styrene, ethyl acrylate at 10 atm., 80-120°.

III.E.1-5   R. H. Crabtree, H. Felkin, and G. E. Morris, J. Organomet. Chem., 141, 205 (1977).

$[Ir(COD)L_2]PF_6$ and $[Ir(COD)L(py)]PF_6$ in CH_2Cl_2 are very active hydrogenation catalysts for olefins.

III.E.1-6   F. Sato et al., J. Organomet. Chem., 142, 71 (1977).

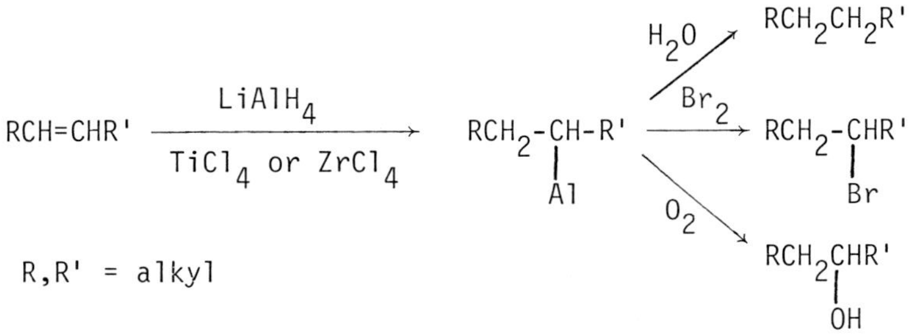

$$RCH=CHR' \xrightarrow[TiCl_4 \text{ or } ZrCl_4]{LiAlH_4} RCH_2\text{-}CH\text{-}R' \begin{array}{c} \xrightarrow{H_2O} RCH_2CH_2R' \\ \xrightarrow{Br_2} RCH_2\text{-}CHR' \\ | \\ Br \\ \xrightarrow{O_2} RCH_2CHR' \\ | \\ OH \end{array}$$
(middle intermediate: Al)

R,R' = alkyl

III.E.1-7   J. Tsuji and H. Suzuki, Chem. Lett., 1083 (1977).

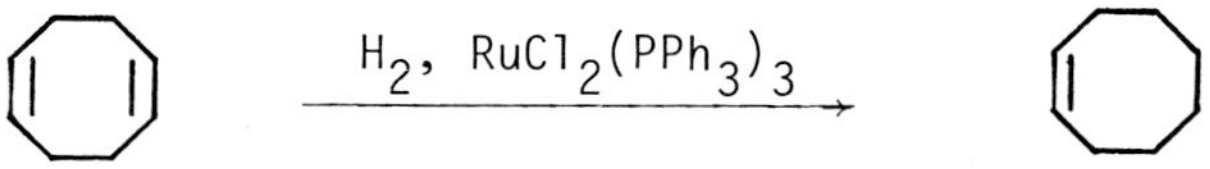

cyclooctatetraene $\xrightarrow{H_2, RuCl_2(PPh_3)_3}$ cyclooctene

89%

Several additional examples of hydrogenation of dienes and trienes to monoenes.

REDUCTIONS

III.E.1-8   E. Yoshii et al., Chem. Pharm. Bull., **25**, 1468 (1977).

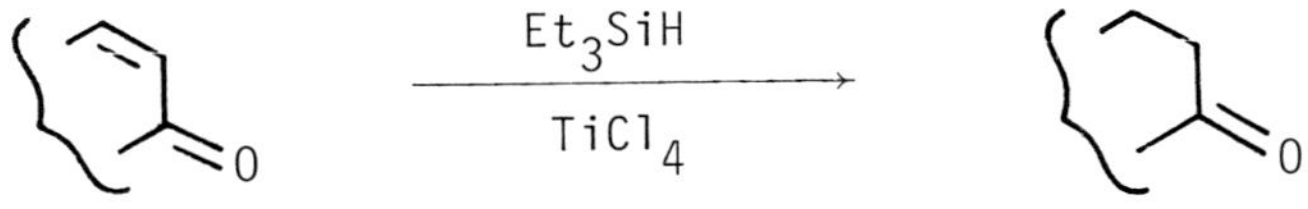

cyclohexenone $\xrightarrow{\text{Et}_3\text{SiH} / \text{TiCl}_4}$ cyclohexanone

~50-75%

III.E.1-9   E. Ucciani and L. Tanguy, Comptes Rendus (C), **284**, 577 (1977).

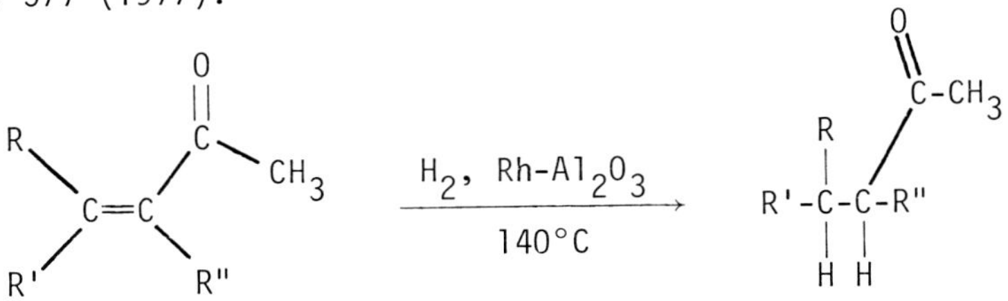

$$\underset{R'}{\overset{R}{>}}C=C\underset{R''}{\overset{C(O)CH_3}{<}} \xrightarrow[140°C]{H_2,\ Rh-Al_2O_3} R'-\underset{\underset{H}{|}}{\overset{\overset{R}{|}}{C}}-\underset{\underset{H}{|}}{\overset{\overset{C(O)CH_3}{|}}{C}}-R''$$

R,R',R" = H, alkyl   24-100%

III.E.1-10   T. W. Russell, D. M. Duncan, and S. C. Hansen, J. Org. Chem., **42**, 551 (1977).

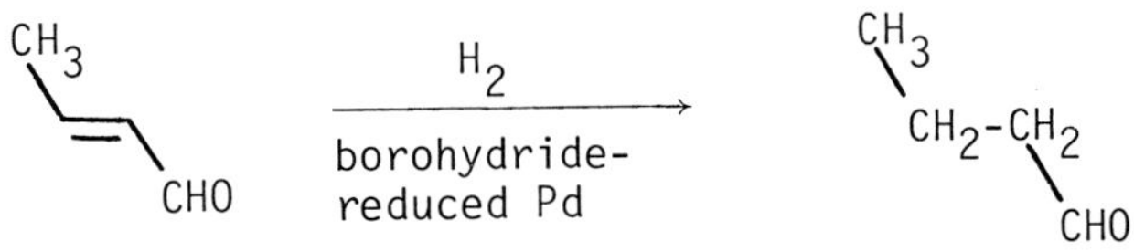

$$CH_3-CH=CH-CHO \xrightarrow[\text{reduced Pd}]{H_2,\ \text{borohydride-}} CH_3-CH(CH_2-CH_2)CHO$$

No yields cited

III.E.1-11 M. F. Semmelhack, R. D. Stauffer, and A. Yamashita, J. Org. Chem., 42, 3180 (1977).

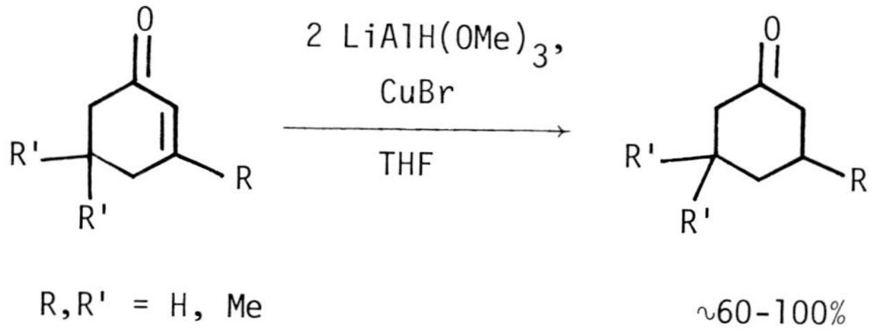

$$\text{R',R' = H, Me} \quad \xrightarrow[\text{THF}]{2\ \text{LiAlH(OMe)}_3,\ \text{CuBr}} \quad \sim 60\text{-}100\%$$

III.E.1-12 V. Z. Sharf et al., Bull. Akad. USSR Chem., 26, 995 (1977).

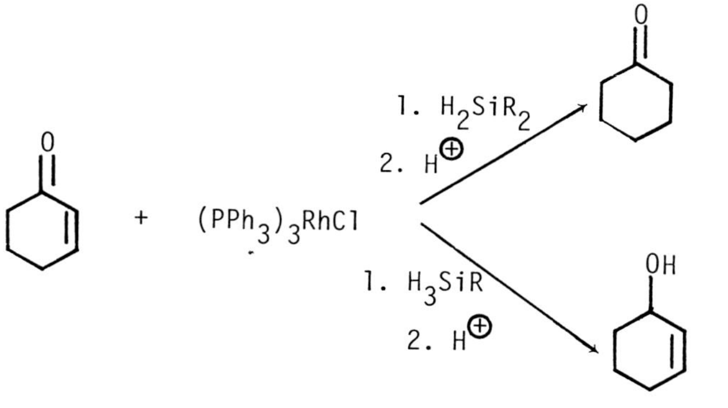

$$\text{cyclohexenone} + (\text{PPh}_3)_3\text{RhCl} \xrightarrow{\substack{1.\ H_2SiR_2 \\ 2.\ H^\oplus}} \text{cyclohexanone}$$

$$\xrightarrow{\substack{1.\ H_3SiR \\ 2.\ H^\oplus}} \text{cyclohexenol}$$

Widely varying yields.

III.E.1-13 G. Consiglio and P. Pino, Israel J. Chem., 15, 221 (1977).

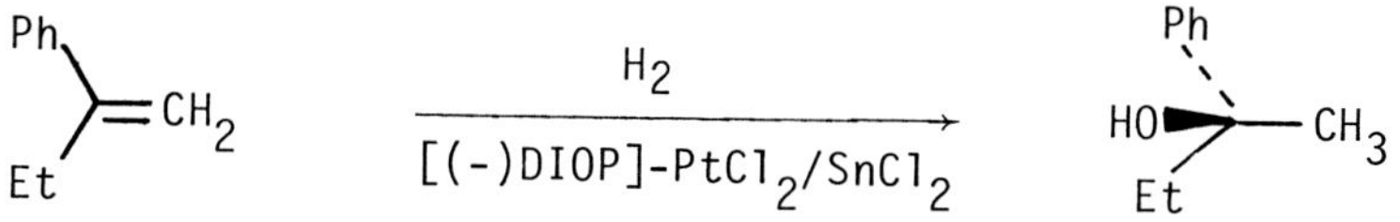

$$\text{Ph(Et)C=CH}_2 \xrightarrow[{[(-)\text{DIOP}]\text{-PtCl}_2/\text{SnCl}_2}]{H_2} \text{Ph(Et)(HO)C-CH}_3$$

90% yield; 37% ee

III.E.1-14 K. Nanjo, K. Suzuki, and M. Sekiya, Chem. Pharm. Bull., 25, 2396 (1977).

$$\underset{R'}{\overset{R}{>}}C=C\underset{X}{\overset{CN}{<}} \xrightarrow[DMF]{Et_3N,\ HCOOH,} \underset{R'}{\overset{R}{>}}CH-CH\underset{X}{\overset{CN}{<}}$$

R,R' = H, Me, Et, Ph, Bz, cyclic   ~70-90%

X = CN, COOEt, SO$_2$Ph

III.E.1-15 B. R. James and R. S. McMillan, Can. J. Chem., 55, 3927 (1977).

$$\overset{\diagdown}{\underset{\diagup}{\overset{C}{\underset{C}{\|}}}}\diagdown \xrightarrow[H_2]{Ru(II)\ chiral\ catalyst} \begin{array}{c} \overset{*}{|} \\ -C-H \\ | \\ -C-H \\ | \end{array}$$

Itaconic acid, atropic   Up to 100% chemical yields,
acid, etc.   up to 25% ee.

III.E.1-16 M. D. Fryzuk and B. Bosnich, J. Am. Chem. Soc., 99, 6262 (1977).

$$\underset{HOOC}{\overset{R}{>}}C=CH_2\ \text{NHCOMe} \xrightarrow[H_2,\ THF]{Chiral\ Rh(I)phosphine} HOOC-\overset{*}{\underset{|}{C}}H-NHCOMe \atop |CH_2R$$

R = H, Ph, alkyl, etc.   74-100% ee

III.E.1-17   W. S. Knowles et al., J. Am. Chem. Soc., 99, 5946 (1977).

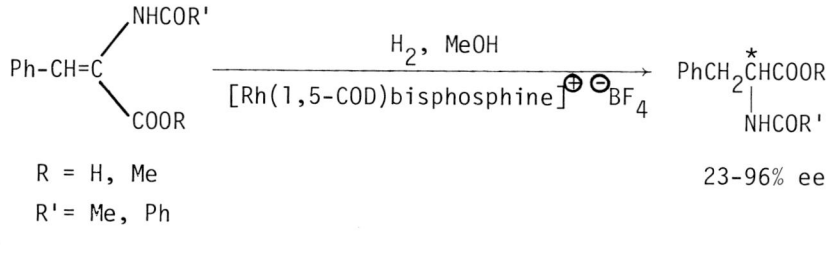

R = H, Me
R'= Me, Ph

23-96% ee

III.E.1-18   C. Fisher and H. S. Mosher, Tetrahedron Lett., 2487 (1977).

Use of a Rhodium(I)-MMPP (MMPP = ℓ-menthylmethylphenylphosphine) catalyst in the asymmetric hydrogenation of α,β-unsaturated carboxylic acids. Chemical yields ∼50-90%, optical yields up to 70%.

III.E.1-19   T. Hayashi, M. Tanako, and I. Ogata, Tetrahedron Lett., 295 (1977).

Use of d-trans-BDPCP as a chiral ligand for use in rhodium-catalyzed asymmetric hydrogenation. Optical yields up to 60% (for α-ethylstyrene) are observed.

d-trans-BDPCP

III.E.1-20   G. Pracejus and H. Pracejus, Tetrahedron Lett., 3497 (1977).

Use of three aminophosphines similar to the one below as chiral Rh(I) ligands for asymmetric hydrogenation catalysts. Using α-acetamidocinnamic acid and its ester as a substrate, yields are up to 99% chemical yield with 80% ee.

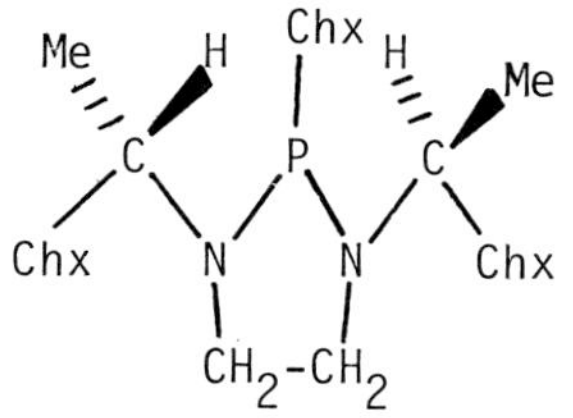

(Chx = cyclohexyl)

III.E.1-21/III.E.2-1   J. J. Brunet and P. Caubere, Tetrahedron Lett., 3947 (1977).

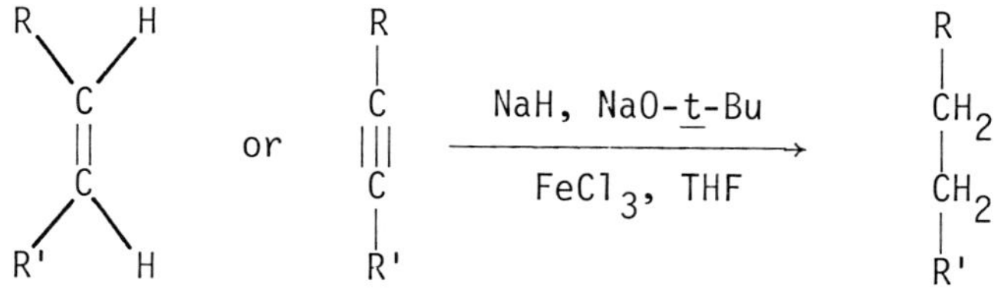

$$\underset{R'}{\overset{R}{C}}=\underset{H}{\overset{H}{C}} \quad \text{or} \quad \underset{R'}{\overset{R}{C}}\equiv C \quad \xrightarrow[FeCl_3, \text{ THF}]{NaH, NaO-\underline{t}-Bu} \quad \underset{R'}{\overset{R}{C}H_2-CH_2}$$

R,R' = n-alkyl, Ph

45-95% for olefins
>95% for acetylenes

III.E.1-22/III.E.2-2   E. C. Ashby and J. J. Lin, Tetrahedron Lett., 4481 (1977).

$$\underset{R'}{\overset{R}{>}}C=C\underset{H}{\overset{H}{<}} \quad \xrightarrow[\text{or } LiAlH_4\text{-}NiCl_2]{LiAlH_4\text{-}CoCl_2} \quad \underset{R'}{\overset{R}{>}}CH_2\text{—}CH_2$$

R,R' = H, alkyl, Ph, cyclic   ~90%

$$R\text{-}C\equiv C\text{-}R' \quad \xrightarrow{\text{same conditions}} \quad \underset{R'}{\overset{R}{>}}C=C\underset{H}{\overset{H}{<}}$$

R,R' = H, alkyl, Ph   70-96%

III.E.1-23/III.E.2-3   J. J. Brunet, P. Gallois, and P. Caubere, Tetrahedron Lett., 3955 (1977).

$$R\text{-}C\equiv C\text{-}R' \quad \xrightarrow[\text{THF/EtOH}]{H_2,\ NaH,\ Ni(OAc)_2} \quad \underset{R}{\overset{H}{>}}C=C\underset{R'}{\overset{H}{<}}$$

R = Ph, n-alkyl
R' = H, Me, Et   78-98%

[cyclohexenyl-vinyl] → [cyclohexenyl-ethyl]   same conditions   98%

240

III.E.2-4 R. Rossi and A. Carpita, Synthesis, 561 (1977).

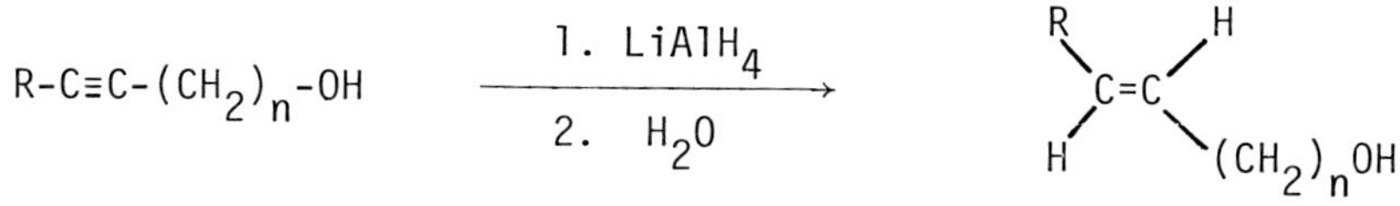

R-C≡C-(CH2)n-OH
1. LiAlH4
2. H2O
R
H
C=C
H
(CH2)nOH

R = 1° alkyl
n = 2-7

>85%

III.E.2-5 G. Zweifel, R. A. Lynd, and R. E. Murray, Synthesis, 52 (1977).

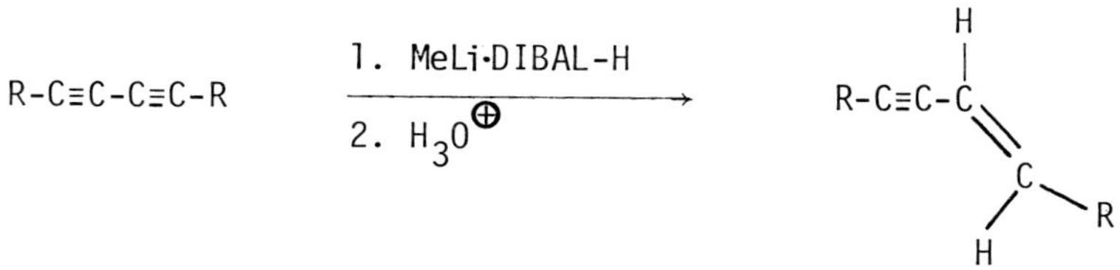

R-C≡C-C≡C-R
1. MeLi·DIBAL-H
2. H3O⊕
H
R-C≡C-C
C
R
H

R = n-Bu, i-Pr, t-Bu, c-Hx

~90%

III.E.3-1 N. L. Holy, Tetrahedron Lett., 3703 (1977).

H2 (1 atm), DMF or CH3CN
Rh(I)(N-phenylanthranilate)2

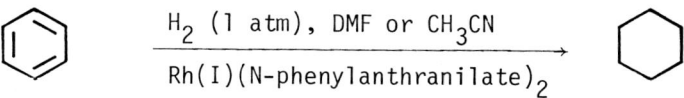

III.E.3-2 P. Patnaik and S. Sarkar, Tetrahedron Lett., 2531 (1977).

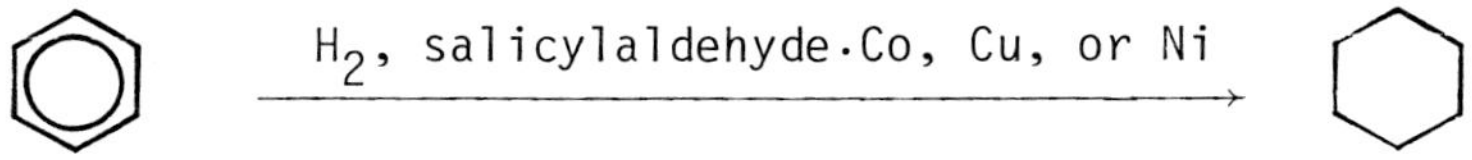

benzene $\xrightarrow{H_2, \text{ salicylaldehyde} \cdot \text{Co, Cu, or Ni}}$ cyclohexane

>95%

III.E.3-3 M. J. Russell, C. White, and P. M. Maitlis, J. Chem. Soc., Chem. Commun., 427 (1977).

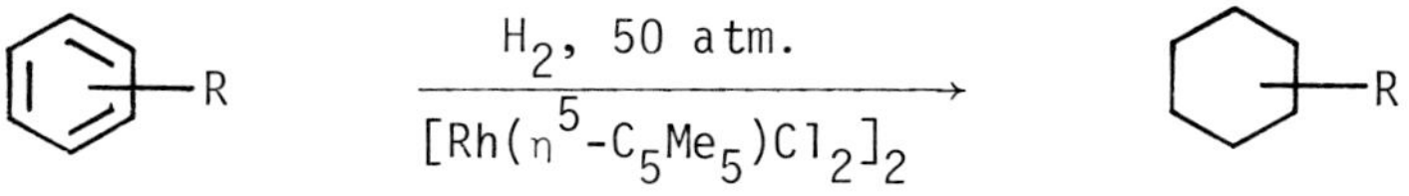

Ar–R $\xrightarrow[{[Rh(\eta^5\text{-}C_5Me_5)Cl_2]_2}]{H_2,\ 50\ \text{atm.}}$ Cy–R

R = alkyl, acyl, ester, amino, etc.

Widely varying yields.

III.E.3-4 P. P. Fu and R. G. Harvey, Tetrahedron Lett., 415 (1977).

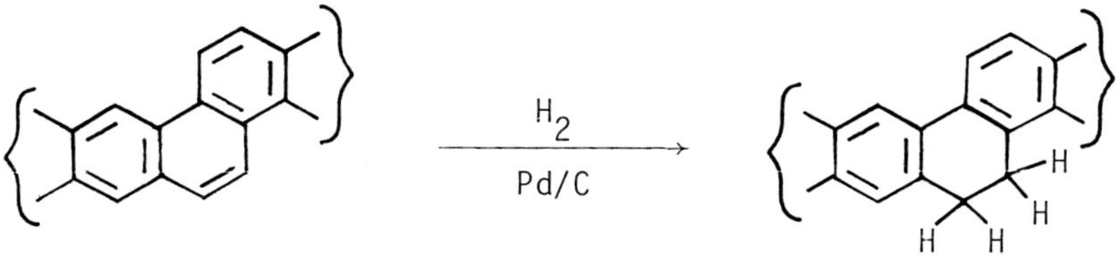

$\xrightarrow[\text{Pd/C}]{H_2}$

70-100%

REDUCTIONS

III.E.3-5   R. E. Donaldson and P. L. Fuchs, J. Org. Chem., 42, 2034 (1977).

$$\text{X-C}_6\text{H}_4\text{-OSiMe}_2\text{-}\underline{t}\text{-Bu} \xrightarrow[\underline{t}\text{-BuOH, -33°}]{\text{Li/NH}_3/\text{THF}} \text{X-C}_6\text{H}_4\text{-OSiMe}_2\text{-}\underline{t}\text{-Bu}$$

X = CH$_3$, OCH$_3$, etc.

∼90%

Can be used to form nonconjugated ketones.

III.E.3-6   H. Minato et al., Chem. Lett., 1091 (1977).

$$\text{Pyridinium-CH}_2\text{-C(=O)-Ph} \; Br^{\ominus} \xrightarrow[\text{H}_2\text{O}]{\text{Na}_2\text{S}_2\text{O}_4} \text{piperidine-N-CH}_2\text{-C(=O)-Ph}$$

64%

$$\text{N-benzylquinolinium} \; Cl^{\ominus} \xrightarrow[\text{DMF, H}_2\text{O}]{\text{Na}_2\text{S}_2\text{O}_4} \text{1-benzyl-1,2,3,4-tetrahydroquinoline}$$

73%

III.F.1-1   R. W. Holder and M. G. Matturro, J. Org. Chem., 42, 2166 (1977).

$$\text{R-OMes} \xrightarrow[\text{THF}]{\text{LiEt}_3\text{BH}} \text{R-H}$$

R = 1°, 2° alkyl, cyclic   65-96%

III.F.1-2 G. W. Gribble, R. M. Leese, and B. E. Evans, Synthesis, 172 (1977).

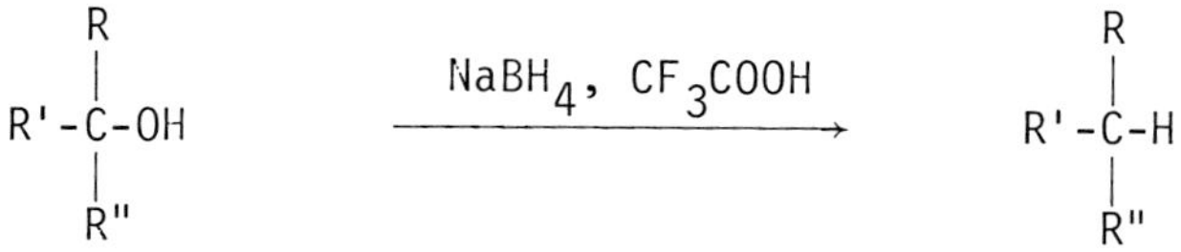

$$\underset{R''}{\overset{R}{\underset{|}{R'-\overset{|}{C}-OH}}} \xrightarrow{NaBH_4, \; CF_3COOH} \underset{R''}{\overset{R}{\underset{|}{R'-\overset{|}{C}-H}}}$$

R's = H, Me, Ph, alkyl ∼90%

III.F.1-3 J.-P. Pete et al., Synthesis, 774 (1977).

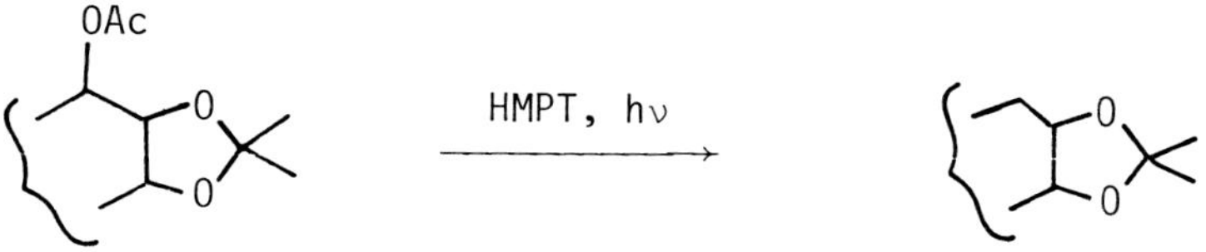

[sugar with OAc group and isopropylidene acetal] →(HMPT, hν)→ [deoxysugar with isopropylidene acetal]

(sugar) (deoxysugar) 60-85%

III.F.1-4 I. E. Felkin and P. Sarda, Tetrahedron, 33, 511 (1977).

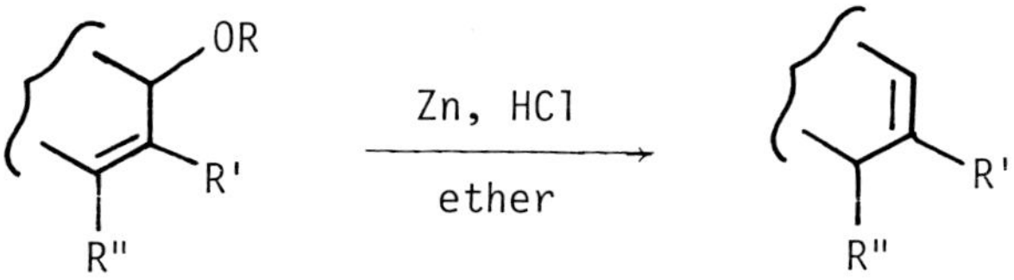

[cyclic enol ether with OR, R', R'' substituents] →(Zn, HCl, ether)→ [cyclic alkene with R', R'' substituents]

R = H, Me, Ac
R', R'' = H, alkyl

>70%

REDUCTIONS

III.F.1-5   L. Mandell, R. F. Daley, and R. A. Day, Jr., J. Org. Chem., 42, 1461 (1977).

$$R\text{-}C_6H_4\text{-}CH_2COOH \xrightarrow[\text{2. NaHCO}_3]{\text{1. electrolysis, DMSO}} R\text{-}C_6H_4\text{-}CHO$$

R =	4-OMe	4-Cl	H	4-NO$_2$
Yield, %	72	62	41	0

III.F.1-6   T. Izawa and T. Mukaiyama, Chem. Lett., 1443 (1977).

$$R\text{-}COOH \longrightarrow \longrightarrow R\text{-}\underset{\underset{O}{\|}}{C}\text{-}N\underset{S}{\overset{}{\bigcirc}} \xrightarrow[\text{2. H}_3O^{\oplus}]{\text{1. DIBAH, toluene}} R\text{-}CHO$$

R = alkyl, benzyl, cinnamyl, etc.   64-93%

III.F.1-7/III.F.2-1   R. O. Hutchins et al., J. Org. Chem., 42, 82 (1977).

$$R\text{-}X \xrightarrow[\text{in HMPA}]{\text{NaBH}_3\text{CN, 9-BBNCN, or polymeric cyanoborane}} R\text{-}H$$

R = 1°,2°,3° alkyl
X = halogen, -OTs

Full paper, wide variety of substrates, full experimental details.

III.F.2-2 E. C. Ashby and J. J. Lin, Tetrahedron Lett., 4481 (1977).

$$\text{R-X} \xrightarrow[\underline{\text{or}}\ \text{LiAlH}_4\text{-NiCl}_2]{\text{LiAlH}_4\text{-CoCl}_2} \text{R-H}$$

X = Cl, Br, I   >90%
R = 1°,2°,3° alkyl, Ph

III.F.2-3 B. Loubinoux, R. Vanderesse, and P. Caubere, Tetrahedron Lett., 3951 (1977).

$$\text{R-X} \xrightarrow[\underline{\text{or}}\ \text{NaH, NaO-}\underline{t}\text{-Am,, ZnCl}_2,\ \text{THF}]{\text{NaH, NaO-}\underline{t}\text{-Am, Ni(OAc)}_2,\ \text{DME}} \text{R-H}$$

X = Cl, Br, I   usually >90%
R = 1°,2° alkyl, cycloalkyl, cinnamyl

III.F.2-4 N. A. Cortese and R. F. Heck, J. Org. Chem., 42, 3491 (1977).

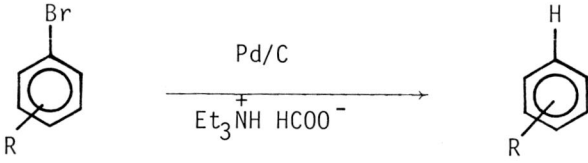

Compatible with $-NO_2$, $-CN$, $-COOR$, etc.

III.F.2-5 M. Tashiro and G. Fukata, J. Org. Chem., 42, 835 (1977).

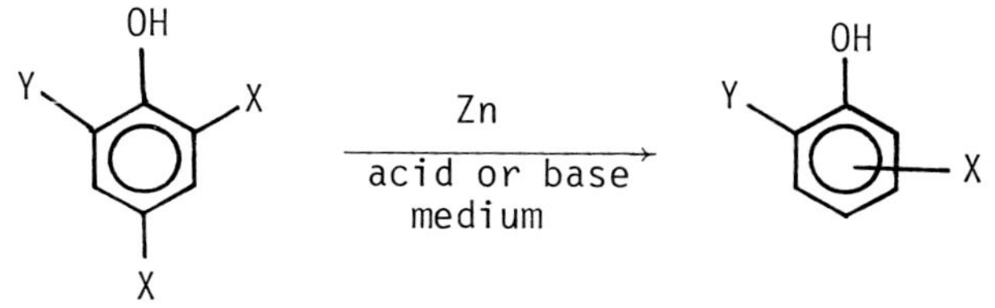

[Structure: 2,4-dihalo-6-Y-phenol] $\xrightarrow[\text{medium}]{\text{Zn, acid or base}}$ [Structure: 2-Y-phenol with X]

X = Cl, Br
Y = alkyl, Cl, Br

Mixtures having lost one or two halogens.

III.F.2-6 T. Kunuda, T. Tamura, and T. Takizawa, Chem. Pharm. Bull., 25, 1749 (1977).

$$R\text{-}CX_3 \xrightarrow[\text{THF}]{Ni(CO)_4} R\text{-}CX_2H$$

X = halogen
R = alkyl, aryl, etc.

~40-80%

III.F.2-7 T. E. Cole and R. Pettit, Tetrahedron Lett., 781 (1977).

$$R\text{-}\underset{\underset{O}{\|}}{C}\text{-}Cl \xrightarrow{3\ NMe_4 \cdot HFe(CO)_4} R\text{-}CHO$$

R = 1°,2°,3° alkyl, Ar, cycloalkyl, heterocyclic

75-100%

III.F.3-1   S. Kurozumi et al., Synth. Commun., 7, 427 (1977).

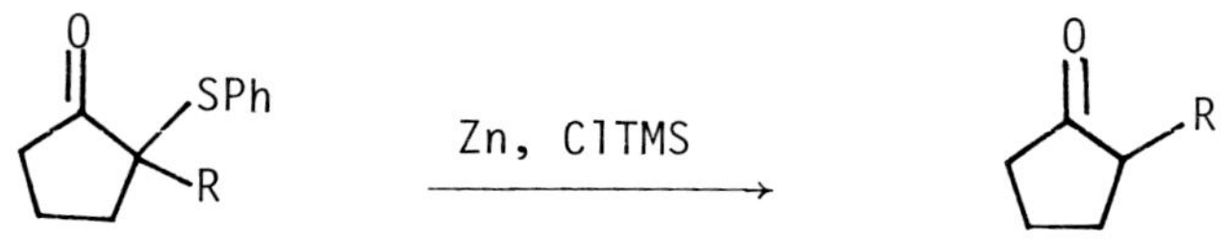

[cyclopentanone with SPh and R substituents at α-carbon] →(Zn, ClTMS)→ [2-alkyl cyclopentanone]

R = alkyl   ~80%

III.F.3-2   H. Alper and H.-N. Paik, J. Org. Chem., 42, 3522 (1977).

$$R_2CS \xrightarrow{Fe(CO)_5, KOH} R_2CH_2$$

(R = Ph, p-CH$_3$Ph, p-OCH$_3$Ph, p-Me$_2$NPh)   60-81%

$$\downarrow Co_2(CO)_8$$

$$R_2C=CR_2$$

$$R-\overset{O}{\underset{\|}{C}}-NHPh \xrightarrow[KOH]{Fe(CO)_5} R-CH_2-NHPh$$

38%, R=Ph

51%, R=Me

REDUCTIONS

III.F.4-1   M. P. Doyle et al., J. Org. Chem., **42**, 3494 (1977).

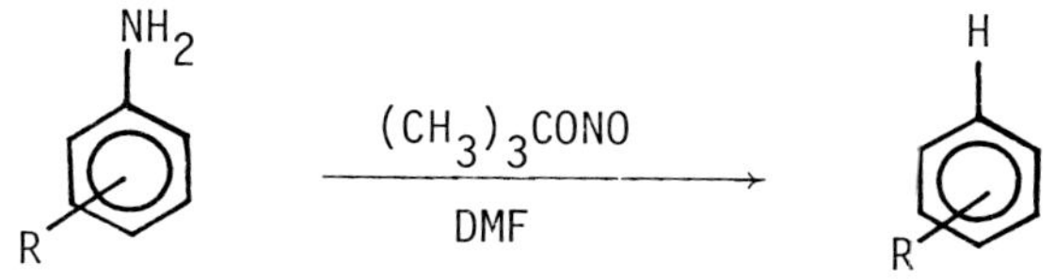

Ar-NH$_2$ → Ar-H, reagent: $(CH_3)_3CONO$ / DMF

R = NO_2, F, Cl, $COCH_3$, CH_3 ∼65-80%

III.F.4-2   A. V. Chebyshev et al., J. Org. Chem. (USSR), **13**, 1081 (1977).

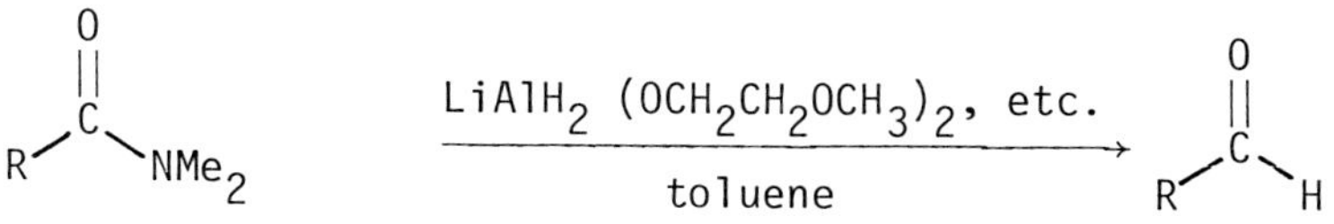

$$R-C(=O)-NMe_2 \xrightarrow{\text{LiAlH}_2(OCH_2CH_2OCH_3)_2, \text{ etc.}}_{\text{toluene}} R-C(=O)-H$$

(palmitic, linoleic, and oleic amides)   41-98%

III.F.4-3   U. K. Pandit, H. van Dam, and J. B. Stevens, Tetrahedron Lett., 913 (1977).

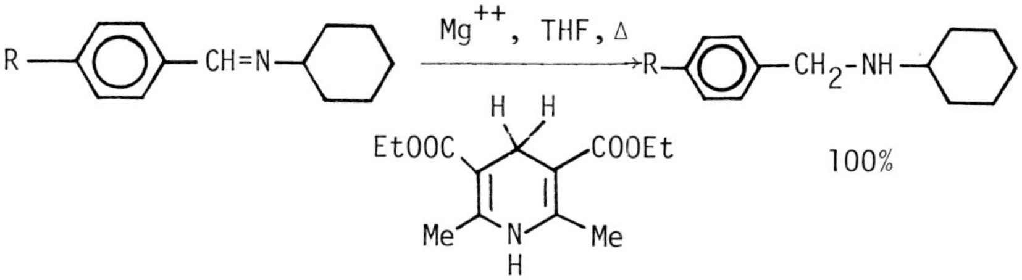

$$R-C_6H_4-CH=N-C_6H_{11} \xrightarrow{Mg^{++}, \text{ THF}, \Delta} R-C_6H_4-CH_2-NH-C_6H_{11}$$

with EtOOC-, COOEt-substituted 2,6-dimethyl-1,4-dihydropyridine

100%

III.G-1 J. A. Marshall and M. E. Lewellyn, J. Org. Chem., 42, 1311 (1977).

$$\text{diol} \xrightarrow{Cl_2PONR_2,\ R = Me, Et} \text{alkene}$$

55-86%

III.G-2 H. Alper and D. Des Roches, Tetrahedron Lett., 4155 (1977).

$$\text{epoxide} \xrightarrow{Fe(CO)_4 TMU} \text{alkene}$$

Widely varying yields.

TMU = tetramethylurea

III.G-3 N. Kornblum and L. Cheng, J. Org. Chem., 42, 2944 (1977).

$$\begin{array}{c} R^1-\underset{\underset{R^4}{|}}{\overset{\overset{R^2}{|}}{C}}-NO_2 \\ R^3-C-NO_2 \end{array} \xrightarrow[0-25°]{Ca/Hg} \begin{array}{c} R^1 \quad R^2 \\ \diagdown \diagup \\ C \\ \| \\ C \\ \diagup \diagdown \\ R^3 \quad R^4 \end{array}$$

R's = alkyl, cyclic, alkyl containing ketones, esters, nitriles

A general method for synthesizing tetrasubstituted olefins.

REDUCTIONS

III.G-4   Y. Gaoni, <u>Tetrahedron Lett.</u>, 947 (1977).

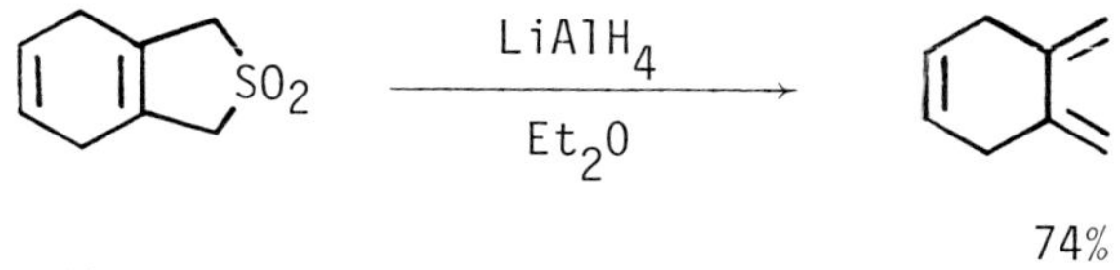

$$\text{[bicyclic sulfone]} \xrightarrow[\text{Et}_2\text{O}]{\text{LiAlH}_4} \text{[diene]} \quad 74\%$$

Several additional examples

III.H-1   R. A. Olofson <u>et al.</u>, <u>Tetrahedron Lett.</u>, 1567 (1977).

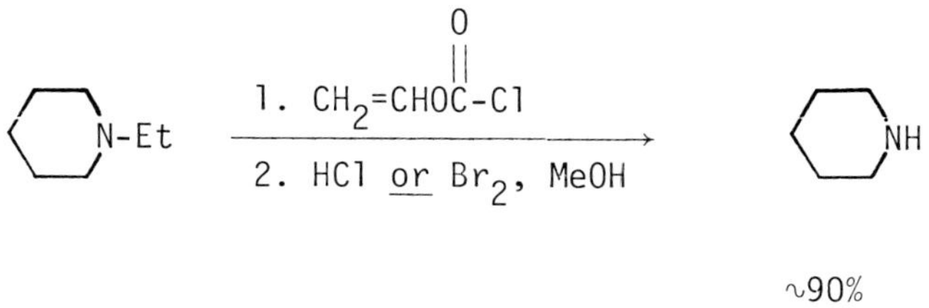

$$\text{N-Et piperidine} \xrightarrow[\text{2. HCl } \underline{\text{or}} \text{ Br}_2, \text{ MeOH}]{\text{1. CH}_2\text{=CHOC(O)-Cl}} \text{NH piperidine} \quad \sim 90\%$$

Several additional examples cited.

III.H-2   J. A. Marshall and R. Bierenbaum, <u>J. Org. Chem.</u>, <u>42</u>, 3309 (1977).

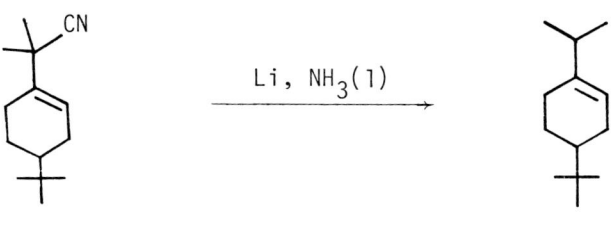

$$\xrightarrow{\text{Li, NH}_3(l)} \quad 80\%$$

III.H-3   N. H. Khan et al., Synth. Commun., 7, 71 (1977).

$$\text{R-CH=NNHPh} \xrightarrow[\text{EtOH, NH}_3]{\text{H}_2, \text{Pd/C}} \text{RCH}_2\text{NH}_2$$

∿50-80%

III.H-4   T.-L. Ho and G. A. Olah, Synthesis, 169 (1977).

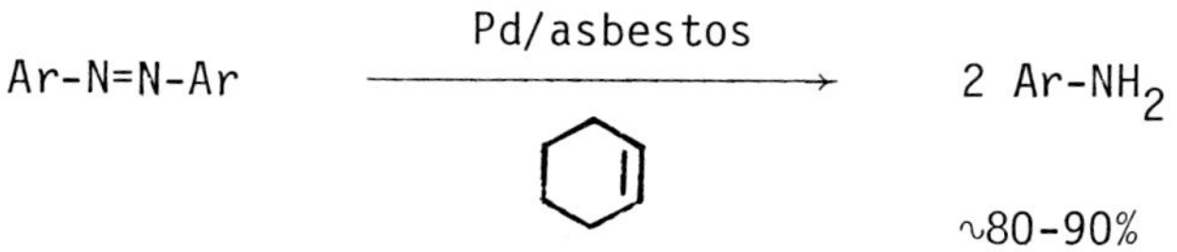

$$\text{Ar-N=N-Ar} \xrightarrow[\bigcirc]{\text{Pd/asbestos}} 2 \text{ Ar-NH}_2$$

∿80-90%

III.I-1   M. M. Midland, A. Tramontano, and S. A. Zderic, J. Am. Chem. Soc., 99, 5211 (1977).

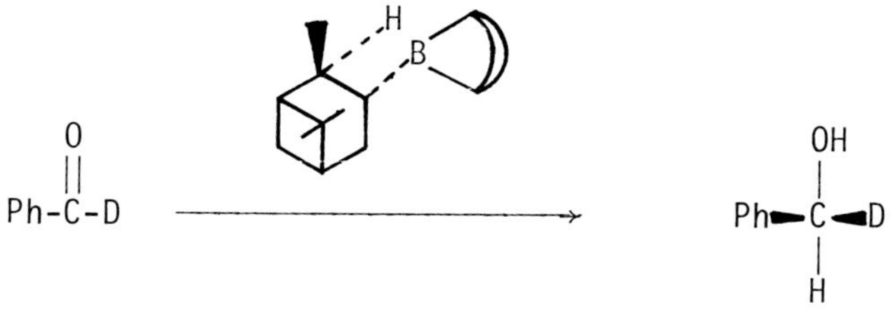

$$\underset{\text{Ph-C-D}}{\overset{\text{O}}{\|}} \longrightarrow \text{Ph}\blacktriangleright\underset{\text{H}}{\overset{\text{OH}}{\text{C}}}\blacktriangleleft\text{D}$$

81.6%
88% ee

REDUCTIONS

III.I-2   S. Krishnamurthy and H. C. Brown, *J. Org. Chem.*, **42**, 1197 (1977).

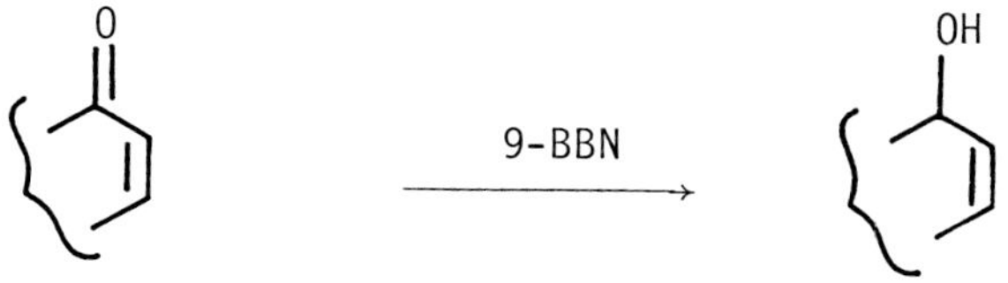

cyclohexenone $\xrightarrow{\text{9-BBN}}$ cyclohexenol

85%

Ph-CH=CHCHO $\xrightarrow{\text{9-BBN}}$ Ph-CH=CHCH$_2$OH

86%

III.I-3   S. Thaisrivongs and J. D. Wuest, *J. Org. Chem.*, **42**, 3243 (1977).

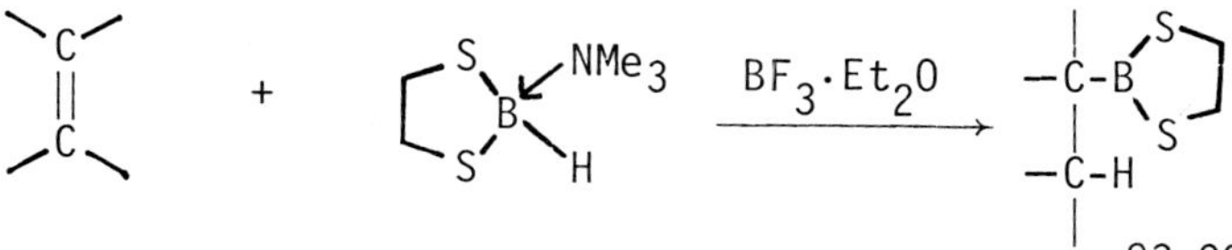

$$\text{C=C} + \text{S-B(NMe}_3\text{)-S with H} \xrightarrow{BF_3 \cdot Et_2O} \text{-C(S-B-S)- -C-H}$$

83-99%

C≡C $\xrightarrow{\text{same conditions}}$ vinyl boronate dithiolane

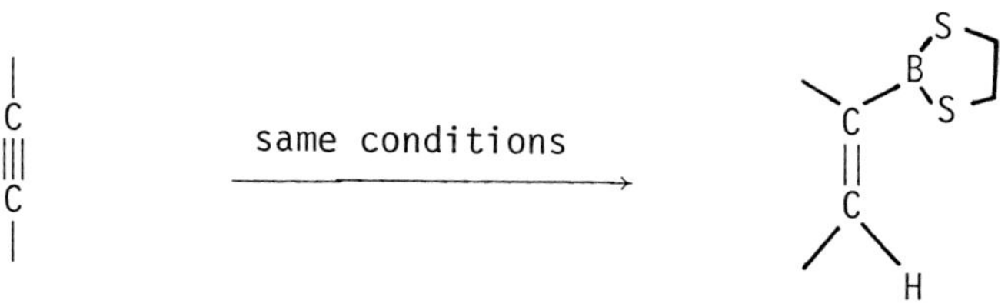

70-91%

III.I-4   C. F. Lane, Aldrichimica Acta, 10, 41 (1977).

Review: "Selective Reductions Using Borane Complexes"

III.I-5   R. Köster, Pure and Appl. Chem., 49, 765 (1977).

Review: "Organoboranes in Synthesis and Analysis"

III.I-6   G. W. Kabalka, J. D. Baker, Jr., and G. W. Neal, J. Org. Chem., 42, 512 (1977).

A study of the uses of catecholborane. Primary alkenes react very sluggishly, while secondary alkenes are unreactive at room temperature, allowing for selective reductions in the presence of alkenes.

III.J-1   G. Büchi and H. Wuest, Tetrahedron Lett., 4305 (1977).

$$\underset{\substack{\text{C}\\||\\\text{C}}}{\diagdown\diagup^{SiMe_3}} \quad \xrightarrow[\text{CH}_3\text{CN, H}_2\text{O (reflux)}]{\underline{p}\text{-toluenesulfinic acid}} \quad \underset{\substack{\text{C}\\||\\\text{C}}}{\diagdown\diagup^{H}}$$

∼80%

IV.A-1   J.-E. Bäckvall, J. Chem. Soc., Chem. Commun., 413 (1977).

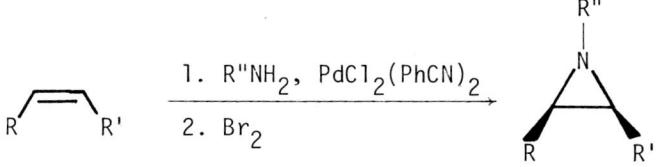

R's = H, alkyl, Ph   ∿40%

IV.A-2   N. DeKimpe et al., Rec. Trav. Chim. Pays-Bas, 96, 242 (1977).

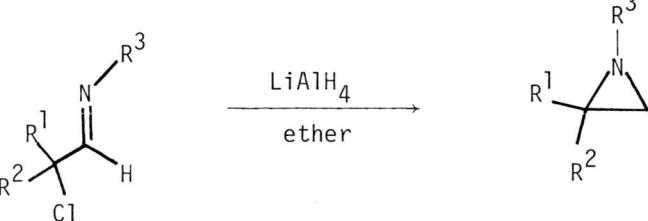

Widely varying yields.

R^1, R^2 = Me, Et, -(CH$_2$)$_5$-

R = Me, Ph, Bz, i-Pr, t-Bu

IV.A-3   P. Carlier, Y. G.-Mialhe, and R. Vessière, Can. J. Chem., 55, 3190 (1977).

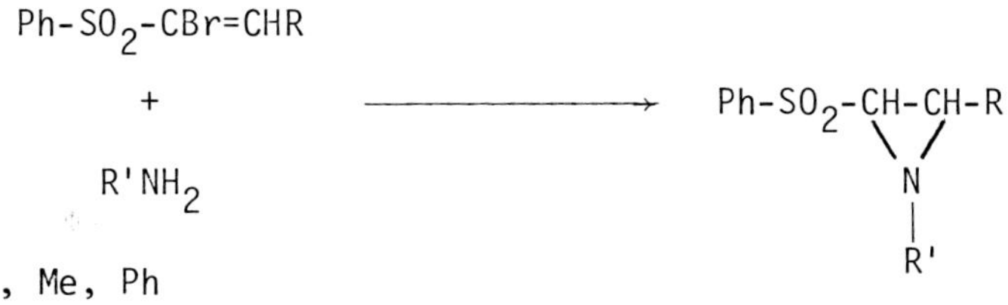

Ph-SO$_2$-CBr=CHR

+   ⟶   Ph-SO$_2$-CH-CH-R
                    \ /
R'NH$_2$ N
                    |
R = H, Me, Ph   R'
R' = Me, Et, i-Pr, t-Bu, Bz   25-81%

IV.B-1 H. Kotake et al., Chem. Lett., 73 (1977).

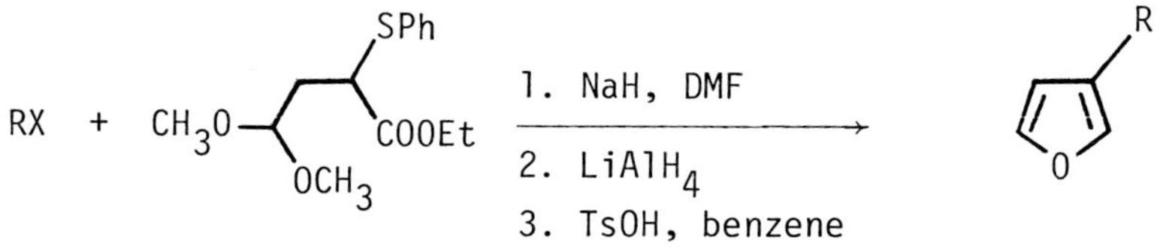

RX + CH₃O–CH(OCH₃)–CH(SPh)–CH₂–COOEt

1. NaH, DMF
2. LiAlH₄
3. TsOH, benzene

→ 3-R-furan

R = alkyl, Bz, allylic
X = Br, I

67-73%

IV.B-2 J. Hambrecht, Synthesis, 280 (1977).

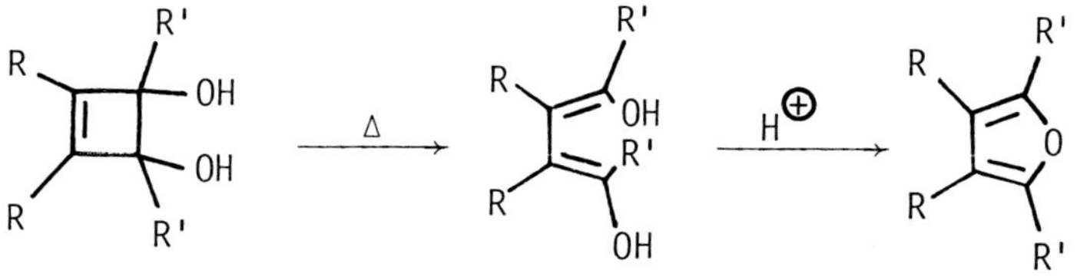

[cyclobutene-diol] →Δ→ [enetriol] →H⊕→ [furan]

R,R' = H, Me, Ph, etc.

50-96%

IV.B-3 B. Harirchian and P. D. Magnus, Synth. Commun., 7, 119 (1977).

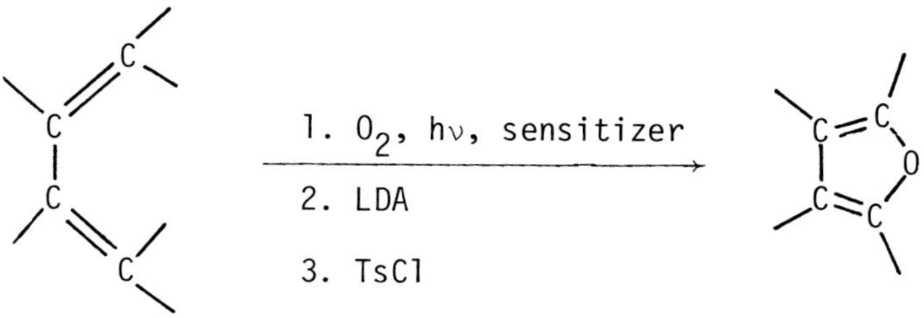

1. O₂, hν, sensitizer
2. LDA
3. TsCl

∼50% overall

IV.B-4   M. Asaoka, N. Sugimura, and H. Takei, Chem. Lett., 171 (1977).

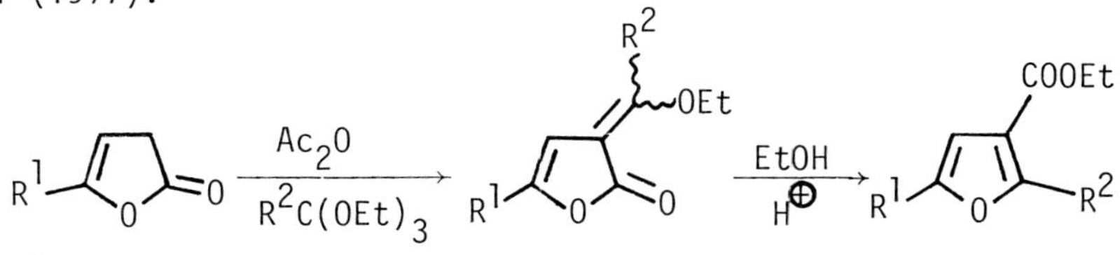

$R^1 = Ph, \underline{p}\text{-}CH_3OPh$

$R^2 = H, Me, Et$

~40-60%

IV.B-5   A. Zamojski and T. Koźluk, J. Org. Chem., 42, 1089 (1977).

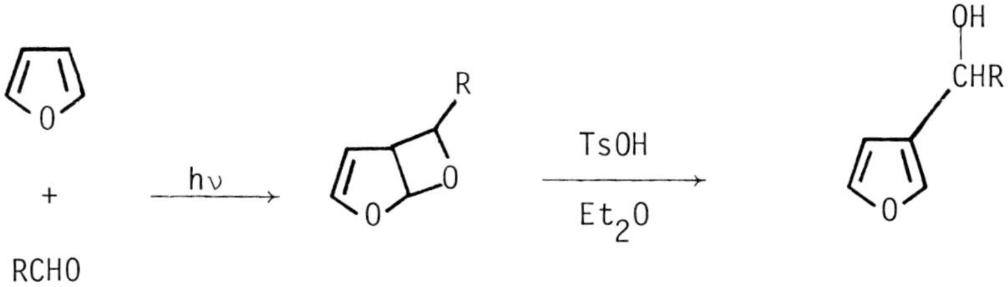

R = alkyl, Ph, COOR

58-73% (second step)

IV.B-6   R. Kada, V. Knoppova, and J. Kovac, Coll. Czech. Chem. Commun., 42, 3333 (1977).

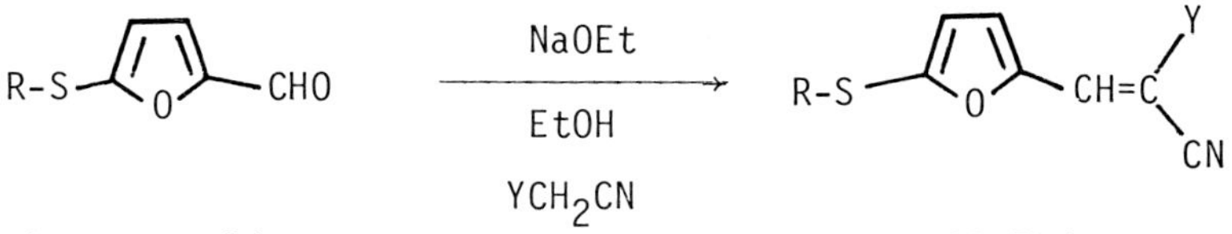

R = heterocyclic
Y = Ph, CN, COOMe

~60-70%

IV.B-7   A. I. Hashem, J. Prakt. Chem., 319, 689 (1977).

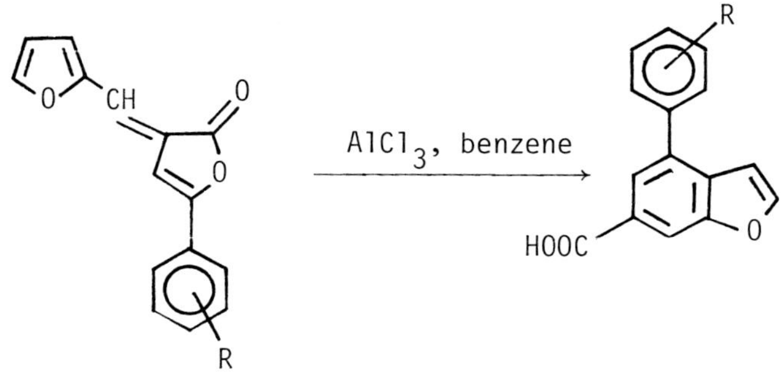

R = H, Me, Cl, OMe   60-76%

IV.B-8   D.L.J. Clive, G. Chittattu, and C. K. Wong, Can. J. Chem., 55, 3894 (1977).

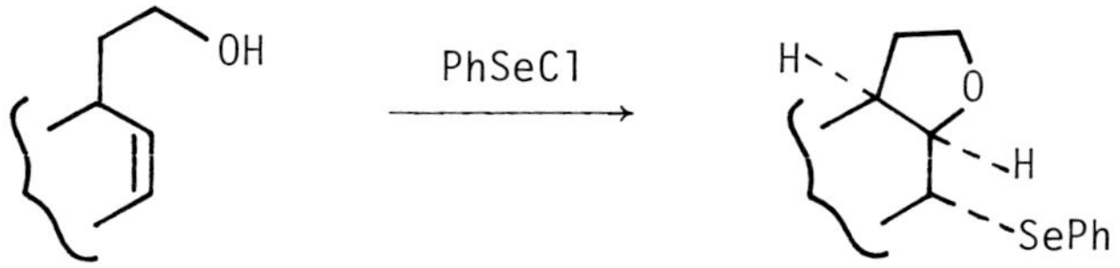

~80%

IV.B-9   Y. Hayakawa et al., Bull. Chem. Soc. Japan, 50, 1990 (1977).

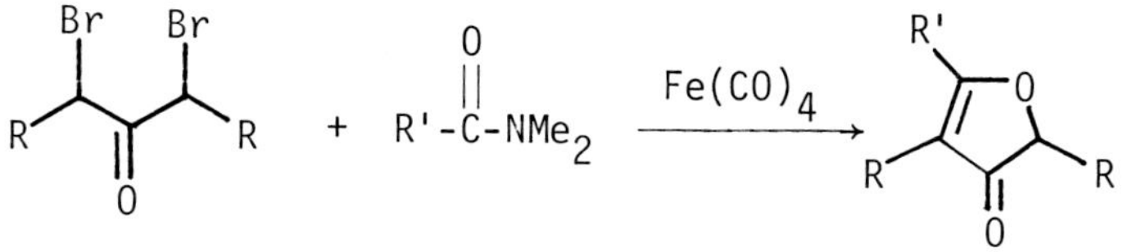

R = Me, Et, i-Pr, t-Bu
R'= H, Me, Ph

30-90%

IV.C-1   Y. Ito, K. Kobayashi, and T. Saegusa, J. Am. Chem. Soc., 99, 3532 (1977).

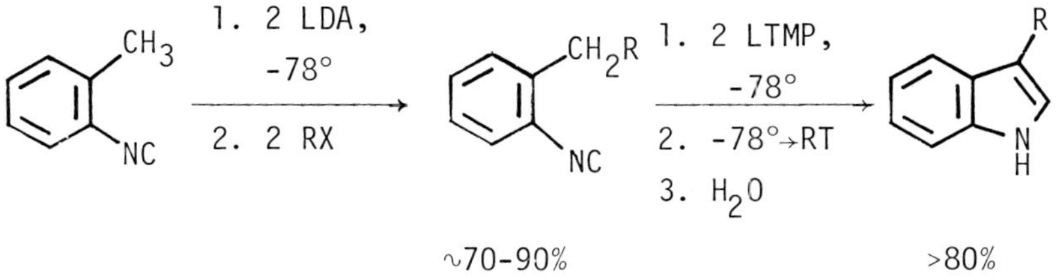

[o-tolyl-NC] — 1. 2 LDA, -78°; 2. 2 RX → [o-(CH$_2$R)-C$_6$H$_4$-NC] — 1. 2 LTMP, -78°; 2. -78°→RT; 3. H$_2$O → 3-R-indole

∼70-90%   >80%

RX = MeI, n-BuBr, i-PrBr, allyl Br, CH$_3$OCOCl, alkylene oxides.

Use of alkylene oxides provides tryptophol derivatives.

IV.C-2   F. DeAngelis, M. Grasso, and R. Nicoletti, Synthesis, 335 (1977).

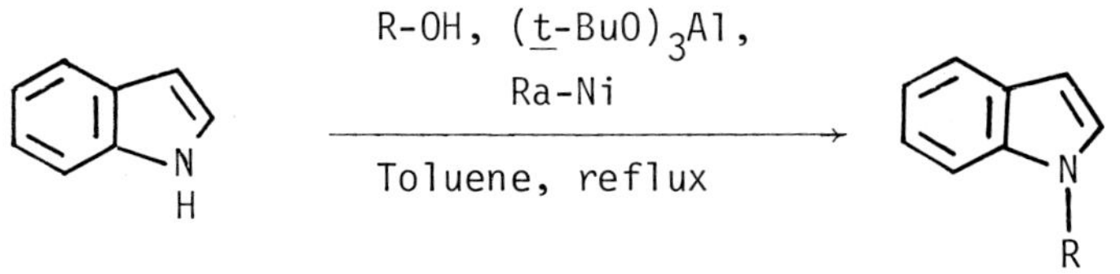

indole (N-H) — R-OH, (t-BuO)$_3$Al, Ra-Ni, Toluene, reflux → N-R indole

R = i-Pr, c-Hx, s-Bu   70-94%

IV.C-3   W. M. Welch, Synthesis, 645 (1977).

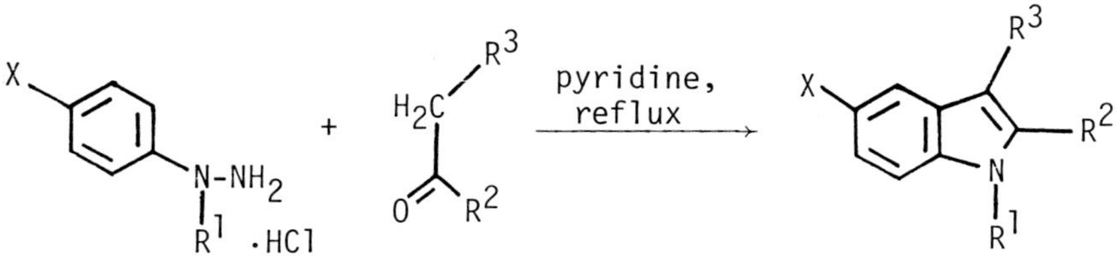

R^1 = H, Ph
R^2, R^3 = alkyl, cyclic, containing carbamates
X = H, Cl, F, OMe

Widely varying yields.

IV.C-4   G. W. Gribble and J. H. Hoffman, Synthesis, 859 (1977).

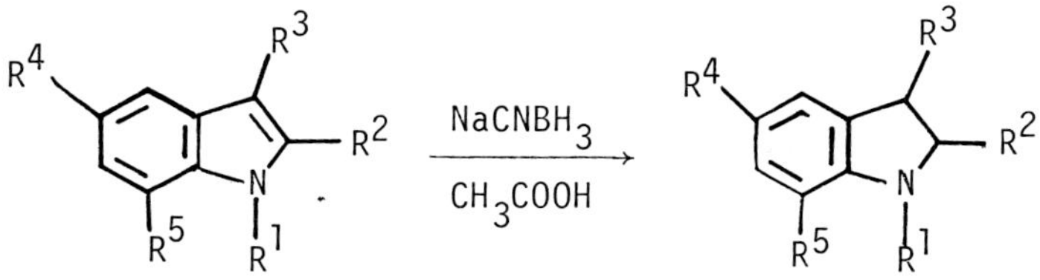

R^1 = H, alkyl, Ph, Bz
R^2, R^3 = H, Me, $-(CH_2)_4-$
R^4, R^5 = H, Me, Br

generally ~80-90%

IV.C-5   K. Isomura, K. Uto, and H. Taniguchi, J. Chem. Soc., Chem. Commun., 664 (1977).

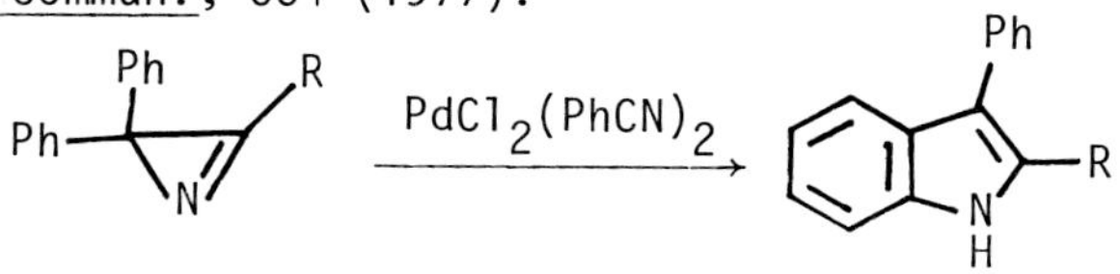

>90% for R = H, Me

IV.C-6   A. Buzas, C. Herisson, and G. Lavielle, Synthesis, 129 (1977).

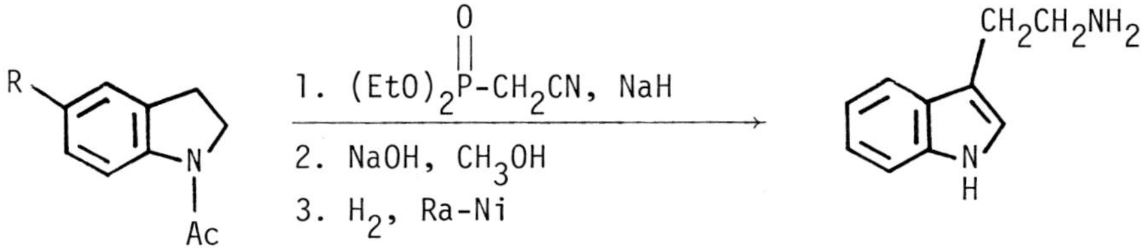

1. $(EtO)_2\overset{O}{\overset{\|}{P}}-CH_2CN$, NaH
2. NaOH, CH_3OH
3. H_2, Ra-Ni

R = H, Me, Cl, OMe   57-65% overall

IV.C-7   M. Mori, K. Chiba, and Y. Ban, Tetrahedron Lett., 1037 (1977).

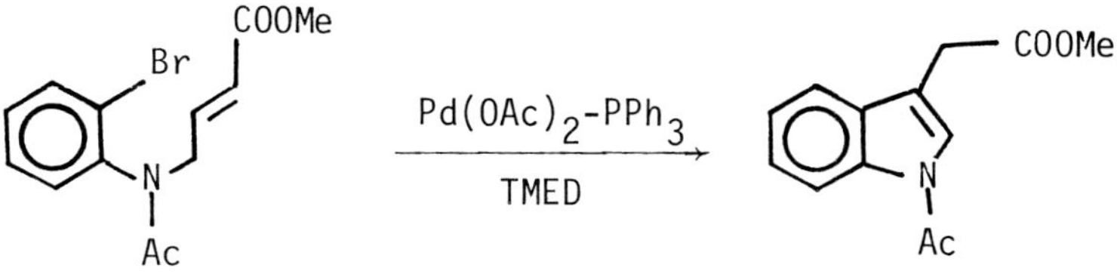

$Pd(OAc)_2-PPh_3$
TMED

37%

IV.C-8   Y. Tamura et al., Tetrahedron Lett., 4417 (1977).

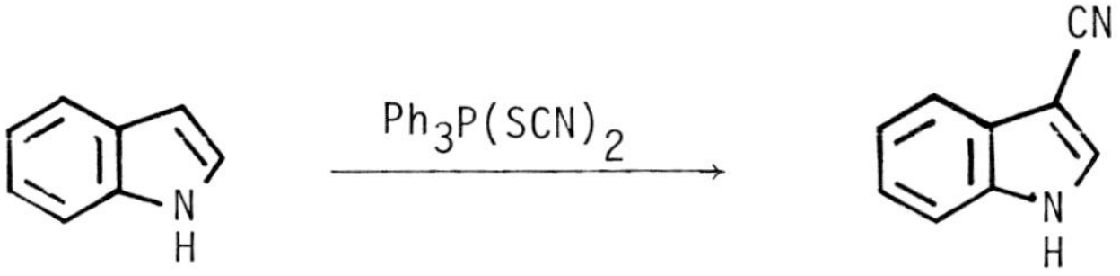

$Ph_3P(SCN)_2$

93%

IV.C-9 R. R. Bard and M. J. Strauss, J. Org. Chem., 42, 435 (1977).

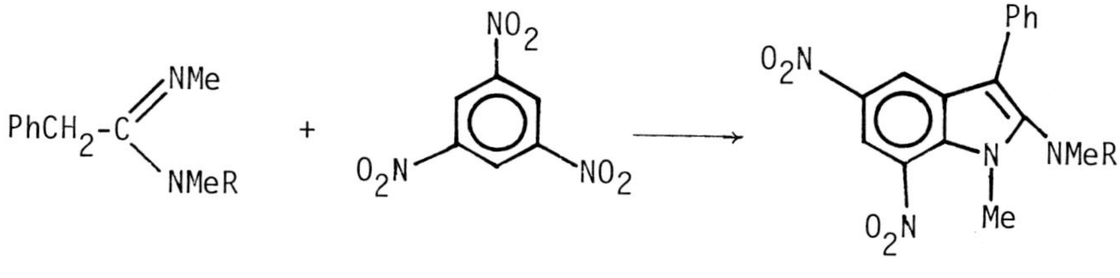

PhCH$_2$-C(=NMe)(NMeR) + 1,3,5-(NO$_2$)$_3$C$_6$H$_3$ ⟶ 3-Ph-2-NMeR-5,7-(NO$_2$)$_2$-1-Me-indole

R = H, Me

~10-20%

IV.C-10 G. Buchi and C.-P. Mak, J. Org. Chem., 42, 1784 (1977).

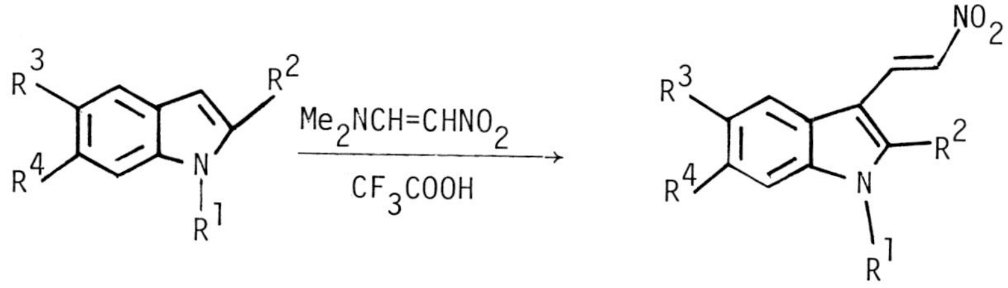

$$\text{indole}(R^1, R^2, R^3, R^4) \xrightarrow[CF_3COOH]{Me_2NCH=CHNO_2} \text{3-(2-nitrovinyl)indole}$$

R^1, R^2 = H, Me
R^3, R^4 = H, Me, OR, OSO$_2$Bz

~90%

IV.C-11 W. O. Siegl, J. Org. Chem., 42, 1872 (1977).

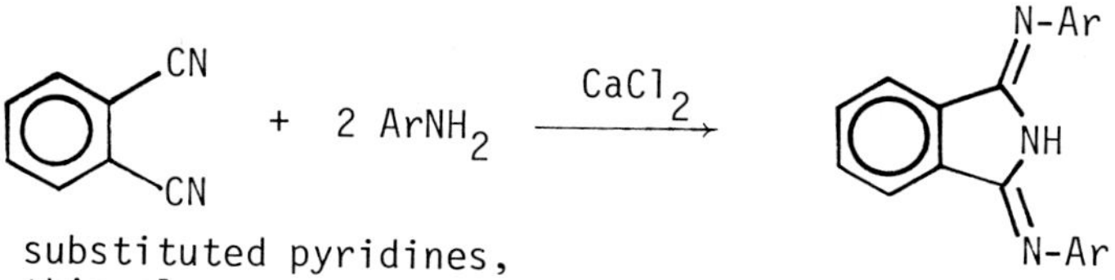

1,2-(CN)$_2$C$_6$H$_4$ + 2 ArNH$_2$ $\xrightarrow{CaCl_2}$ bis(arylimino)isoindoline

Ar = substituted pyridines, thiazoles

~50-80%

IV.C-12/IV.D-1   M. Haimova et al., Comptes Rendus (C), 285, 353 (1977).

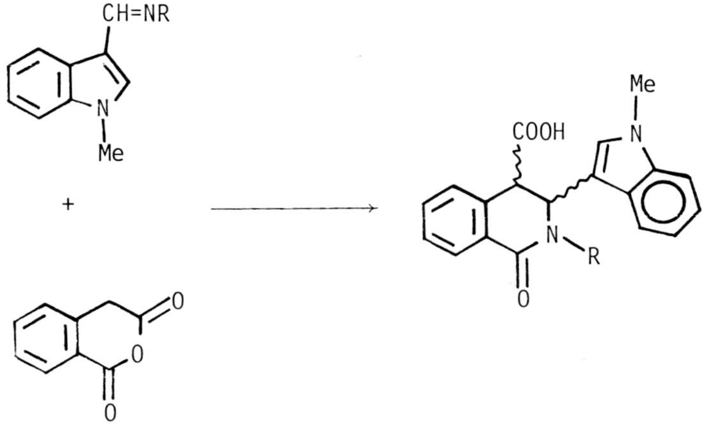

R = Me, Et, Pr, Bz   ~50%

IV.D-2   L. Birkofer and J. Schramm, Liebigs Ann. Chem., 760 (1977).

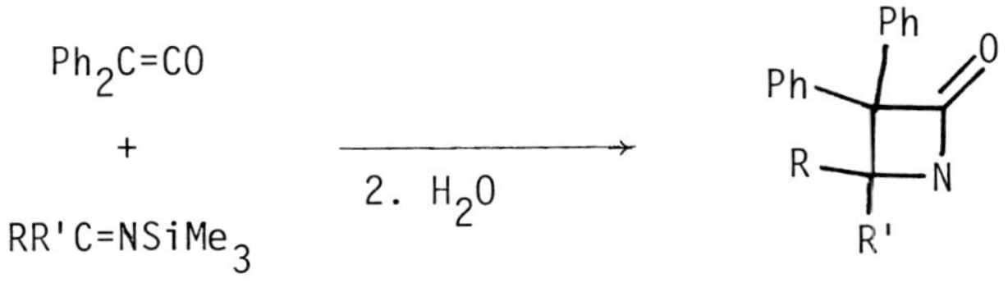

$Ph_2C=CO$

+

$RR'C=NSiMe_3$

2. H_2O

R' = Ph, p-MeOPh   Widely varying yields.

R" = H, Ph

IV.D-3   A. K. Bose et al., Synthesis, 407 (1977).

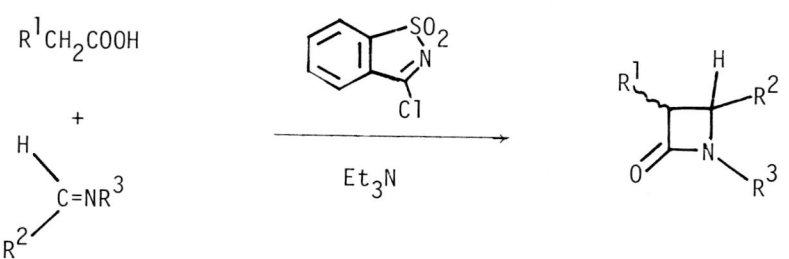

R^1 = OR, Br, N_3, etc.
R^2, R^3 = Ph, p-OMePh

Widely varying yields.

IV.D-4   H. H. Wasserman and A. W. Tremper, Tetrahedron Lett., 1449 (1977).

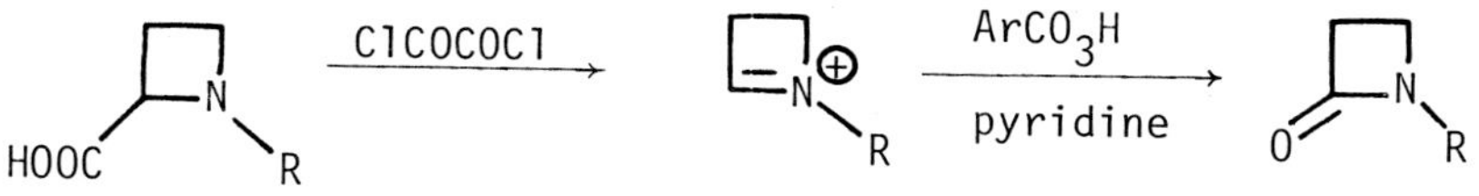

HOOC—[azetidine]—N—R →(ClCOCOCl)→ [azetidine]=N⊕—R →(ArCO$_3$H / pyridine)→ O=[azetidinone]—N—R

R = t-Bu, Bz, cyclohexyl, etc.   70-80%

IV.D-5   M. Rosenblum et al., J. Am. Chem. Soc., 99, 2823 (1977).

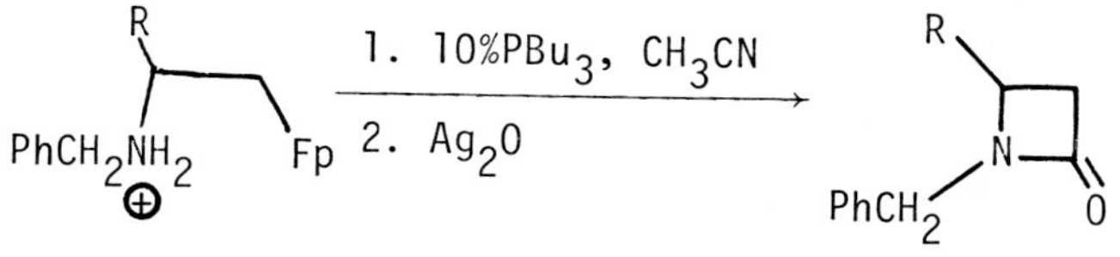

PhCH$_2$NH$_2$⊕—CH(R)—CH$_2$—Fp →(1. 10%PBu$_3$, CH$_3$CN; 2. Ag$_2$O)→ [azetidinone with R, N-CH$_2$Ph]

59%, R = H
82%, R = Me

IV.D-6   I. Ojima, S. Inaba, and K. Yoshida, Tetrahedron Lett., 3643 (1977).

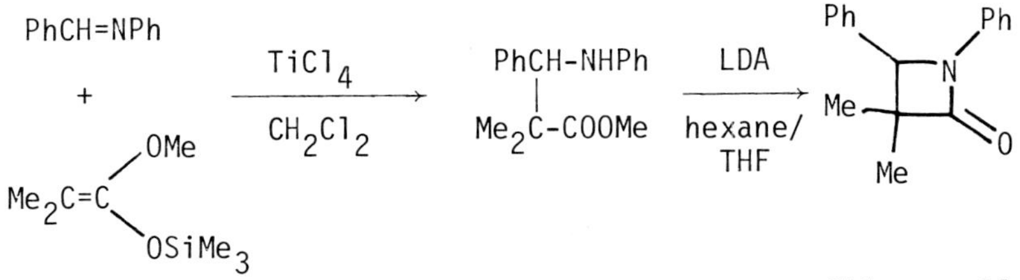

$$PhCH=NPh + Me_2C=C(OMe)(OSiMe_3) \xrightarrow{TiCl_4, CH_2Cl_2} PhCH-NHPh | Me_2C-COOMe \xrightarrow{LDA, hexane/THF} $$

[β-lactam: Ph, N-Ph, Me, Me, C=O ring]

81% overall

IV.D-7   C. Belzecki and Z. Krawczyk, J. Chem. Soc., Chem. Commun., 302 (1977).

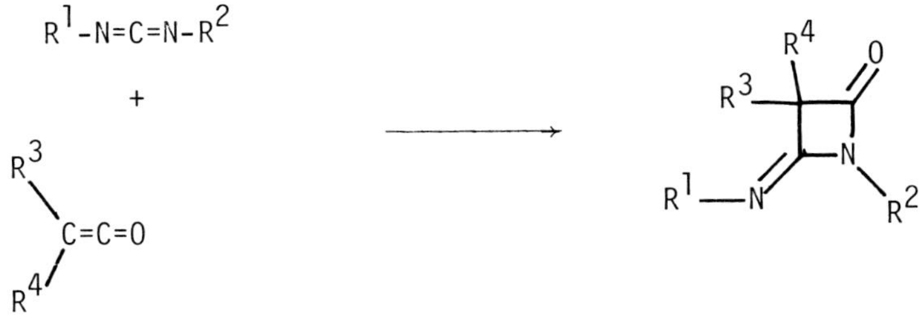

$$R^1-N=C=N-R^2 + R^3R^4C=C=O \longrightarrow$$

[β-lactam product with R^4, R^3, C=O, N-R^2, R^1-N= ring]

R^1, R^2 = alkyl
R^3, R^4 = alkyl, Ph, COOR

28-64%

IV.D-8   P. Tomasik and A. Woszczyk, Tetrahedron Lett., 2193 (1977).

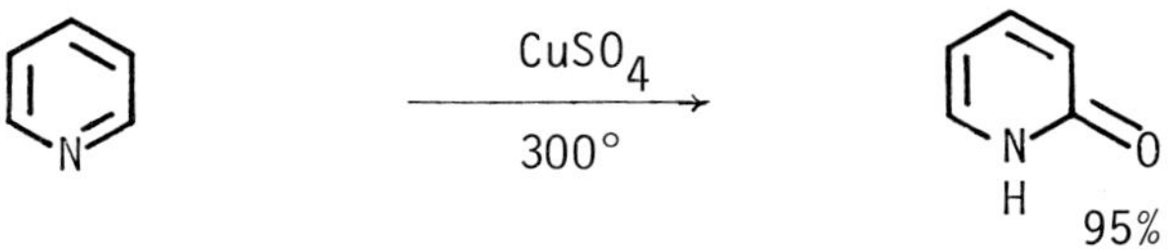

$$\text{pyridine} \xrightarrow{CuSO_4, 300°} \text{2-pyridone}$$

95%

IV.D-9 T. Kobayashi and T. Hiraoka, Chem. Lett., 1399 (1977).

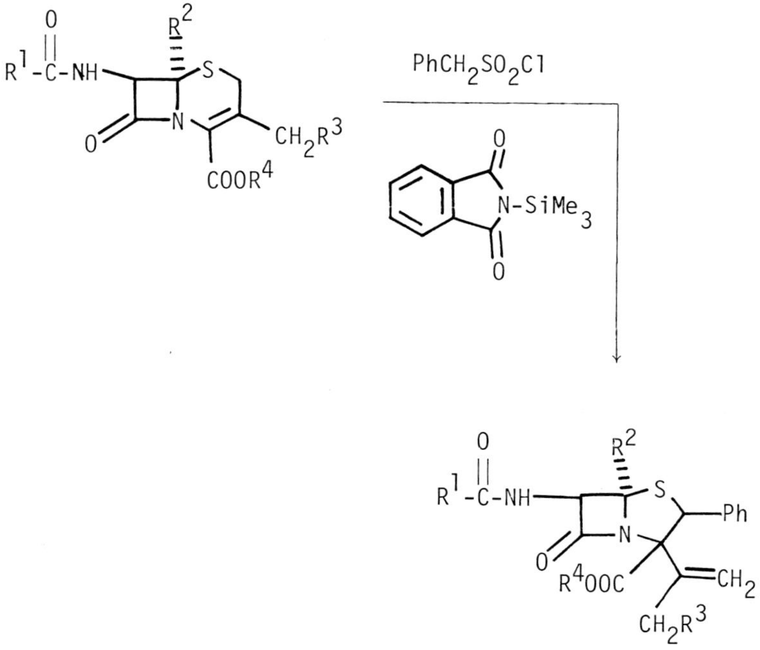

[Reaction scheme: cephem-type β-lactam with $R^1-C(O)-NH$, R^2, S, N, CH_2R^3, $COOR^4$ substituents reacts with $PhCH_2SO_2Cl$ and phthalimide-$N-SiMe_3$ to give rearranged β-lactam product with Ph, CH_2R^3, $=CH_2$, R^4OOC groups]

37%

IV.D-10 B. G. Chatterjee and D. P. Sahu, J. Org. Chem., 42, 3162 (1977).

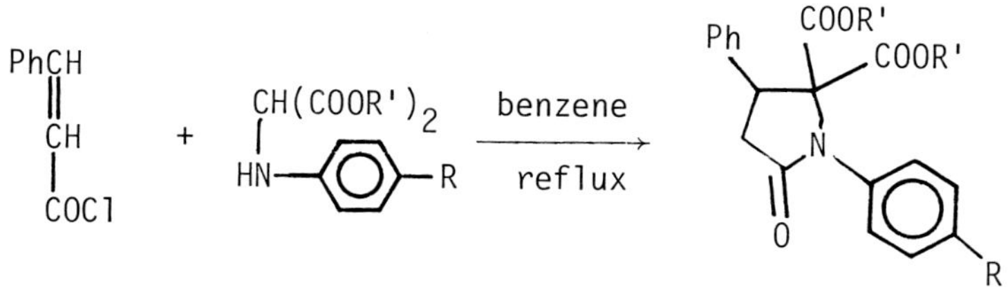

[Reaction scheme: $PhCH=CH-COCl$ + $CH(COOR')_2$ with $HN-C_6H_4-R$, benzene reflux → pyrrolidinone with Ph, $COOR'$, $COOR'$, N-aryl-R substituents]

R = H, Me, Hal, COOR
R' = Me, Et

~40-80%

IV.D-11   K. Saito et al., Synthesis, 841 (1977).

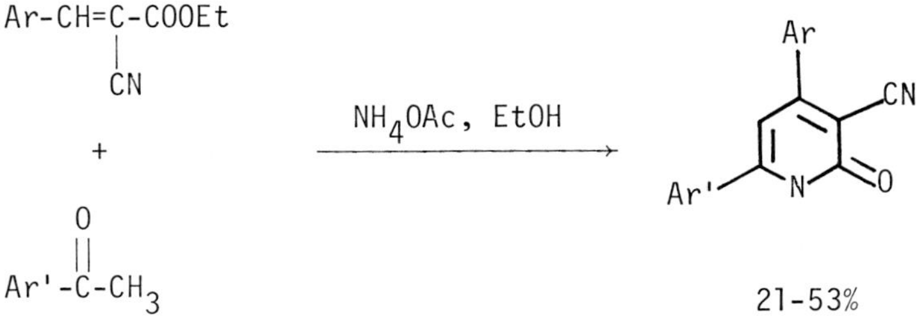

$$Ar-CH=C(CN)-COOEt + Ar'-C(=O)-CH_3 \xrightarrow{NH_4OAc, EtOH} \text{pyridinone (Ar, Ar', CN)}$$

21-53%

IV.D-12   L. E. Overman and S. Tsuboi, J. Am. Chem. Soc., 99, 2814 (1977).

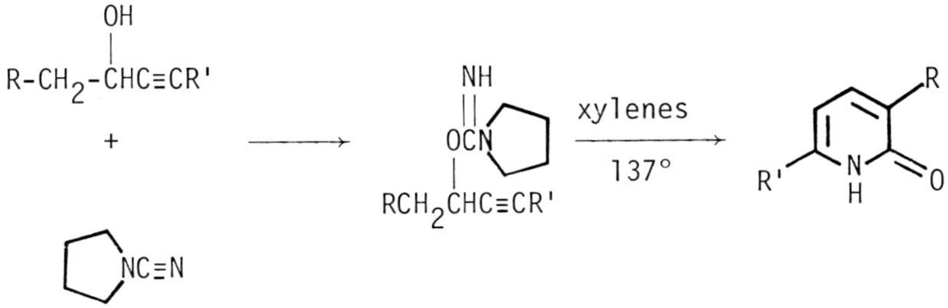

$$R-CH_2-CH(OH)-C\equiv CR' + \text{pyrrolidine-NC}\equiv N \longrightarrow \text{carbamate intermediate} \xrightarrow{\text{xylenes}, 137°} \text{pyridinone}$$

R,R' = H, 1°, 2°, 3° alkyl, Ph   ~60-70%

IV.D-13   P. Hong and H. Yamazaki, Tetrahedron Lett., 1333 (1977).

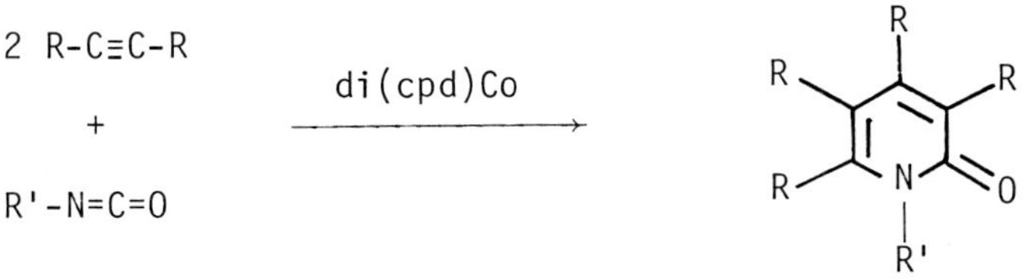

$$2\ R-C\equiv C-R + R'-N=C=O \xrightarrow{di(cpd)Co} \text{N-substituted pyridinone}$$

R,R' = H, Ph, Me, COOMe   ~20-30%

IV.D-14   D.H.R. Barton et al., J. Chem. Soc., Perkin I, 1107
(1977).

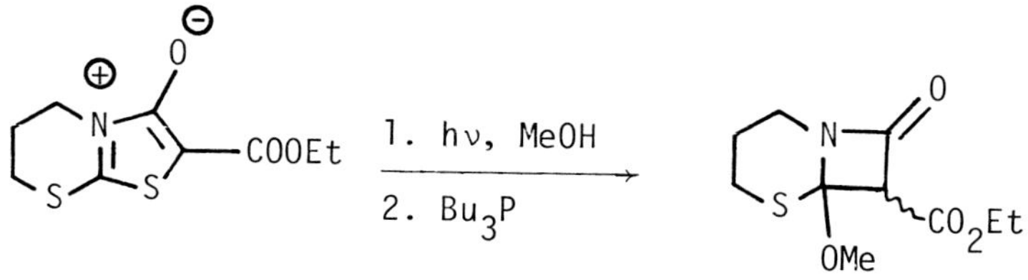

[Structure: bicyclic thiazolium-3-olate with -COOEt] → 1. hν, MeOH; 2. Bu$_3$P → [β-lactam structure with OMe and CO$_2$Et]

9%

Many additional examples.

IV.D-15   P. Hong and H. Yamazaki, Synthesis, 50 (1977).

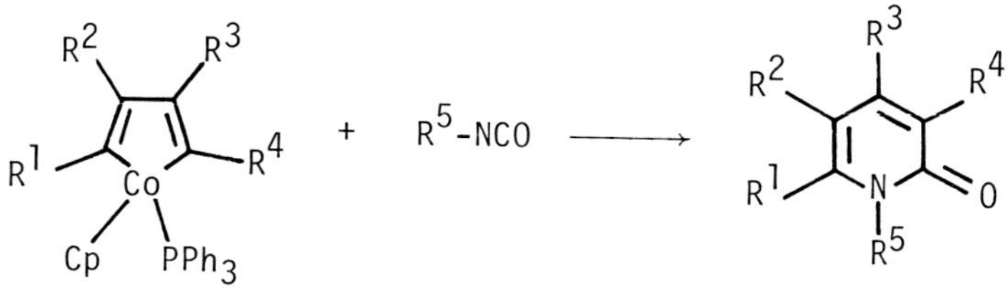

[Cobalt cyclopentadienone complex with R^1, R^2, R^3, R^4, Cp, PPh$_3$] + R^5-NCO → [2-pyridone with R^1, R^2, R^3, R^4, R^5]

R^1,...,R^4 = Me, Ph, COOMe   ∼70%
R^5 = Me, Bu, Ph

IV.D-16   J. T. Boyers and E. E. Glover, J. Chem. Soc., Perkin I, 1960 (1977).

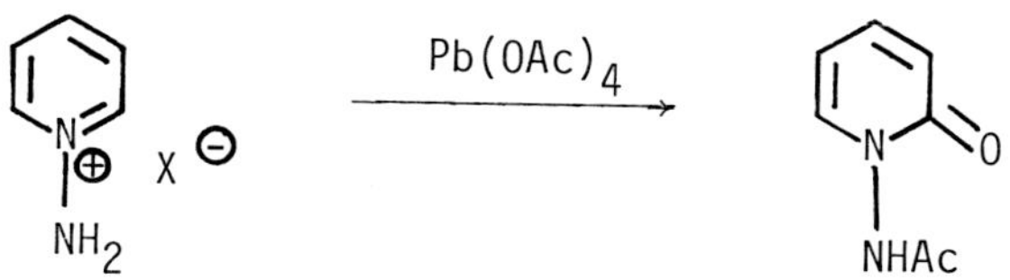

[1-aminopyridinium X$^⊖$] → Pb(OAc)$_4$ → [1-(NHAc)-2-pyridone]   40%

Several additional examples.

268

IV.D-17 J. E. Douglas and D. A. Hunt, J. Org. Chem., 42, 3974 (1977).

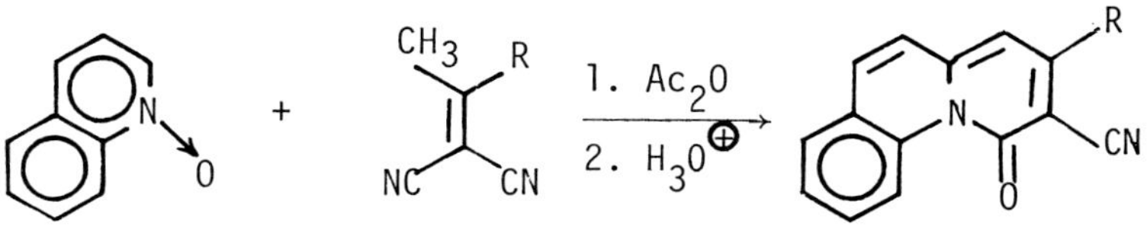

R = t-Bu, Ph   60-70%

IV.D-18 M. A. Haimova et al., Tetrahedron, 33, 331 (1977).

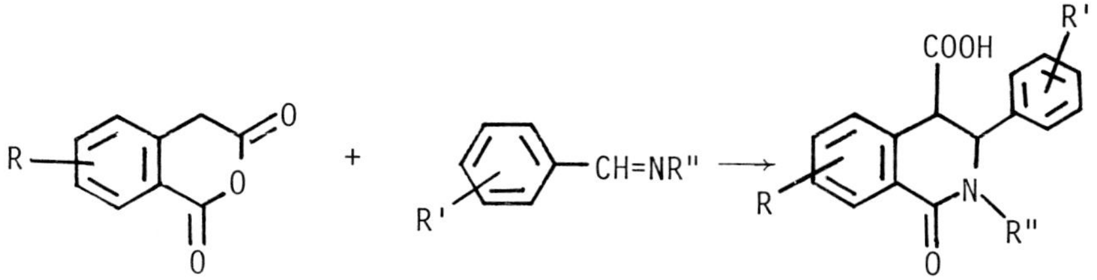

R,R' = H, OMe   ~80-90%
R" = H, alkyl, Ph

IV.D-19 E. Schmitz et al., J. Prakt. Chem., 319, 274 (1977).

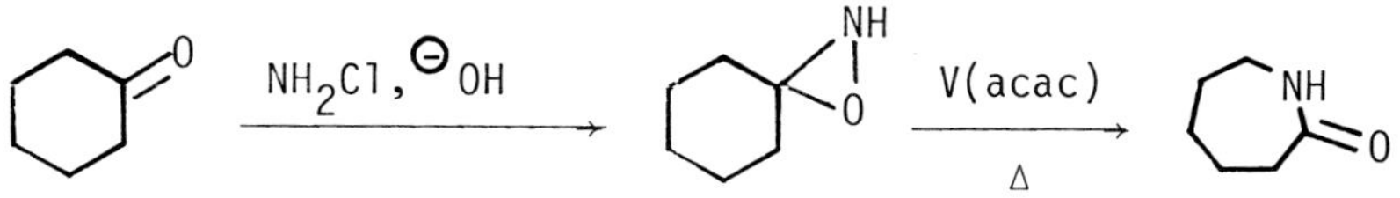

83-88%

IV.D-20   A. Gossauer, R.-P. Hinze, and H. Zilch, Angew. Chem. Int. Ed., 16, 418 (1977).

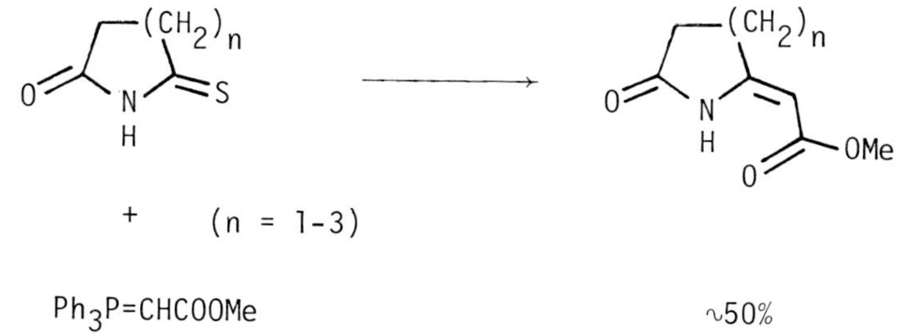

[Structure: pyrrolidinone-thione with $(CH_2)_n$ ring + $Ph_3P=CHCOOMe$ → pyrrolidinone with exocyclic =CHCOOMe, $(n = 1-3)$, ~50%]

IV.D-21   W. N. Speckamp et al., Tetrahedron Lett., 939 (1977).

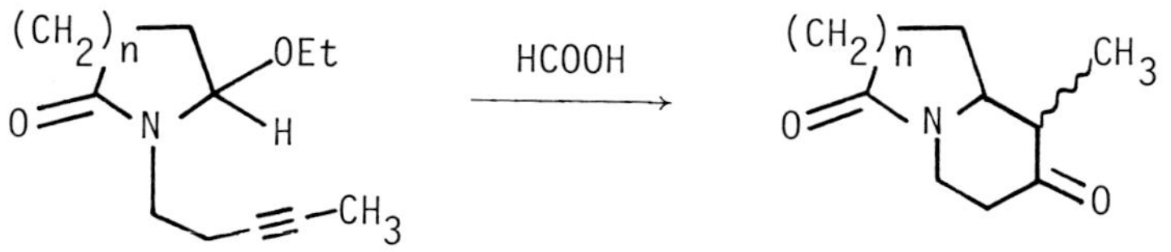

[Structure: N-substituted lactam with OEt, H, and CH$_2$C≡C-CH$_3$ substituents, with HCOOH → bicyclic lactam ketone with CH$_3$ group, ~90%]

IV.D-22/IV.E-1   A. E. Greene et al., Tetrahedron Lett., 2365 (1977).

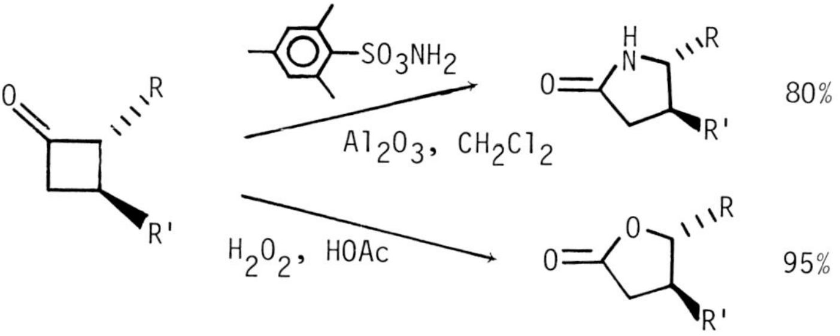

[Cyclobutanone with R, R' substituents:
- with mesitylene-SO$_3$NH$_2$, Al$_2$O$_3$, CH$_2$Cl$_2$ → pyrrolidinone (R, R' substituted), 80%
- with H$_2$O$_2$, HOAc → γ-butyrolactone (R, R' substituted), 95%]

IV.E-2   G. D. Annis and S. V. Ley, J. Chem. Soc., Chem. Commun., 581 (1977).

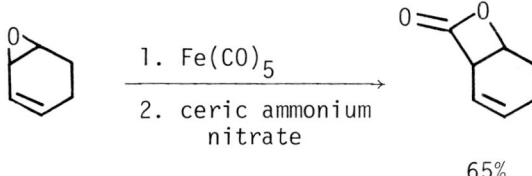

65%

IV.E-3   P. Brownbridge and S. Warren, J. Chem. Soc., Chem. Commun., 465 (1977).

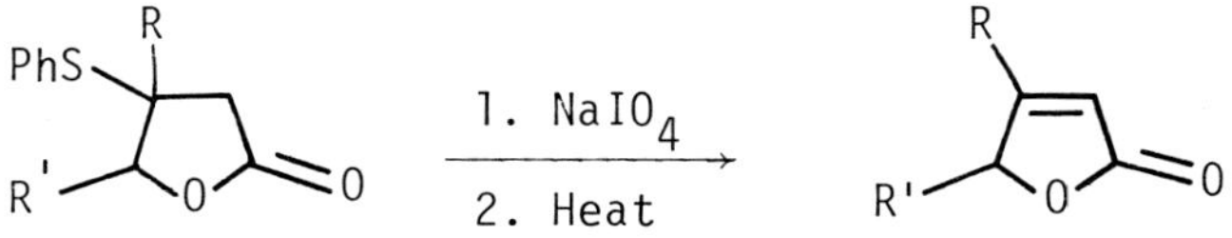

[Structure: PhS-substituted tetrahydrofuranone with R, R' groups] →(1. NaIO$_4$; 2. Heat)→ [butenolide with R, R' groups]

R,R' = H, Me

IV.E-4   K. Iwai et al., Bull. Chem. Soc. Japan, 50, 242 (1977).

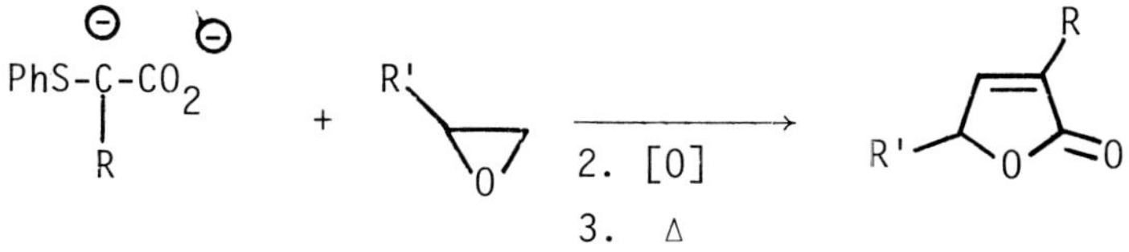

$PhS-\underset{R}{\overset{\ominus}{C}}-CO_2^{\ominus}$ + [epoxide with R'] →(2. [O]; 3. Δ)→ [butenolide with R, R']

R = H, Me, Et
R'= Me, Bu, Ph

IV.E-5 A. Kasahara et al., Bull. Chem. Soc. Japan, 50, 1899 (1977).

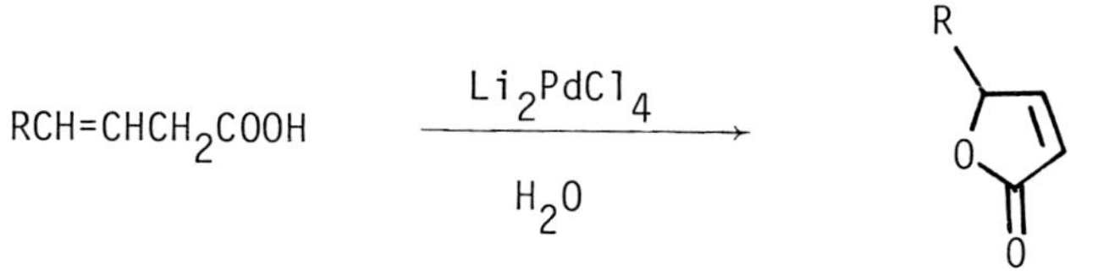

RCH=CHCH$_2$COOH $\xrightarrow{\text{Li}_2\text{PdCl}_4}{\text{H}_2\text{O}}$ [butenolide with R substituent]

∼30-40%

[cyclohexenyl-CH$_2$-COOH] $\xrightarrow{\text{Li}_2\text{PdCl}_4}{\text{H}_2\text{O}}$ [bicyclic lactone]

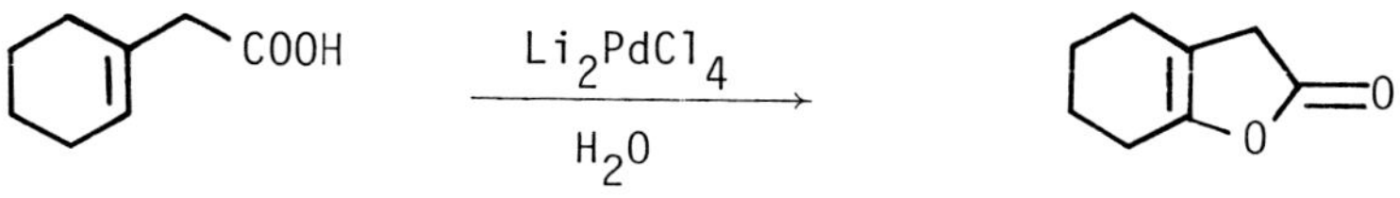

38%

IV.E-6 A. Mosterd and H.J.T. Bos, Rec. Trav. Chim. Pays-Bas, 96, 275 (1977).

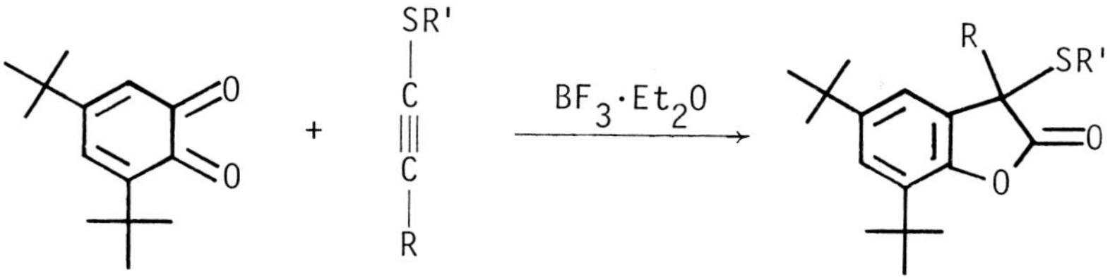

[di-t-butyl-o-benzoquinone] + R'S–C≡C–R $\xrightarrow{\text{BF}_3\cdot\text{Et}_2\text{O}}$ [benzofuranone product with R, SR' substituents]

R = Me, Et
R' = Me, Ph, t-Bu

50-90%

IV.E-7   A. Dobrev and C. Ivanov, <u>Synthesis</u>, 562 (1977).

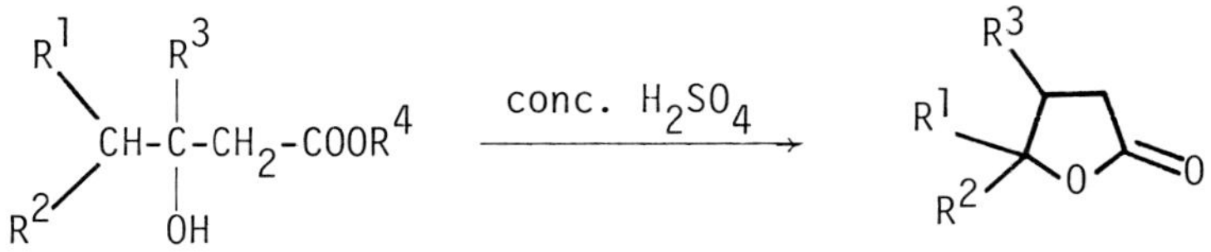

$$\underset{R^2}{\overset{R^1}{>}}CH-\underset{OH}{\overset{R^3}{\underset{|}{C}}}-CH_2-COOR^4 \xrightarrow{\text{conc. } H_2SO_4} \text{[lactone with } R^1, R^2, R^3\text{]}$$

R^1, R^2, R^3 = alkyl, cyclic   65-94%

R^4 = Et, <u>t</u>-Bu

IV.E-8   U. Ravid and R. M. Silverstein, <u>Tetrahedron Lett.</u>, 423 (1977).

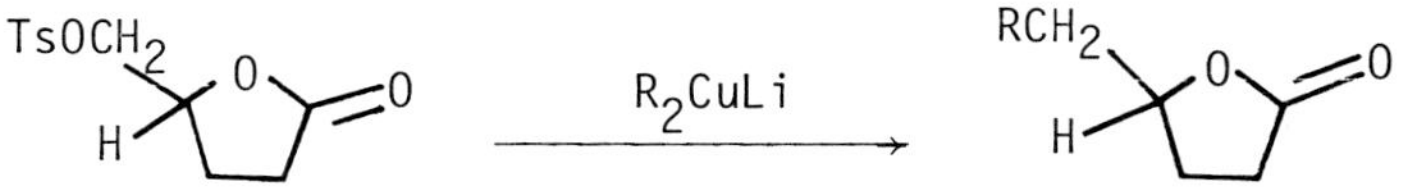

$$\text{TsOCH}_2\text{-[lactone]-H} \xrightarrow{R_2\text{CuLi}} \text{RCH}_2\text{-[lactone]-H}$$

R = Me, <u>n</u>-Bu, -CH=CH-C_5H_{11}

yields 66%, 41%, 12%

IV.E-9   K. C. Nicolaou and Z. Lysenko, <u>J. Chem. Soc., Chem. Commun.</u>, 293 (1977).

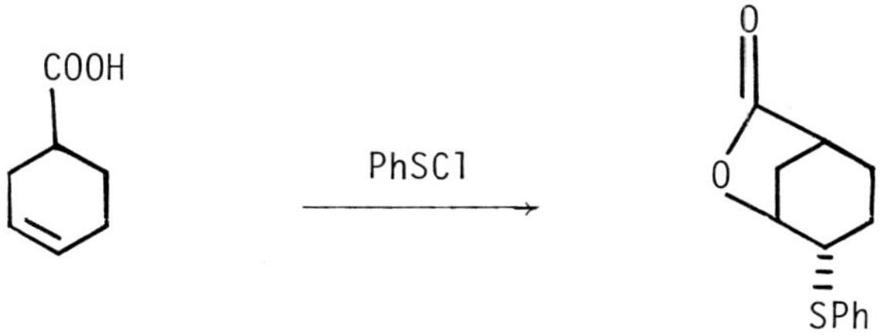

$$\text{cyclohex-3-ene-COOH} \xrightarrow{\text{PhSCl}} \text{bicyclic lactone-SPh}$$

70%

IV.E-10 J. A. Marshall and G. A. Flynn, Synth. Commun., 7, 417 (1977).

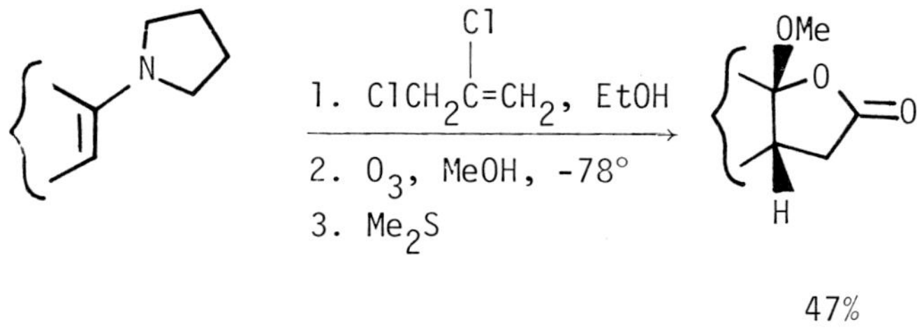

1. ClCH$_2$C(Cl)=CH$_2$, EtOH
2. O$_3$, MeOH, -78°
3. Me$_2$S

47%

IV.E-11 D.L.J. Clive and G. Chittattu, J. Chem. Soc., Chem. Commun., 484 (1977).

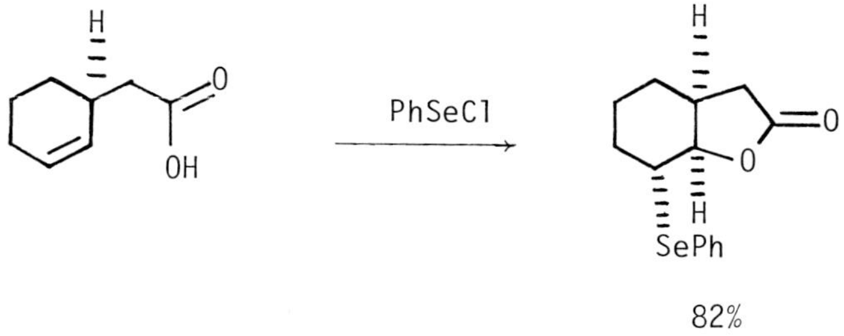

PhSeCl

82%

IV.E-12 N. Minami and I. Kuwajima, Tetrahedron Lett., 1423 (1977).

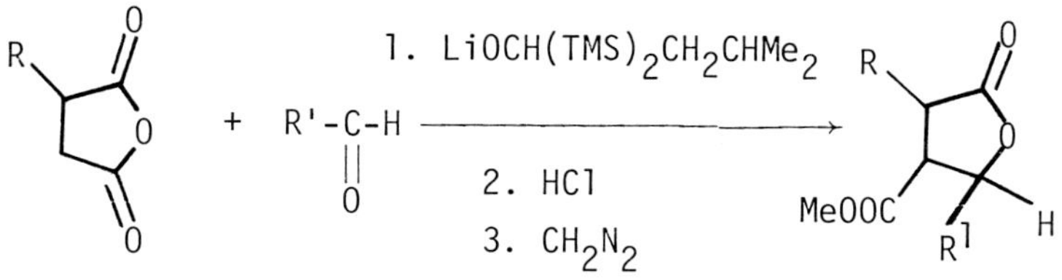

1. LiOCH(TMS)$_2$CH$_2$CHMe$_2$
2. HCl
3. CH$_2$N$_2$

R = H, Me
R' = Ph, alkyl, vinyl

51-85%

IV.E-13   V. P. Sendrik, O. Paleta, and V. Dedek, Coll. Czech. Chem. Commun., 42, 2530 (1977).

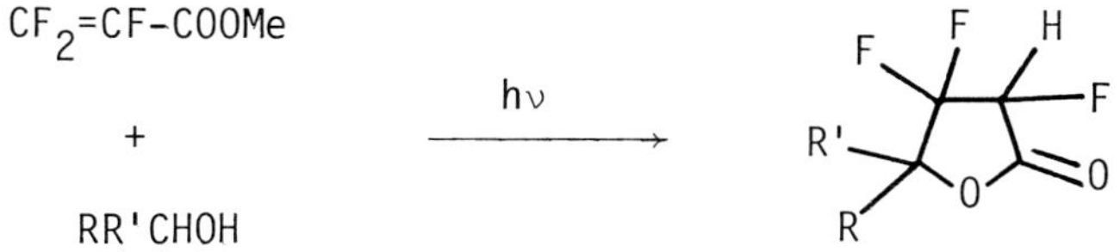

$CF_2=CF-COOMe$

\+

RR'CHOH

$\xrightarrow{h\nu}$

[β-fluoro-γ-butyrolactone product]

R,R' = Me, Et, $-(CH_2)_5-$ 32-58%

IV.E-14   V. Jäger and H. J. Gunther, Tetrahedron Lett., 2543 (1977).

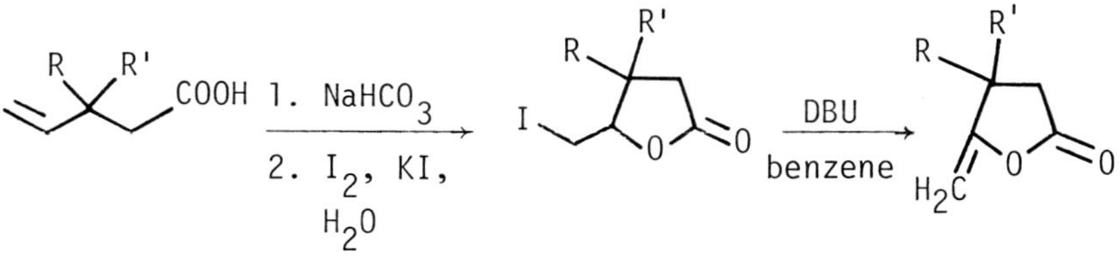

[alkene-COOH] $\xrightarrow[\text{2. } I_2, KI, H_2O]{\text{1. } NaHCO_3}$ [iodo-lactone] $\xrightarrow[\text{benzene}]{DBU}$ [methylene-lactone]

R,R' = H, Me, Et   ~70%

IV.E-15   M. Kato, M. Kageyama, and A. Yoshikoshi, J. Chem. Soc., Perkin I, 1305 (1977).

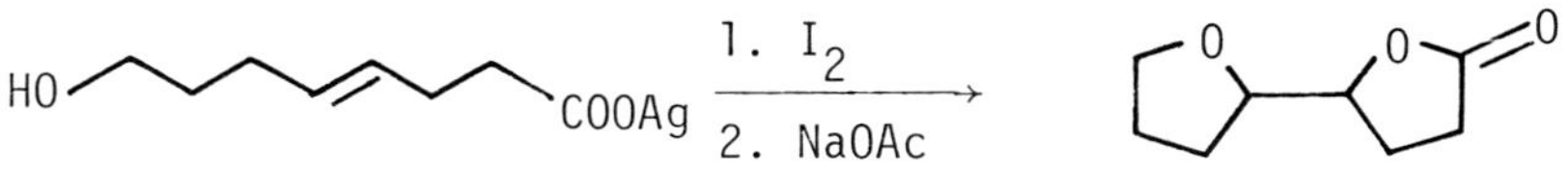

HO~~~~~=~~~~COOAg $\xrightarrow[\text{2. NaOAc}]{\text{1. } I_2}$ [bis-tetrahydrofuran lactone]

58%

Several additional examples.

IV.E-16   D. E. Korte, L. S. Hegedus, and R. P. Wirth, J. Org. Chem., 42, 1329 (1977).

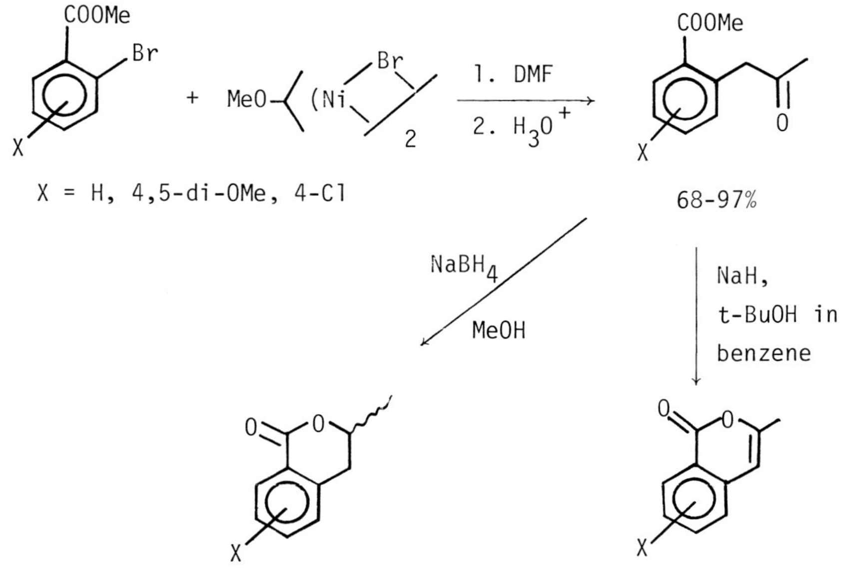

X = H, 4,5-di-OMe, 4-Cl

68-97%

IV.E-17   R. Aumann and H. Ring, Angew. Chem. Int. Ed., 16, 50 (1977).

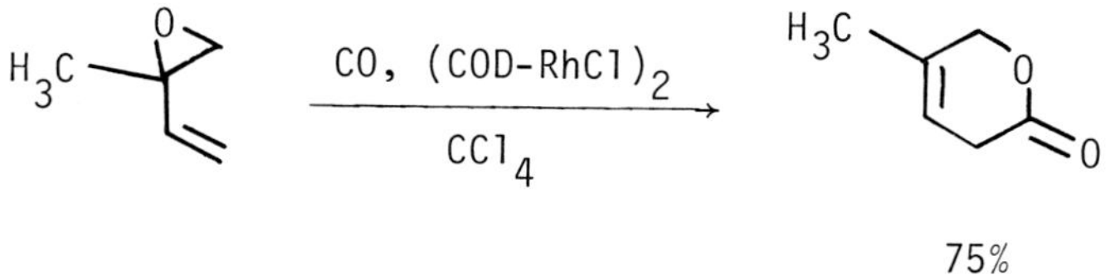

75%

IV.E-18 P. A. Grieco et al., J. Chem. Soc., Chem. Commun., 870 (1977).

$$(CH_2)_n \text{-lactone-R} \xrightarrow{\text{PhSe-OH, } H_2O_2} (CH_2)_n \text{-unsaturated lactone-R}$$

n = 3, 4

R = Me, H, ring juncture

50-83%

IV.E-19 C. W. Bird and P. Thorley, Chem. and Ind., 872 (1977).

$$\begin{array}{c} R-\overset{O}{\underset{\|}{C}}-CH_2-C\diagdown^{O} \\ C\diagup \\ R-\overset{\|}{\underset{O}{C}}-CH_2 ^{O} \end{array} \xrightarrow{Pb(OAc)_4} \text{4-hydroxy-2H-pyran-2-one with COR substituent}$$

No details.

IV.E-20 A. Chatterjee et al., Indian J. Chem., 15B, 214 (1977).

indanone $\xrightarrow{\text{MCPBA}, Ac_2O, H_2SO_4}$ chromanone $\xrightarrow[\substack{1.\text{Pd/C} \\ 2.\text{KOH} \\ 3.\text{HCl}}]{}$ coumarin

~60% overall

IV.E-21   B. K. Sarkhel and J. N. Srivastava, Indian J. Chem., 15B, 103 (1977).

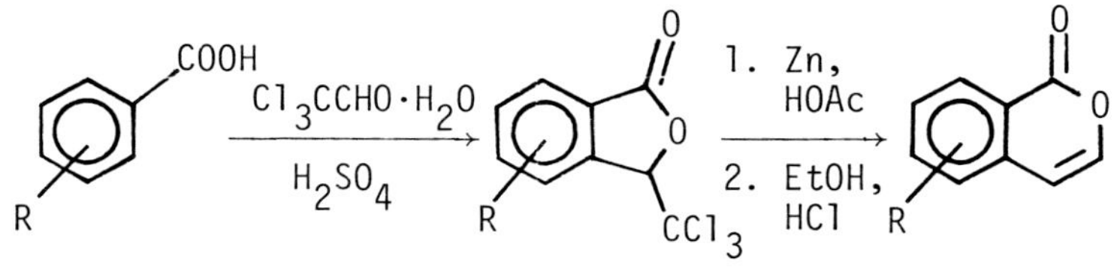

R = OH, OMe   ~30% overall

IV.E-22   V. K. Ahluwalia and D. Kumar, Indian J. Chem., 15B, 514 (1977).

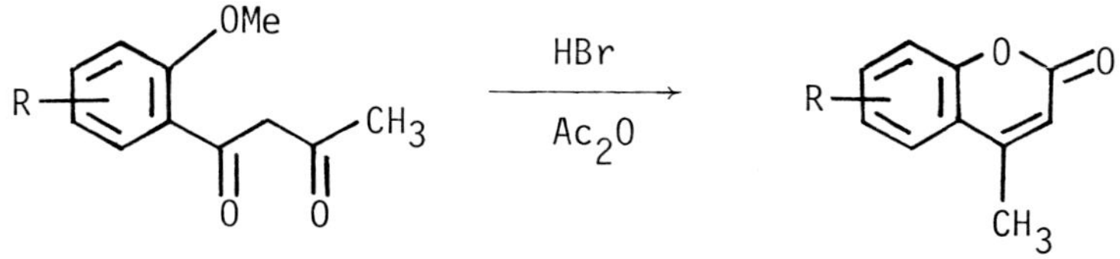

R = H, OMe   ~20-90%

IV.E-23   R. S. Mali and V. J. Yadav, Synthesis, 464 (1977).

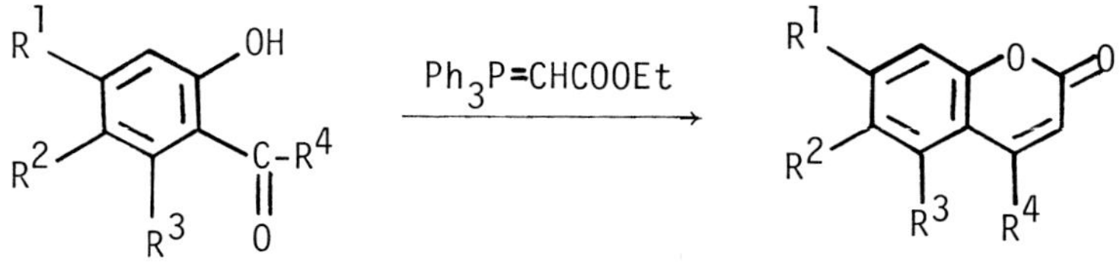

R's = H, Me, OMe

42-90%
35%, R^1 = OH

IV.E-24 K. Narasaka, T. Masui, and T. Mukaiyama, <u>Chem. Lett.</u>, 763 (1977).

HO-(CH$_2$)$_n$-COOH $\xrightarrow[\text{Et}_3\text{N}]{\text{CH}_3\text{CN}}$ cyclic (CH$_2$)$_n$-O-C=O

n =	14	11	10
% yield =	84	69	61

IV.E-25 T. Mukaiyama, K. Narasaka, and K. Kikuchi, <u>Chem. Lett.</u>, 441 (1977).

HO-(CH$_2$)$_n$—COOH + Et$_3$N 1. [N-methyl-2-chloropyridinium iodide], [2-phenyl-pyridone] 2. TsOH, CH$_2$Cl$_2$, rfx → cyclic (CH$_2$)$_n$ lactone

n = 10, 73%
n = 11, 99%
n = 14, 100%

IV.E-26 E. J. Corey, D. J. Brunelle, and K. C. Nicolaou, <u>J. Am. Chem. Soc.</u>, <u>99</u>, 7359 (1977).

(CH$_2$)$_n$ lactone with CH(CH$_2$)$_3$OH substituent $\xrightarrow{\text{TsOH, CH}_2\text{Cl}_2}$ (CH$_2$)$_n$ lactone with HO ~65%

n = 4,5

279

IV.E-27   E. Shalom, J.-L. Zenou, and S. Shatzmiller, J. Org. Chem., 42, 4213 (1977).

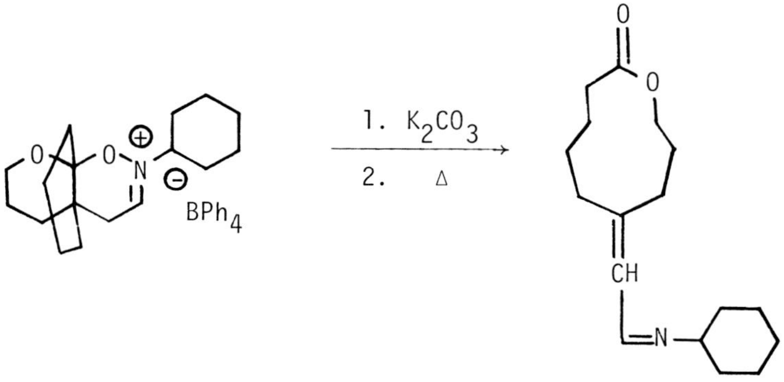

IV.E-28   T. G. Beck, Tetrahedron, 33, 3041 (1977).

Review: "The Synthesis of Macrocyclic Lactones. Approaches to Complex Macrolide Antibiotics"

IV.E-29   S. S. Newaz, Aldrichimica Acta, 10, 64 (1977).

Review: "Recent Methods for the Synthesis of Conjugated Lactones"

IV.F-1 C. Botteghi, G. Caccia, and S. Gladiali, Synth. Commun., 6, 549 (1976).

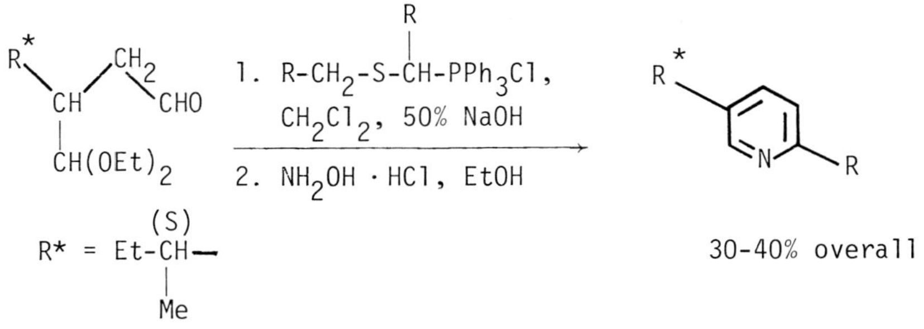

$R^*\text{-CH(CH(OEt)}_2\text{)-CH}_2\text{-CHO}$

1. $R\text{-CH}_2\text{-S-CH(R)-PPh}_3\text{Cl}$, CH_2Cl_2, 50% NaOH
2. $NH_2OH \cdot HCl$, EtOH

→ R^*-pyridine-R

$R^* = \text{Et-CH(Me)}—$ (S)

R = H, Me

30-40% overall

IV.F-2 C.-S. Giam and K. Ueno, J. Am. Chem. Soc., 99, 3166 (1977).

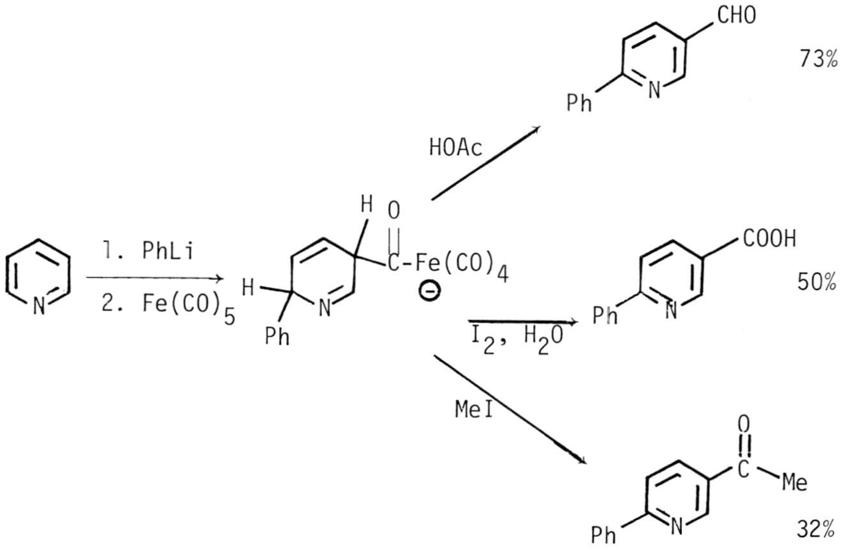

pyridine
1. PhLi
2. $Fe(CO)_5$

→ H-(dihydropyridine-Ph)-C(=O)-Fe(CO)$_4^{\ominus}$

HOAc → 5-CHO-2-Ph-pyridine 73%

I_2, H_2O → 5-COOH-2-Ph-pyridine 50%

MeI → 5-C(=O)Me-2-Ph-pyridine 32%

IV.F-3   R. Faragher and T. L. Gilchrist, J. Chem. Soc., Chem. Commun., 252 (1977).

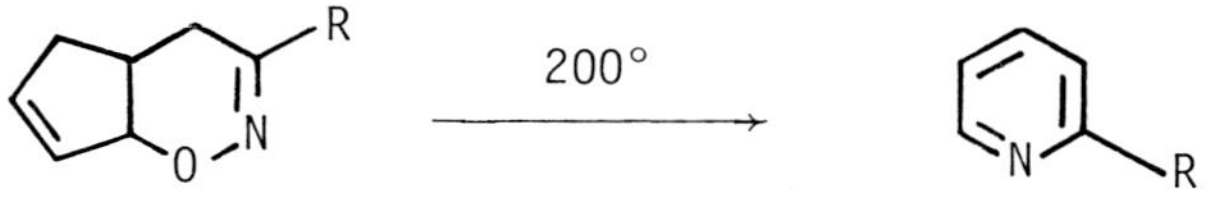

R = Ph   4-BrPh   2-Furyl
   80%   78%   40%

IV.F-4   G. Dauphin, et al., Tetrahedron, 33, 1129 (1977).

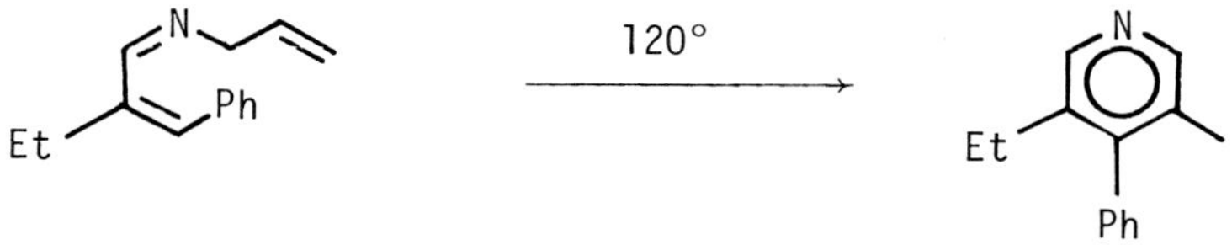

60%

IV.F-5   H. Greuter and D. Bellus, J. Het. Chem., 14, 203 (1977).

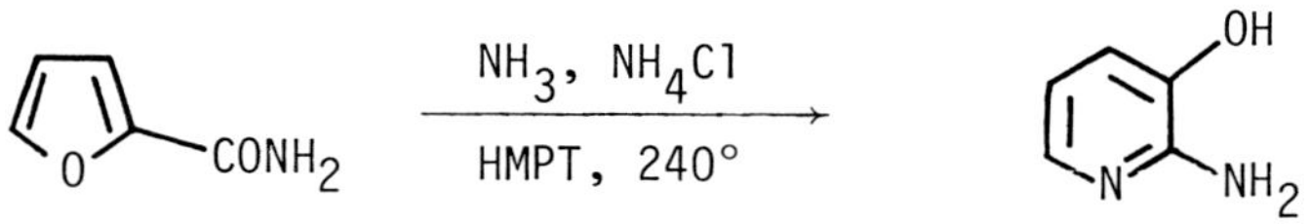

55%

IV.F-6   A. Naiman and K.P.C. Vollhardt, Angew. Chem. Int. Ed., 16, 708 (1977).

$$(CH_2)_n\diagup^{C\equiv CH}_{C\equiv CH} + \underset{N}{\overset{R}{\underset{|||}{\underset{C}{|}}}} \xrightarrow{CpCo(CO)_2} (CH_2)_n\text{-pyridine ring with }R$$

n = 3,4,5   ~40-70%

R = alkyl, Ph, ester, etc.

IV.F-7   C. Jutz, H.-G. Löbering, and K.-H. Trinkl, Synthesis, 326 (1977).

$$\underset{H_2N}{\overset{H\diagdown\diagup Z}{\underset{CH_3}{\overset{C}{\underset{||}{C}}}}} \xrightarrow[\text{2. }\Delta]{1.\ Me_2N\text{-}CH=CRCH(OEt)_2,\ NaOMe} \quad R\text{-pyridine-}Z,CH_3$$

Z = COOMe, CN   R = H, Me, OMe, Ph   52-95%

IV.F-8   D. E. Korte, L. S. Hegedus, and R. P. Wirth, J. Org. Chem., 42, 1329 (1977).

$$\text{2-allyl-benzamide} + PdCl_2\cdot 2CH_3CN \xrightarrow[THF]{NaH} \text{3-methyl-1-hydroxyisoquinoline}$$

68%

IV.F-9   A. G. Anastassiou et al., J. Org. Chem., 42, 2651
(1977).

[Cyclopentadienyl-N(COOEt)-CH=CHCHO] →(Al₂O₃) [cyclopenta-fused pyridine with N-COOEt] →(LiAlH₄) [cyclopenta[b]pyridine] + [cyclopenta[c]pyridine isomer]

No yields given.

IV.F-10   L. B. Davies, P. G. Sammes, and R. A. Watt, J. Chem. Soc., Chem. Commun., 663 (1977).

[pyrimidinone with (CH₂)₃-allyl side chain, R substituent, HN, N, Me] →(Δ, CH₃CN or DMF) [cyclopenta-fused pyridine with Me and R]

R = Me, Ph   51-65%

IV.F-11   S. B. Kadin and C. H. Lamphere, Synthesis, 500 (1977).

[R-substituted benzoxazine-2,4-dione] →(1. H₂C(CN)₂, Et₃N; 2. 48% HBr or NaOH) [R-substituted 2-amino-4-hydroxyquinoline]

R = H, Cl, Me, OMe   69-86%

IV.F-12   R. F. Abdulla et al., Synth. Commun., 7, 305 and 313 (1977).

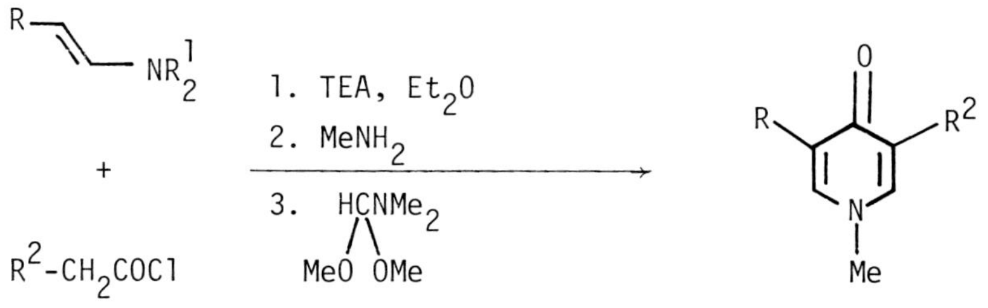

$R-CH=CH-NR^1_2$ + R^2-CH_2COCl

1. TEA, Et_2O
2. $MeNH_2$
3. $HCNMe_2$ / MeO OMe

→ 3,5-disubstituted-1-methyl-4(1H)-pyridinone (R at 5, R^2 at 3, Me on N)

R, R^2 = H, aryl, furyl, COOH, COOR, etc.

Widely varying yields.

IV.F-13   E. B. Pedersen and D. Carlsen, Synthesis, 890 (1977).

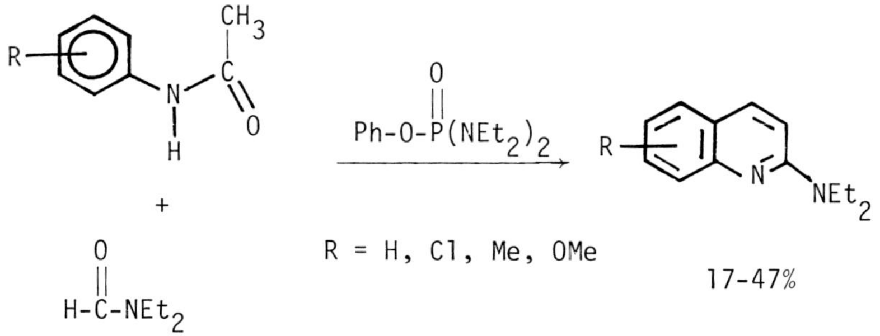

$R-C_6H_4-N(H)-C(CH_3)=O$ + $H-C(=O)-NEt_2$

$\xrightarrow{Ph-O-\overset{O}{\underset{\|}{P}}(NEt_2)_2}$

$R-$quinoline$-2-NEt_2$

R = H, Cl, Me, OMe

17-47%

IV.F-14   C. M. Leir, J. Org. Chem., 42, 911 (1977).

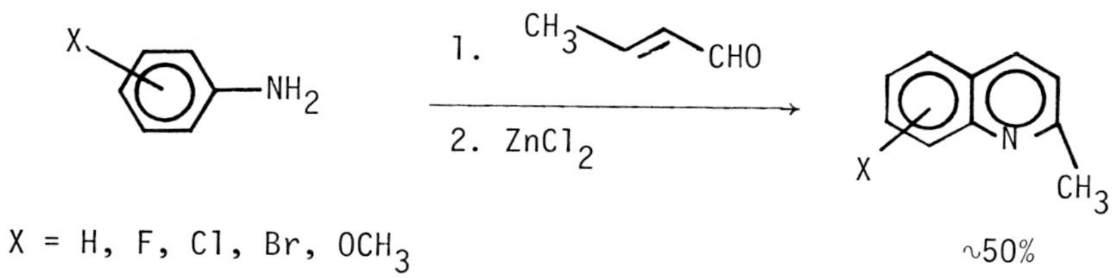

X = H, F, Cl, Br, OCH$_3$

~50%

IV.F-15   H. Meyer, F. Bossert, and H. Horstman, Liebigs Ann. Chem., 1888 and 1895 (1977).

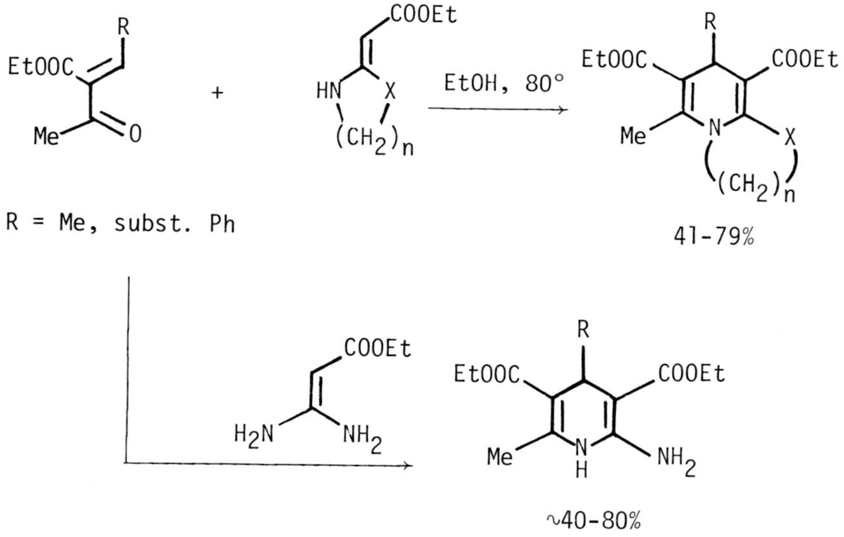

R = Me, subst. Ph

41-79%

~40-80%

IV.F-16 A. Gnanasekaran, N. Soundararajan, and P. Shanmugam, Synthesis, 612 (1977).

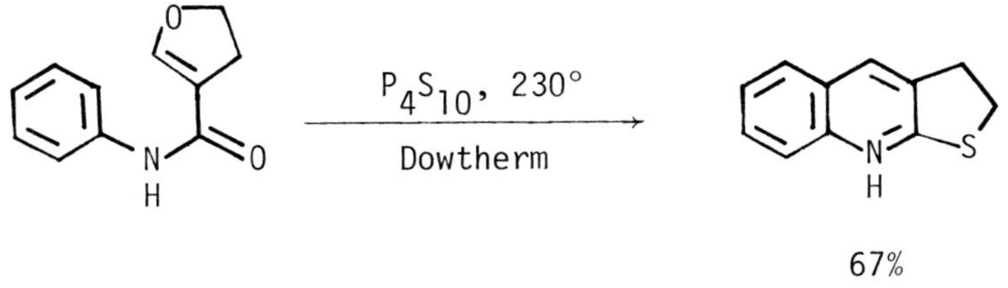

[Reaction: N-phenyl dihydrofuran-3-carboxamide → P_4S_{10}, 230°, Dowtherm → 2,3-dihydrothieno[2,3-b]quinoline, 67%]

IV.F-17 D. D. Chapman et al., J. Org. Chem., 42, 2474 (1977).

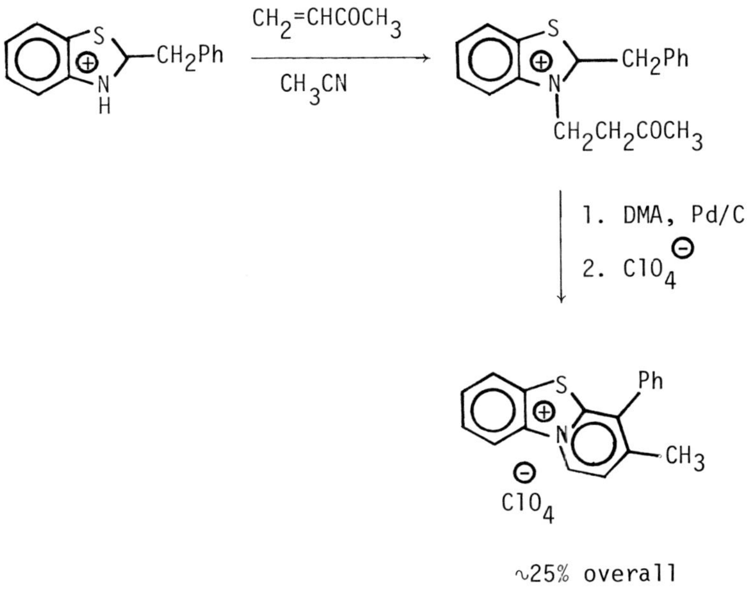

[Reaction: 2-benzyl-2,3-dihydrobenzothiazole + $CH_2=CHCOCH_3$ / CH_3CN → N-(3-oxobutyl) benzothiazolium intermediate → 1. DMA, Pd/C; 2. $ClO_4^{\ominus}$ → benzothiazolo-pyridinium perchlorate with Ph and CH_3 substituents, ~25% overall]

287

IV.F-18 M. Mori, K. Chiba, and Y. Ban, Tetrahedron Lett., 1037 (1977).

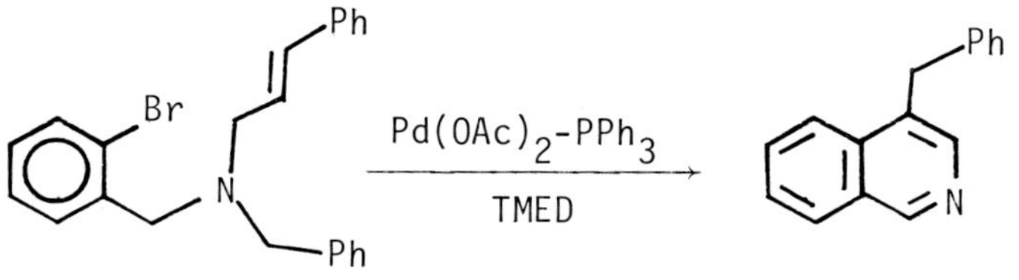

[Reaction: 2-bromobenzyl-N-(cinnamyl)-N-benzylamine → 4-benzylisoquinoline, Pd(OAc)$_2$-PPh$_3$, TMED, 27%]

IV.F-19 A. McKillop and D. P. Rao, Synthesis, 760 (1977).

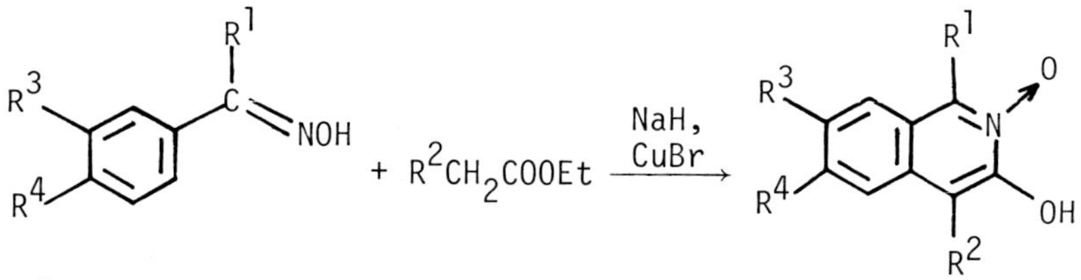

[Reaction: aryl ketoxime + R^{2}CH$_2$COOEt, NaH, CuBr → isoquinoline N-oxide derivative, 50-79%]

R^1 = H, Me
R^2 = -COR, -COOR, -CN
R^3, R^4 = H, -O-CH$_2$-O-

IV.F-20 M. Makosza et al., Synthesis, 56 (1977).

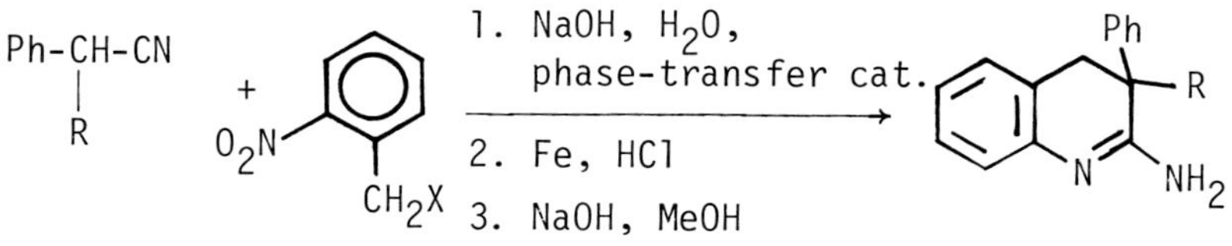

Ph-CH-CN
   |
   R

+ o-O$_2$N-C$_6$H$_4$-CH$_2$X

1. NaOH, H$_2$O, phase-transfer cat.
2. Fe, HCl
3. NaOH, MeOH

→ 3-Ph-3-R-2-amino-3,4-dihydroquinoline

R = alkyl, Ph, Bz
X = Cl, Br

∼30% overall

SYNTHESIS OF HETEROCYCLES

IV.F-21 C. A. Hergrueter et al., Tetrahedron Lett., 4145 (1977).

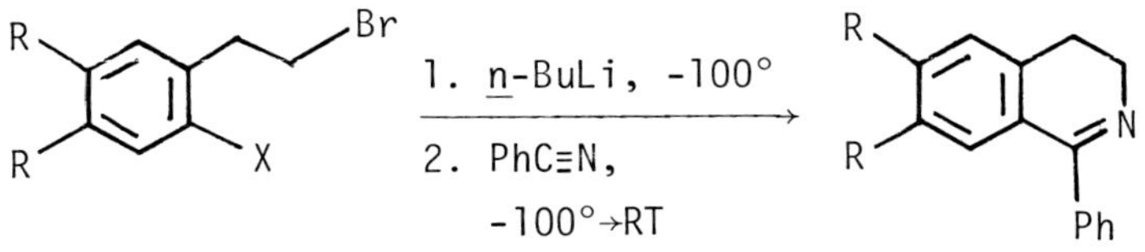

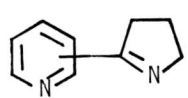

1. n-BuLi, -100°
2. PhC≡N, -100°→RT

R = H, OMe, -OCH$_2$O-

X = Br, I

80-96%

IV.F-22/IV.G-1 D. Spitzner, Synthesis, 242 (1977).

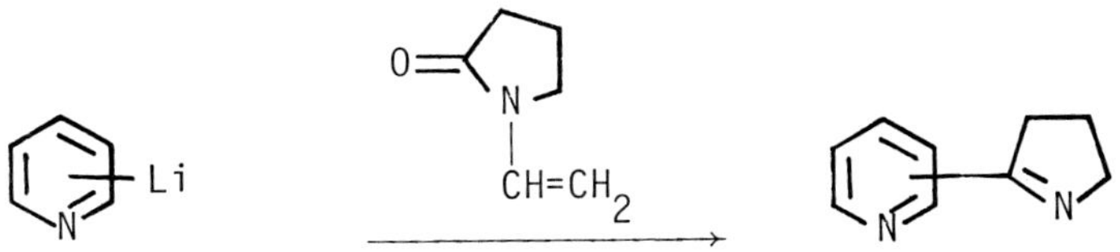

25-28%

IV.F-23/IV.G-2 M. Sainsbury, Synthesis, 437 (1977).

Review: "The Synthesis of 6H-Pyrido[4,3-b]-carbazoles"

IV.G-3   K. T. Potts and S. J. Chen, J. Org. Chem., 42, 1639 (1977).

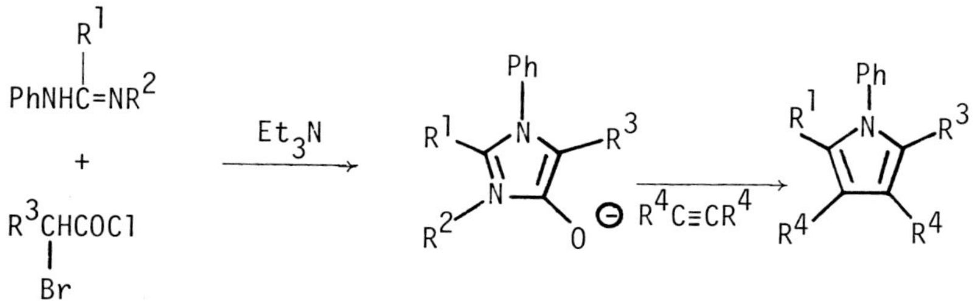

R^1 = Ph, MeS
R^2 = Me, Ph
R^3 = Ph, COOEt
R^4 = COOH, COOMe, COPh

32-77%   60-87%

IV.G-4   J. Hambrecht, Synthesis, 280 (1977).

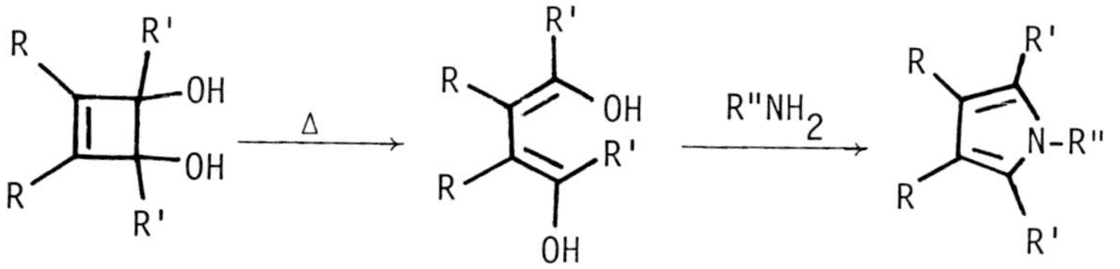

R,R' = H, Me, Ph, etc.   R" = H, Me   30-91%

SYNTHESIS OF HETEROCYCLES

IV.G-5   T. Severin and H. Poehlmann, Chem. Ber., 110, 491 (1977).

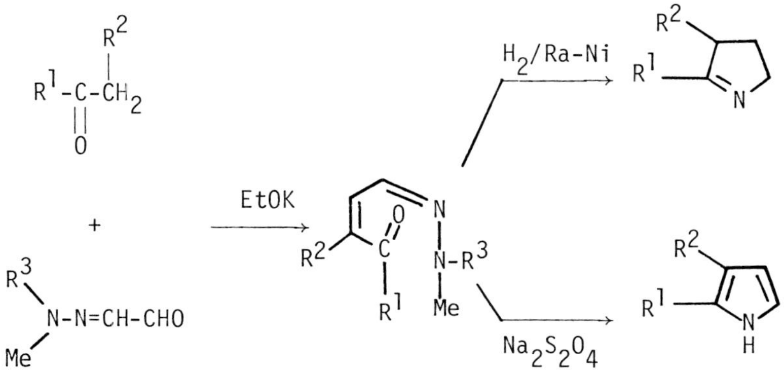

R^1 = H, Et, Ar
R^2, R^3 = H, Me, Ph

IV.G-6   P.F.S. Filho and U. Schuchardt, Angew. Chem. Int. Ed., 16, 647 (1977).

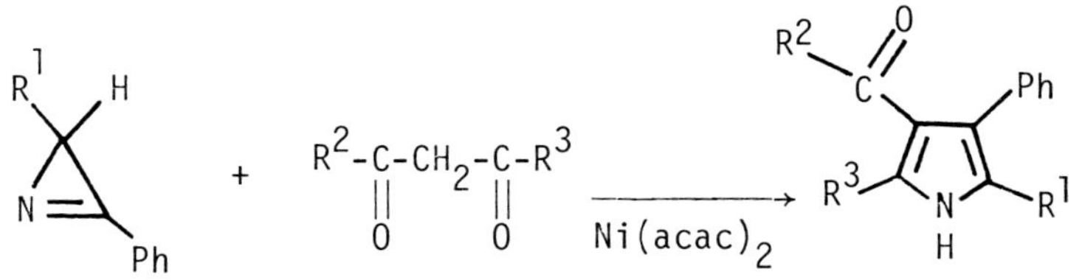

R^1 = H, Me, Ph
R^2 = Me, Et, Ph ⎫
R^3 = Me, Ph   ⎭ cyclic

generally 50-100%

IV.G-7 G. Schulz and W. Steglich, Angew. Chem. Int. Ed., 16, 252 (1977).

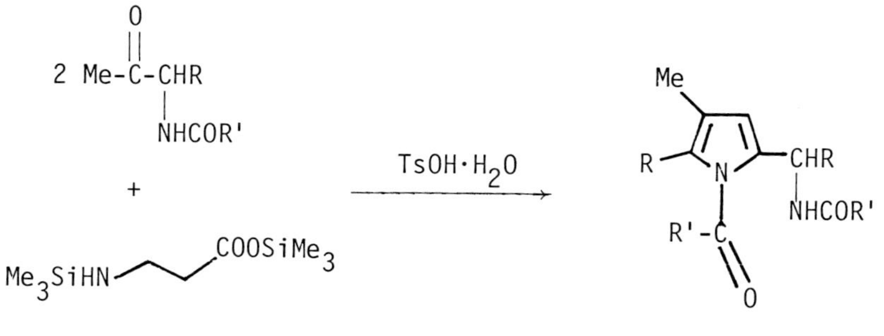
2 Me-C-CHR
O
NHCOR'
+
Me_3SiHN
$COOSiMe_3$
$TsOH \cdot H_2O$
Me
R
N
CHR
NHCOR'
R'-C
O

R = H, Me, i-Pr

R'= Me, Ph

40-75%

IV.G-8 Y. Tamura et al., Tetrahedron Lett., 4417 (1977).

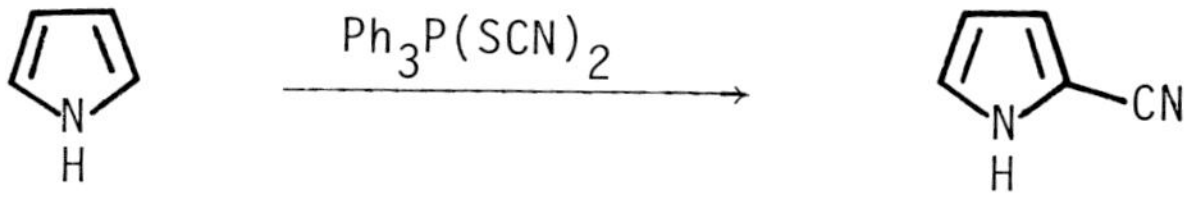
N
H
$Ph_3P(SCN)_2$
CN

50%

IV.G-9 M. T. Pizzorno and S. M. Albonico, J. Org. Chem., 42, 909 (1977).

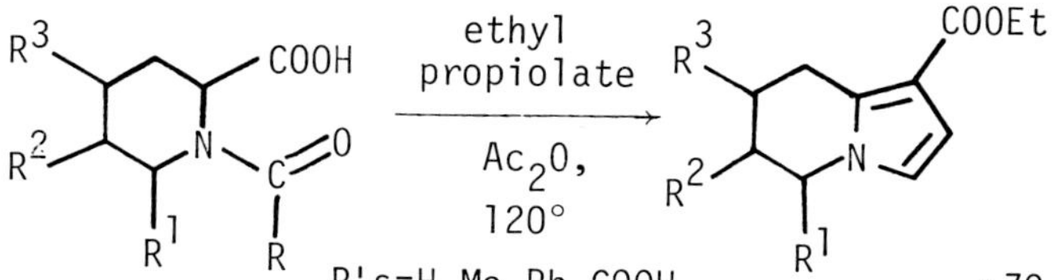
R^3
R^2
R^1
COOH
N
C
O
R
ethyl
propiolate
Ac_2O,
120°
COOEt
N

R's=H,Me,Ph,COOH

~70-80%

IV.G-10   H. Pauls and F. Kröhnke, Chem. Ber., 110, 1294 (1977).

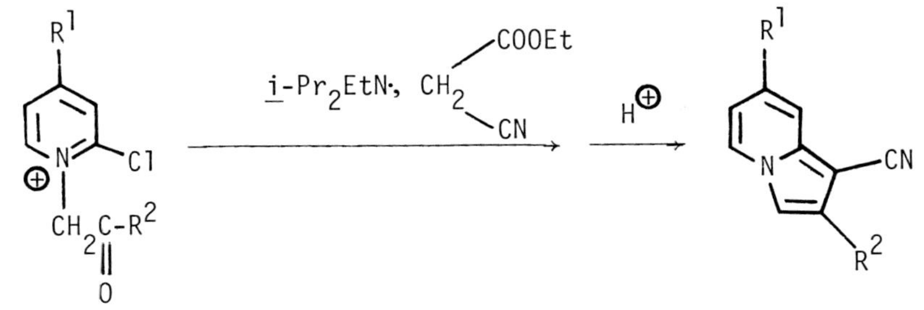

R^1 = H, Me
R^2 = Me, Ph, p-tolyl

IV.G-11   J.C.L. Armande and U. K. Pandit, Tetrahedron Lett., 897 (1977).

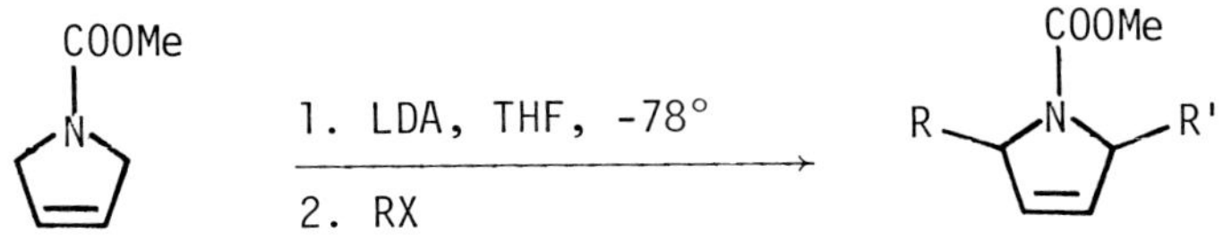

1. LDA, THF, -78°
2. RX

R = n-Bu, R' = H, 62%

R = n-C_7H_{14}OTr, R' = H, 35%

R = R' = n-Bu, 21%

IV.H-1   J. J. Eisch and K. R. Im, *J. Organomet. Chem.*, **139**, C51 (1977).

$$\text{(dibenzothiophene-like with S, E)} \xrightarrow[\text{THF, 50°}]{2(\text{COD})_2\text{NiBipy}} \text{(fluorene-like with E)}$$

E = O, NH, S   50-70%

IV.H-2   G. Piancatelli, A. Scettri, and M. D'Auria, *Tetrahedron Lett.*, 2199 (1977).

$$\underset{\text{furan-CH(OH)-R}}{} \xrightarrow[\text{CH}_2\text{Cl}_2]{\text{C}_5\text{H}_5\text{NHCrO}_3\text{Cl}} \underset{\text{pyranone with HO, R}}{}$$

R = 1° alkyl   >90%

IV.H-3   T. Watanabe et al., *J. Chem. Soc., Chem. Commun.*, 493 (1977).

$$R^1\text{-C=C(R}^2\text{)-OLi} + \underset{\text{o-AcO-C}_6\text{H}_4\text{-COCl}}{} \longrightarrow \text{chromone with } R^1, R^2$$

R^1, R^2 = H, Me, $-(\text{CH}_2)_n-$

IV.H-4   R. Graham and J. R. Lewis, Chem. and Ind., 798 (1977).

[Scheme: substituted salicylic acid methyl ether + methyl 4-hydroxybenzoate → 1. Condensation, 2. hν → benzophenone intermediate → KOH, H_2O, rfx → xanthone-COOK, ~20% overall; R = H, Cl]

IV.H-5   S. Ueda and K. Kurosawa, Bull. Chem. Soc. Japan, 50, 193 (1977).

[Scheme: 2-hydroxybenzophenone derivative $\xrightarrow{Mn(OAc)_3}$ xanthone; R,R' = H, OH, OMe; 24-65%]

IV.H-6   R. H. Everhardus, H. G. Eeuwhorst, and L. Brandsma, J. Chem. Soc., Chem. Commun., 801 (1977).

[Scheme: 2-RS-5-SMe-thiophene (H at 5-position) → 1. BuLi, THF; 2. MeI; 3. H_2O → 2-Me-5-SMe-thiophene; R = Ph, alkyl; 75-80%]

IV.H-7   S. Rajappa and R. Sreenivasan, Indian J. Chem., 15B, 301 (1977).

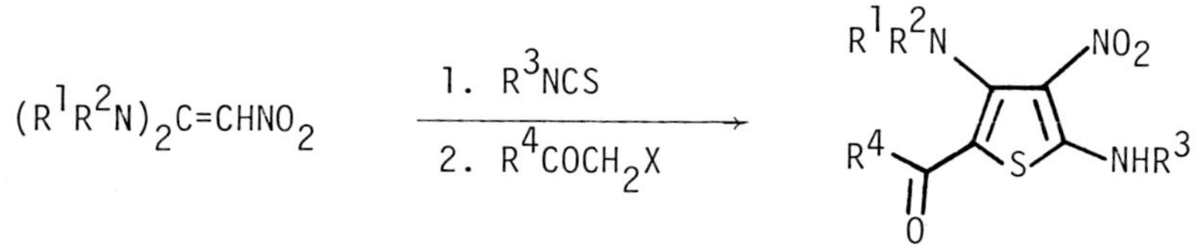

$(R^1R^2N)_2C=CHNO_2$ $\xrightarrow[\text{2. }R^4COCH_2X]{\text{1. }R^3NCS}$ [thiophene with R^1R^2N, NO_2, R^4CO, NHR^3 substituents]

R^1R^2 = HMe, Me$_2$, -(CH$_2$)$_4$-
R^3 = H, alkyl, aryl, benzoyl
R^4 = Me, Ph
X = Cl, Br

Widely varying yields.

IV.H-8   J. M. McIntosh and R. S. Steevensz, Can. J. Chem., 55, 2442 (1977).

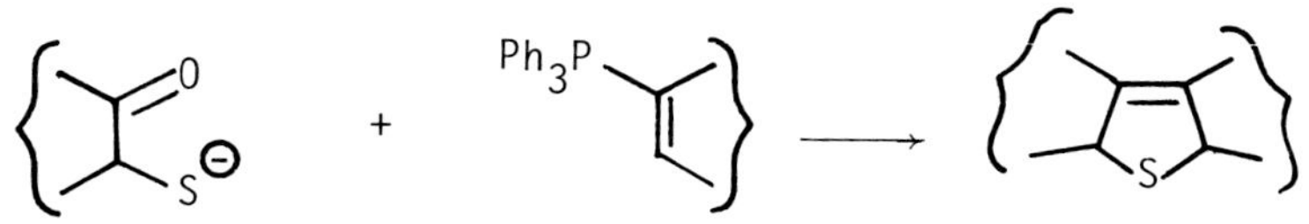

[cyclic ketone with S⁻ + Ph$_3$P-diene → fused thiophene]

Widely varying yields.

IV.H-9   Y. Tamaru, Y. Yamada, and Z. Yoshida, Tetrahedron Lett., 3365 (1977).

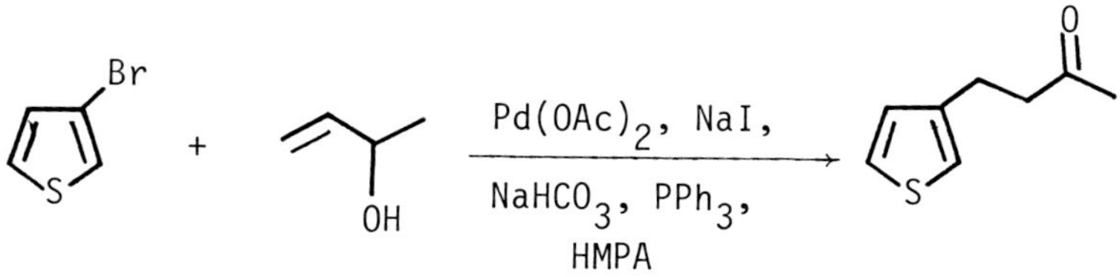

[3-bromothiophene] + [CH$_2$=C(CH$_3$)-OH, methallyl alcohol] $\xrightarrow[\text{NaHCO}_3\text{, PPh}_3\text{, HMPA}]{Pd(OAc)_2, \text{NaI,}}$ [3-thienyl-CH$_2$CH$_2$-C(=O)CH$_3$]

90%

IV.H-10 F. M. Benitez and J. R. Grunwell, Tetrahedron Lett., 3413 (1977).

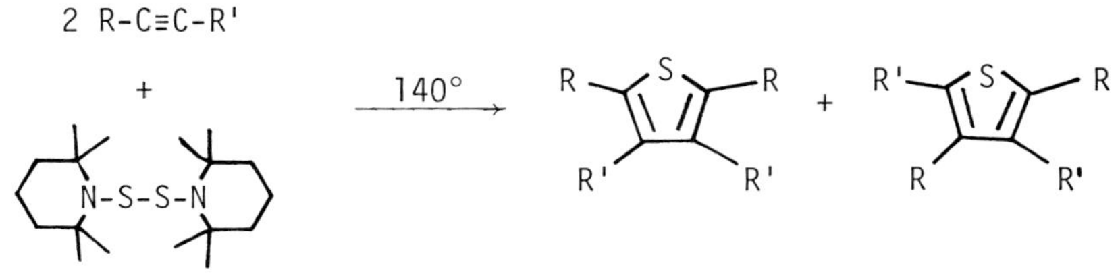

2 R-C≡C-R' + [piperidine-N-S-S-N-piperidine] →(140°) thiophene products, 25-82%

R = Ph, COOEt
R' = H, Ph

IV.H-11 D. Nasipuri, I. D. Dalal, and S. K. Ghosh, Synthesis, 59 (1977).

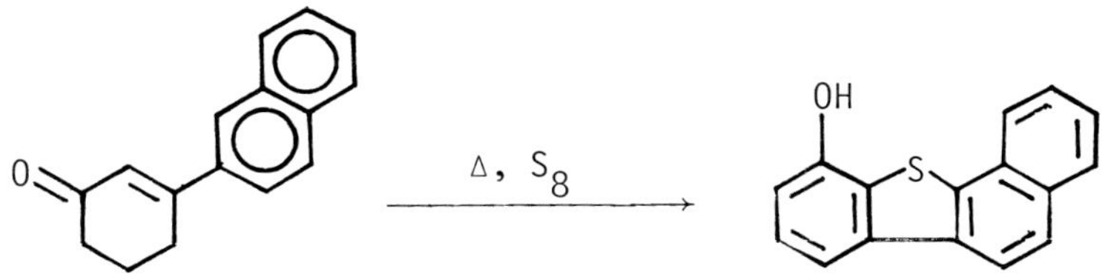

→ (Δ, S_8) 50%

IV.H-12 R. Pellicciari and B. Natalini, J. Chem. Soc., Perkin I, 1822 (1977).

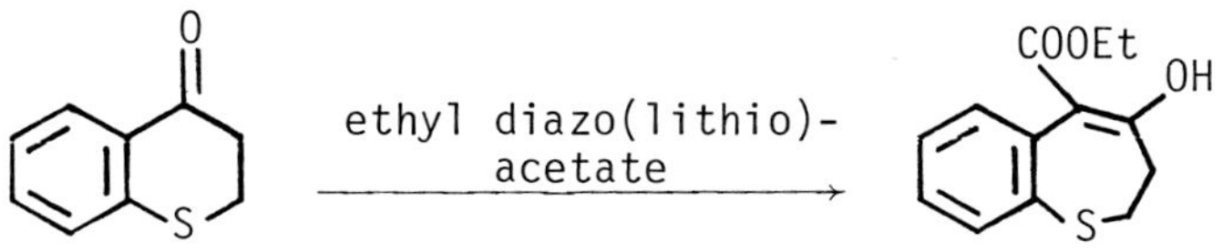

→ (ethyl diazo(lithio)acetate) 33%

IV.H-13 D. C. Palmer and M. J. Strauss, Chem. Rev., **77**, 1 (1977).

Review: "Benzomorphans: Synthesis, Stereochemistry Reactions, and Spectroscopic Characterizations"

IV.I.1.a-1 A. M. van Leusen, J. Wildeman, and O. H. Oldenziel, J. Org. Chem., **42**, 1153 (1977).

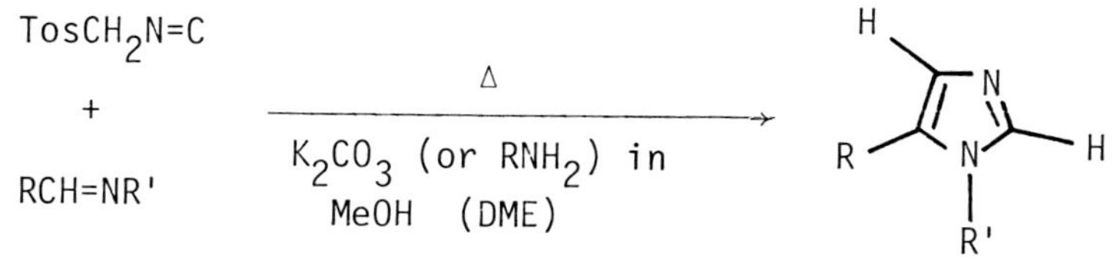

TosCH$_2$N=C
+
RCH=NR'

$\xrightarrow{\Delta}$ K$_2$CO$_3$ (or RNH$_2$) in MeOH (DME)

→ (imidazole product with R, H, R' substituents)

R,R' = Me, t-Bu, Ar   ~50-80%

IV.I.1.a-2 L. Citerio and R. Stradi, Tetrahedron Lett., 4227 (1977).

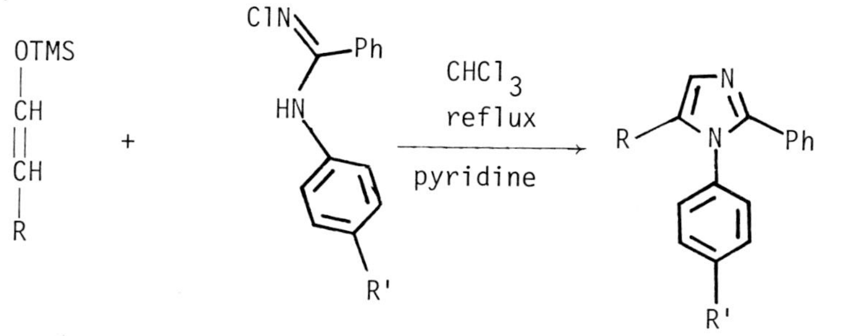

OTMS
|
CH
||
CH
|
R

+ (ClN=C(Ph)-NH-Ar-R')

$\xrightarrow[\text{pyridine}]{\text{CHCl}_3 \text{ reflux}}$

→ (imidazole product, R, Ph, N-aryl-R')

R = H, Me, Et, n-alkyl   55-75%
R' = H, Me, Br

SYNTHESIS OF HETEROCYCLES

IV.I.1.a-3   E. Regel and K.-H. Büchel, Liebigs Ann. Chem., 145 and 159 (1977).

$$\underset{R}{\text{imidazole}} \xrightarrow[Et_3N]{R'COCl} \underset{R}{\text{imidazole}}-\underset{O}{\overset{\|}{C}}R'$$

R = Me, Et, Ph, Bz   Widely varying yields.
R'= aryl, heterocyclic, OEt, etc.

IV.I.1.a-4   E. P. Papadopoulos, J. Org. Chem., 42, 3925 (1977).

$$\underset{R}{\text{imidazole}} + R'NCO \xrightarrow[reflux]{PhNO_2} \underset{R}{\text{imidazole}}-\underset{O}{\overset{\|}{C}}-NHR'$$

R = H, Me
R'= Bu, 1-naphthyl, Ph   Widely varying yields.

IV.I.1.a-5   V. E. Gunn, M. Koga, and J.-P. Anselme, J. Org. Chem., 42, 754 (1977).

$$ArCCH_2Br + Me_2NNH_2 \longrightarrow Ar-\text{imidazole}-COAr$$
(with C=O on the left reagent)

No details (reference to thesis).

IV.I.1.a-6   S. R. Landor et al., Tetrahedron Lett., 3743 (1977).

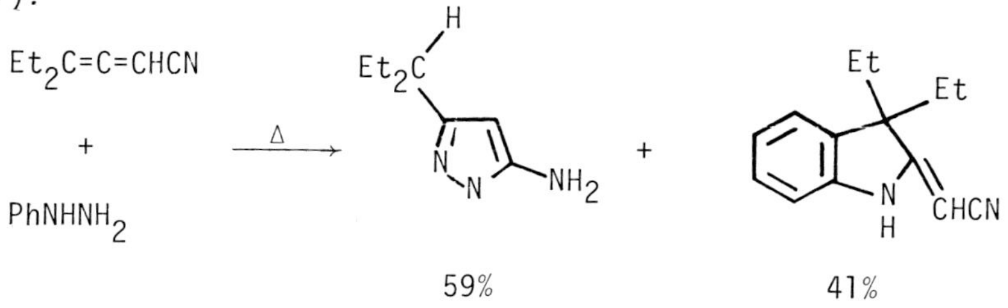

Et$_2$C=C=CHCN + PhNHNH$_2$ $\xrightarrow{\Delta}$

Et$_2$CH-[pyrazole]-NH$_2$ 59%

+ [3,3-diethylindolin-2-ylidene]=CHCN   41%

IV.I.1.a-7   B. George and E. Papadopoulos, J. Org. Chem., 42, 441 (1977).

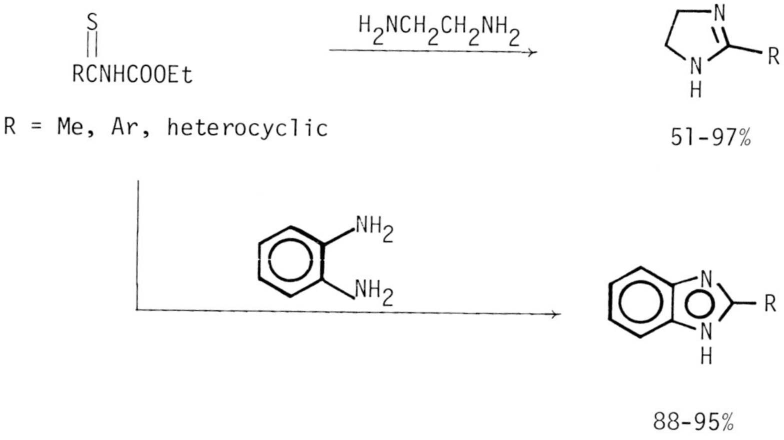

$\underset{\text{RCNHCOOEt}}{\overset{\text{S}}{\|}}$ $\xrightarrow{\text{H}_2\text{NCH}_2\text{CH}_2\text{NH}_2}$ [imidazoline]-R

R = Me, Ar, heterocyclic   51-97%

[o-phenylenediamine] $\longrightarrow$ [benzimidazole]-R   88-95%

IV.I.1.a-8 R. Stradi and G. Verga, Synthesis, 688 (1977).

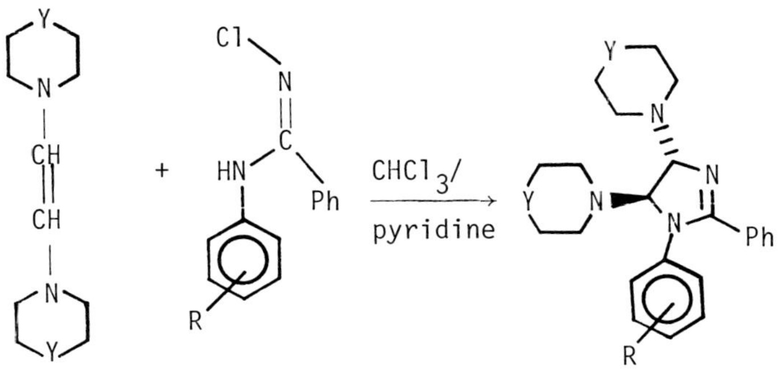

R = H, Me, F, Br   75-80%

IV.I.1.a-9 F. Yoneda, M. Higuchi, and S. Matsumoto, J. Chem. Soc., Perkin I, 1754 (1977).

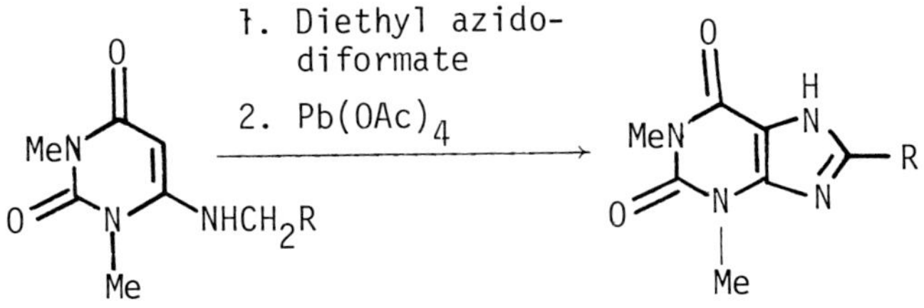

1. Diethyl azido-
   diformate
2. Pb(OAc)$_4$

R = Et, Pr, Ph, Bz, etc.   ~70-80%

IV.I.1.a-10   K. Senga et al., Synthesis, 264 (1977).

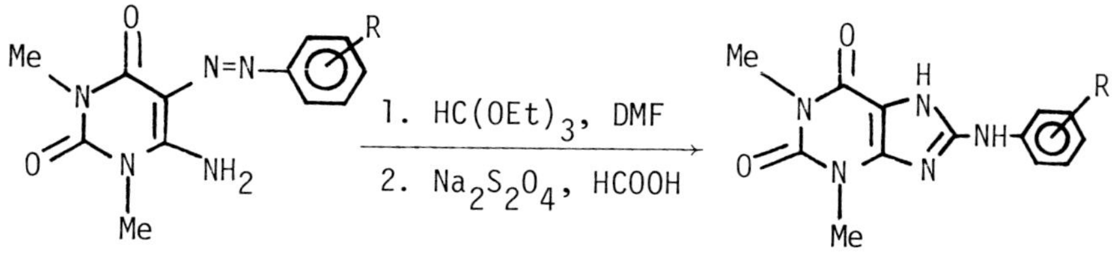

[Structure: 1,3-dimethyl-5-(arylazo)-6-aminouracil]

1. HC(OEt)$_3$, DMF
2. Na$_2$S$_2$O$_4$, HCOOH

[Structure: 8-(arylamino)-7H-3-methylxanthine]

R = H, Br, Cl, Me, OMe   ~30% overall

IV.I.1.a-11   Y. Ito, T. Hiras, and T. Saegusa, J. Organomet. Chem., 131, 121 (1977).

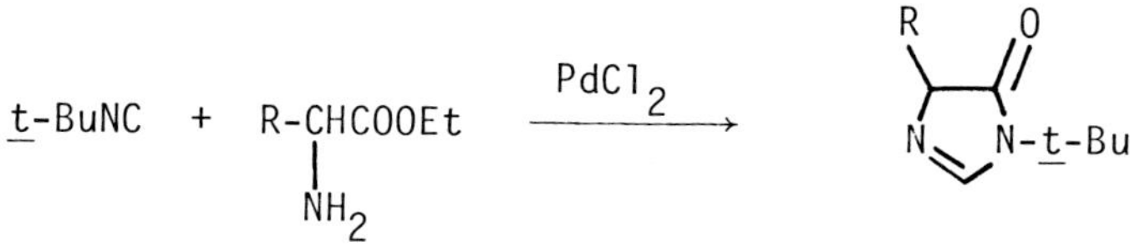

t-BuNC + R-CHCOOEt $\xrightarrow{PdCl_2}$ [imidazole structure with N-t-Bu]
         |
         NH$_2$

R = 1°,2° alkyl, Bz   ~70-80%

IV.I.1.a-12   K. Matsumoto et al., Synthesis, 249 (1977).

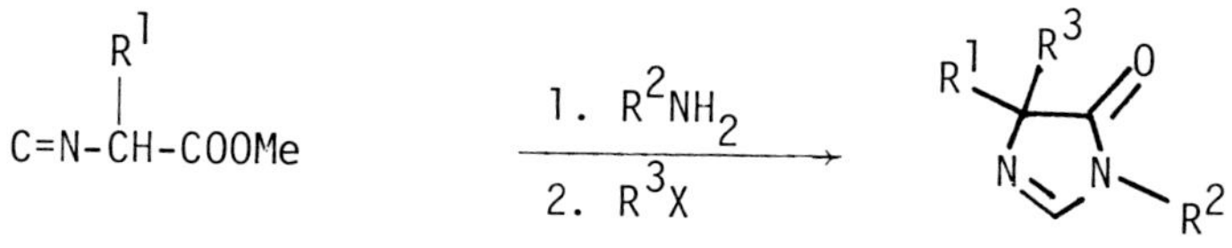

R^1
|
C=N-CH-COOMe   1. R^{2}NH$_2$ [imidazolone structure with R^1, R^3, R^2]
                  2. R^{3}X

R's = Me, i-Pr, Bz, etc.   ~50-60% overall

IV.I.1.a-13   O. S. Wolfbeis, Synthesis, 136 (1977).

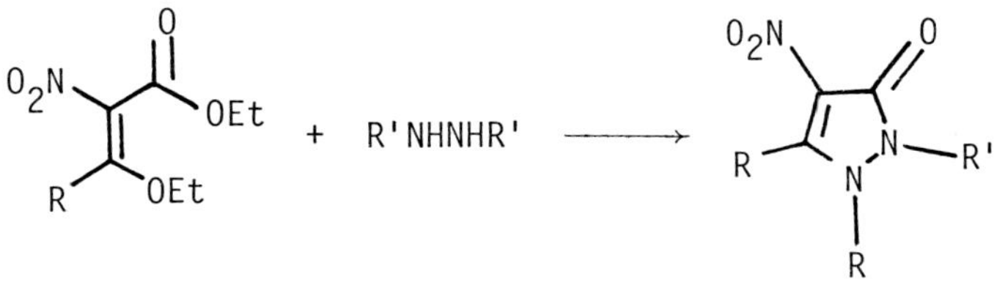

R = H, Me   ~70%
R'= H, Me, Ph

IV.I.1.a-14   M. Lange et al., Z. Chem., 17, 94 (1977).

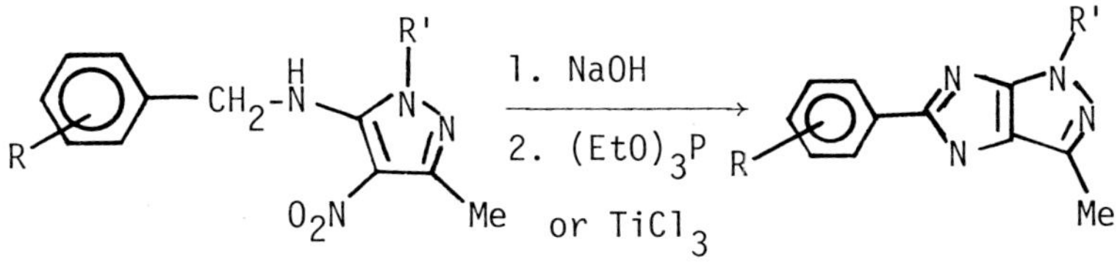

R = H, Me, Hal, OH, NH$_2$, COOH   ~40-60% overall
R'= p-nitrophenyl

IV.I.1.a-15   Y. Tamaru, T. Harada, and Z. Yoshida, Tetrahedron Lett., 4323 (1977).

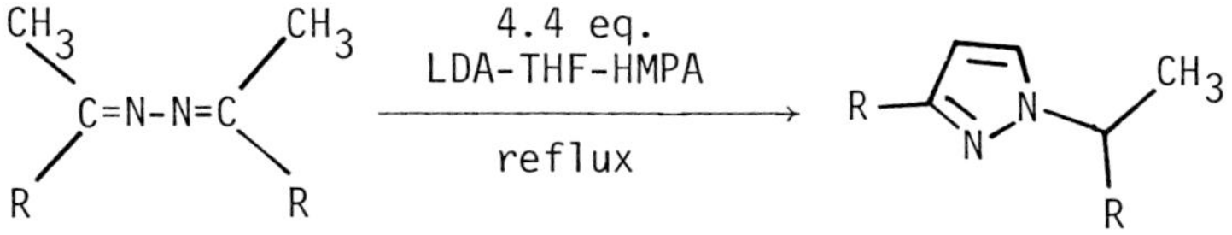

R = 2°,3° alkyl, aryl   25-46%

IV.I.1.a-16   K. Gewald and O. Calderon, Monatsh. Chem., 108, 611 (1977).

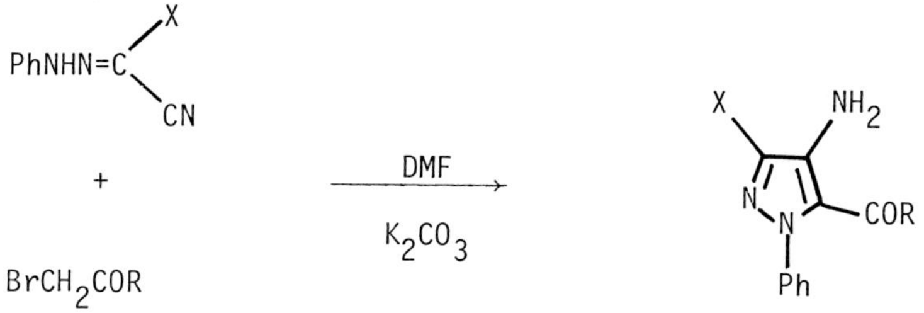

PhNHN=C(X)(CN) + BrCH$_2$COR $\xrightarrow{\text{DMF, } K_2CO_3}$ [pyrazole: X, NH$_2$, COR, N-Ph]

X = COPh, CONH$_2$, COOEt, CN   60-88%
R = Ph, OEt

IV.I.1.a-17   W. Pfeiffer, E. Dilk, and E. Bulka, Synthesis, 196 (1977).

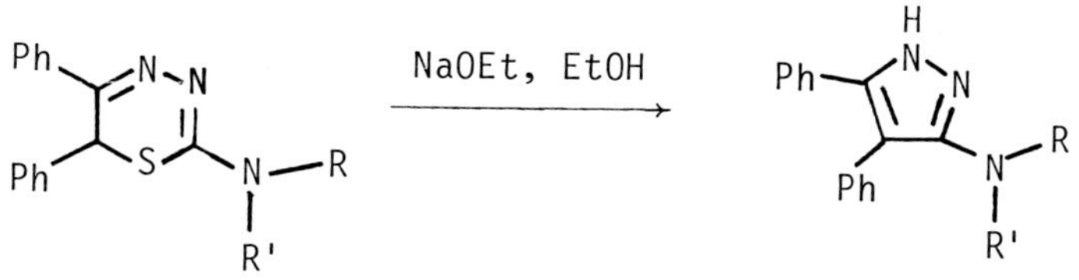

[Ph,Ph-thiadiazine with NRR' substituent] $\xrightarrow{\text{NaOEt, EtOH}}$ [Ph,Ph-pyrazole with NRR' substituent]

R = H, Me, alkyl, Ph, etc.   ~80-90%
R' = H, Me

IV.I.1.b-1   K. Takagi and M. Hubert-Habart, Bull. Soc. Chim. France, 369 (1977).

Review: "Formation of Pyrimidines from Oxygen-Containing Heterocyclic Compounds"

SYNTHESIS OF HETEROCYCLES

IV.I.1.b-2   D. J. Brown and P. Waring, Aust. J. Chem., 30, 621 (1977).

Multistep synthetic routes to 2-(pyrimidin-2'-yl)acetic acids and esters, e.g.:

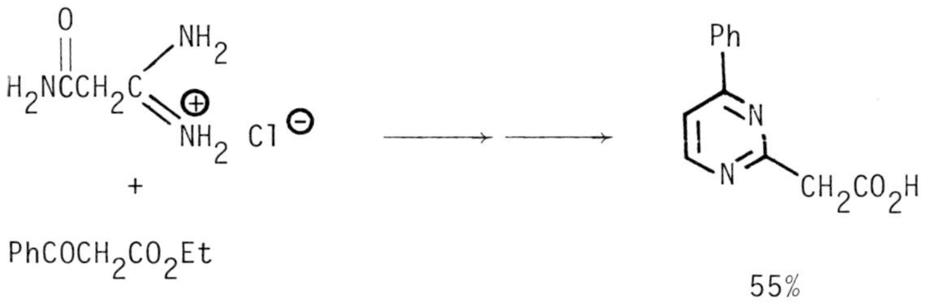

$H_2NCOCH_2C(=NH_2^+)NH_2$ Cl^- + PhCOCH$_2$CO$_2$Et $\longrightarrow$ 4-Ph-pyrimidin-2-yl-CH$_2$CO$_2$H

55%

IV.I.1.b-3   B. Singh and G. Y. Lesher, J. Het. Chem., 14, 1413 (1977).

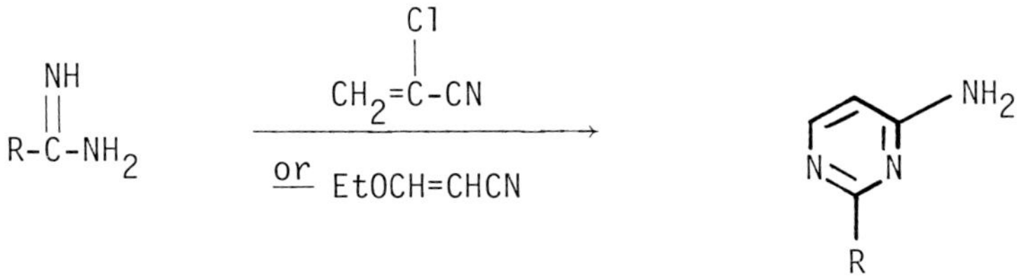

$R-C(=NH)-NH_2$ + CH$_2$=C(Cl)-CN or EtOCH=CHCN $\longrightarrow$ 4-amino-2-R-pyrimidine

R = pyridyl, Me, Ph   50-80%

IV.I.1.b-4   A. Kreutzberger and D. Wiedemann, Liebigs Ann. Chem., 537 (1977).

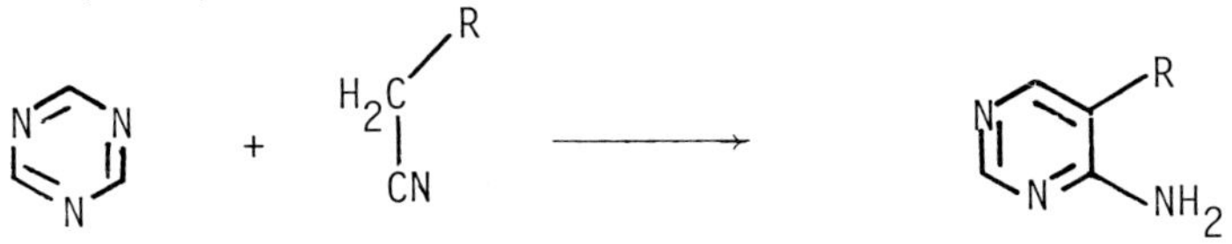

1,3,5-triazine + H$_2$C(R)CN $\longrightarrow$ 5-R-4-amino-pyrimidine

R = aryl, vinyl, heterocyclic   6-58%

IV.I.1.b-5   I. I. Lapkin, V. I. Semenov, and M. I. Belonovich, J. Org. Chem. (USSR), 13, 1217 (1977).

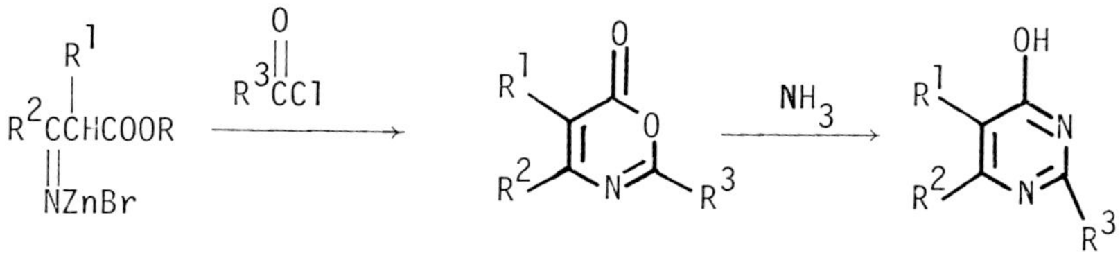

R's = H, Me, Et, Ph   21-62%

IV.I.1.b-6   R.L.N. Harris and J. L. Huppatz, Angew. Chem. Int. Ed., 16, 779 (1977).

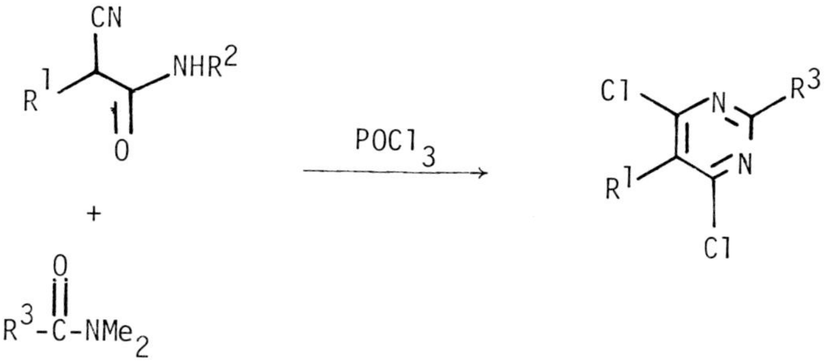

R^1 = Me, Et, Pr   ~20-60%

R^2 = Me, i-Pr, Hx

R^3 = H, Me, Ph

IV.I.1.b-7   A. V. Ivashchenko and V. M. Dziomko, Russ. Chem. Rev., 46, 115 (1977).

Review: "The Reactions of Isatin and its Derivatives with Aromatic and Heterocyclic o-Diamines."

IV.I.1.b-8   B. Stelander and H. G. Viehe, <u>Angew. Chem. Int. Ed.</u>, <u>16</u>, 189 (1977).

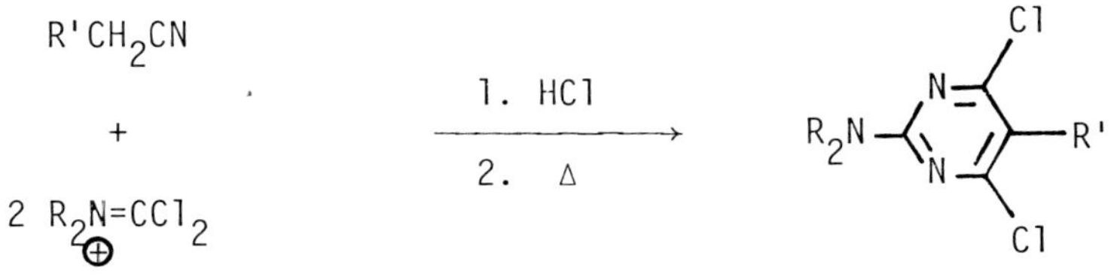

R'CH$_2$CN
+
2 R$_2$N=CCl$_2$ $^\oplus$

$\xrightarrow{\text{1. HCl} \quad \text{2. }\Delta}$

[pyrimidine product: 4,6-dichloro-2-(R_2N)-5-R'-pyrimidine]

R = Me, Et
R' = Me, Ph

~40-60% overall

IV.I.1.b-9   S. Evans and E. E. Schweizer, <u>J. Org. Chem.</u>, <u>42</u>, 2321 (1977).

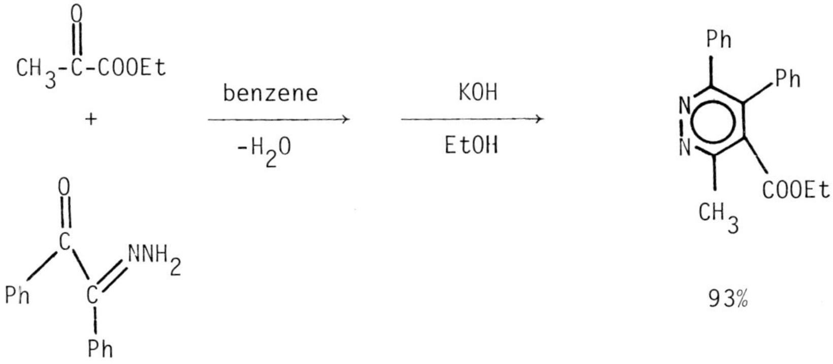

CH$_3$-C(=O)-COOEt
+
Ph-C(=O)-C(=NNH$_2$)-Ph

$\xrightarrow[-H_2O]{\text{benzene}}$ $\xrightarrow{\text{KOH}}{\text{EtOH}}$

[pyridazine product with Ph, Ph, COOEt, CH$_3$ substituents]

93%

IV.I.1.b-10 R. Lantzsch and D. Arlt, Synthesis, 756 (1977).

$$CH_3\text{-}\underset{\underset{O}{\|}}{C}\text{-}CH_2\text{-}\underset{\underset{CH_3}{|}}{\overset{\overset{CH_3}{|}}{C}}\text{-}NCO \quad \xrightarrow{RNH_2, \, \Delta} \quad$$

[pyrimidinone product with R on N, Me substituents, and C=O]

R = Ph, 4-ClBz   70-82%

IV.I.1.b-11 C. Bischoff, H. Herma, and E. Schroder, J. Prakt. Chem., 319, 230 (1977).

[2-R-cyclohexanone] + H_2NCONH_2 → [fused quinazolinedione]

R = 1° alkyl, Ph   25-91%

IV.I.1.b-12 M. Sato and C. H. Stammer, J. Het. Chem., 14, 149 (1977).

[diamine with two NH$_2$] + [pentachlorophenyl ester: Cl$_4$-C$_6$-O-CHNHCPh(=O) with COOEt, OH] → [piperazinone with NHCPh(=O) substituent]

~70-90%

IV.I.1.b-13   J. R. Piper and J. A. Montgomery, J. Org. Chem., 42, 208 (1977).

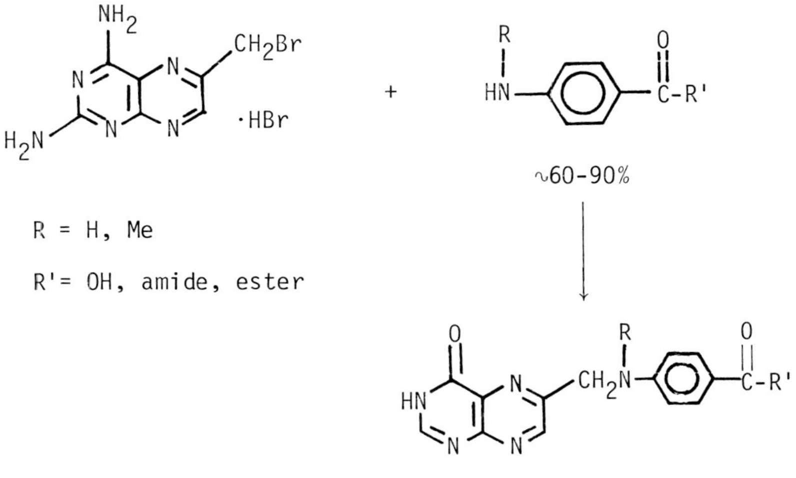

R = H, Me
R' = OH, amide, ester

∼60-90%

A useful route in the synthesis of pteridine derivatives.

IV.I.1.b-14   E. B. Pederson, Synthesis, 180 (1977).

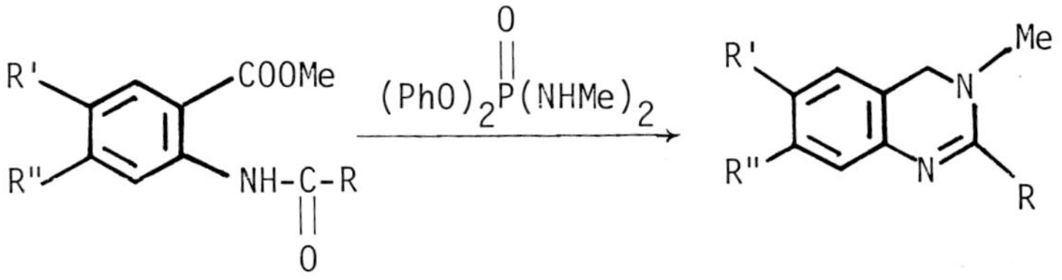

R = H, Me, Et, Ph
R' = H, Me, OEt
R" = H, OMe

32-74%

IV.I.1.c-1   L. Garanti and G. Zecchi, *J. Chem. Soc., Perkin I*, 2092 (1977).

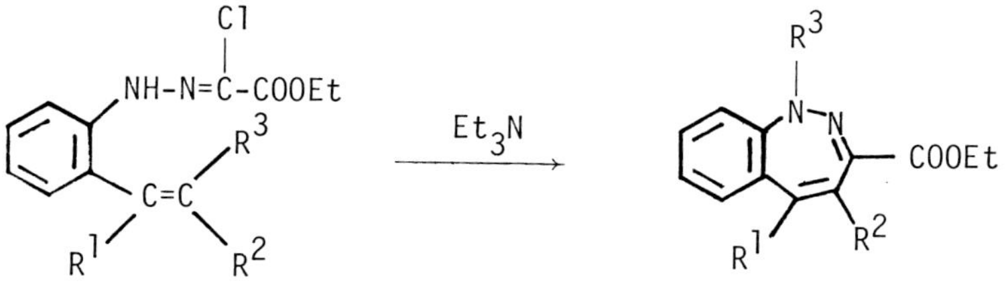

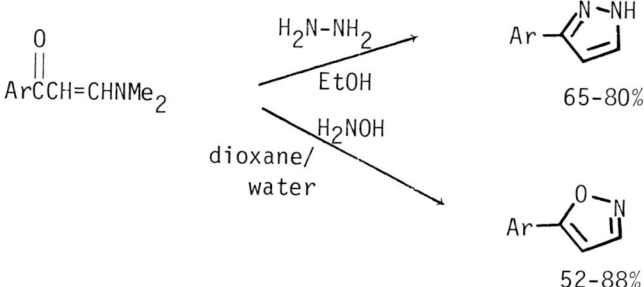

R's = H, Me, Ph, COOEt   17-75%

IV.I.2-1   Y. Lin and S. A. Lang, Jr., *J. Het. Chem.*, **14**, 345 (1977).

ArCOCH=CHNMe$_2$ → (H$_2$N-NH$_2$, EtOH) → Ar—[pyrazole N-NH]   65-80%

→ (H$_2$NOH, dioxane/water) → Ar—[isoxazole O-N]   52-88%

IV.I.2-2   G. P. Pollini et al., *Synthesis*, 837 (1977).

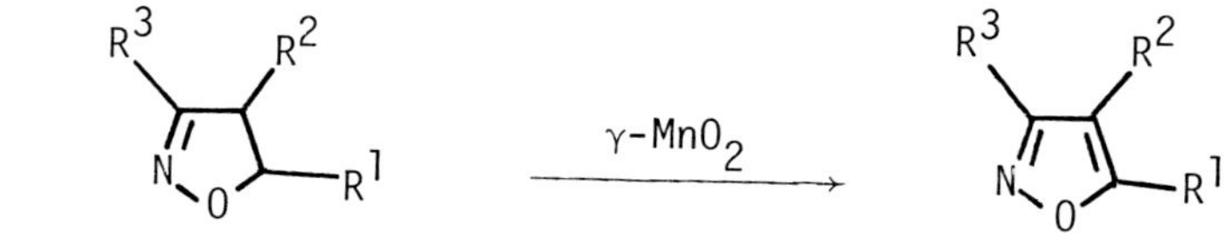

[isoxazoline with R^3, R^2, R^1] —γ-MnO$_2$→ [isoxazole with R^3, R^2, R^1]

R^1, R^3 = alkyl, aryl, COOMe
R^2 = H, NO$_2$

>97%

IV.I.2-3   U. R. Kalkote and D. D. Goswami, Aust. J. Chem., 30, 1847 (1977).

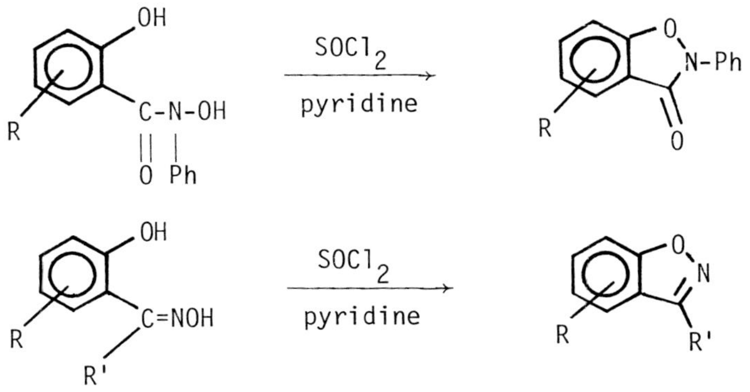

IV.I.2-4   A. Grignard and M. Lamant, Comptes Rendus (C), 285, 203 (1977).

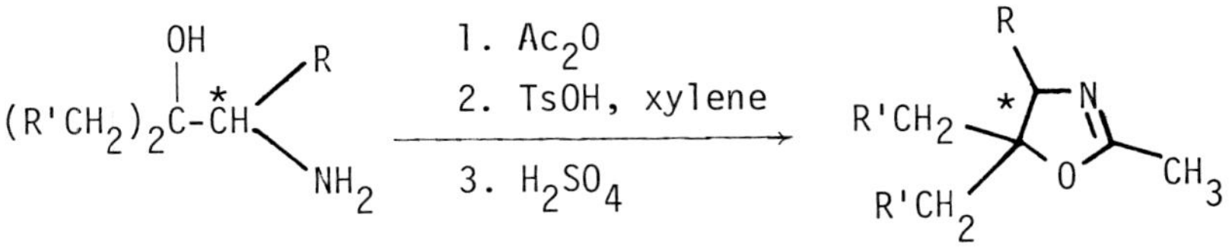

R = n-alkyl
R'= H, Me, i-Pr, Bz, CH$_3$SCH$_2$CH$_2$

Widely varying yields.

IV.I.2-5   K. Kitatani, T. Hiyama, and H. Nozaki, Bull. Chem. Soc. Japan, 50, 1647 (1977).

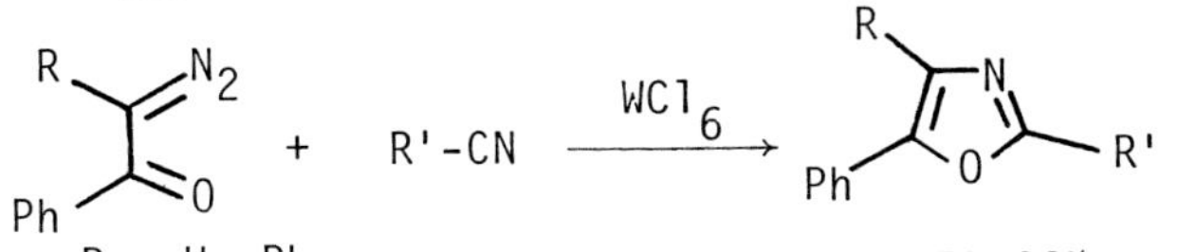

R = H, Ph
R'= Me, Et, Ph, vinyl

50-66%

IV.I.2-6 I. Matsuda, T. Takahashi, and Y. Ishii, <u>Chem. Lett.</u>, 1457 (1977).

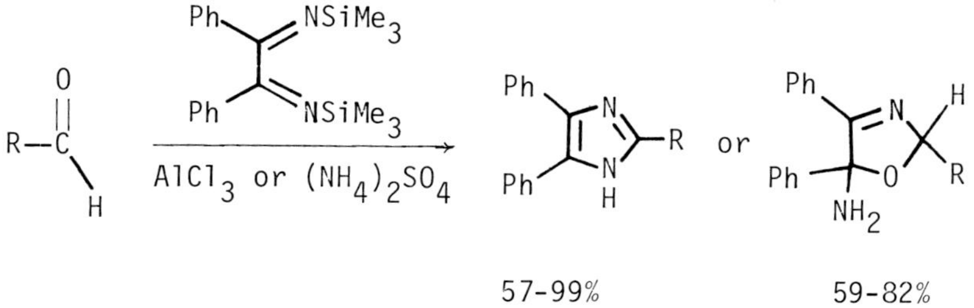

$$R-\overset{O}{\underset{H}{C}}-H \quad \xrightarrow[\text{AlCl}_3 \text{ or } (NH_4)_2SO_4]{Ph-C(=NSiMe_3)-C(Ph)=NSiMe_3}$$

product A (imidazole):
Ph, Ph substituted imidazole with R, 57-99%

or

product B (oxazoline):
Ph, Ph substituted with R, NH$_2$, 59-82%

R = aryl, cinnamyl, trihalo-
  methyl, etc.

(product varies according to conditions)

IV.I.2-7 B. George and E. Papadopoulos, <u>J. Org. Chem.</u>, <u>42</u>, 441 (1977).

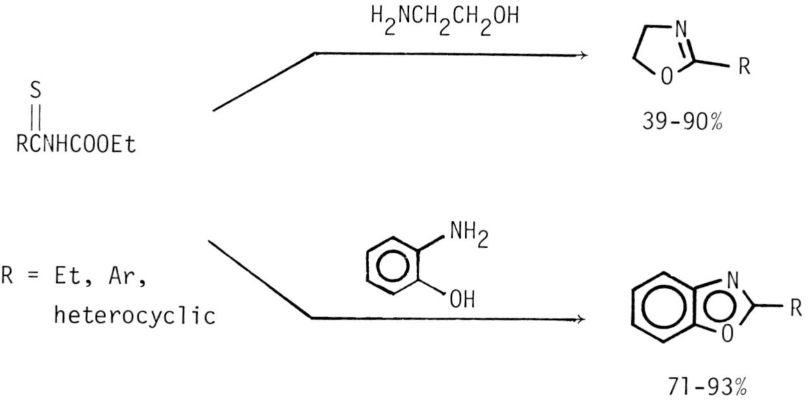

$$\underset{\text{RCNHCOOEt}}{\overset{S}{\|}} \xrightarrow{H_2NCH_2CH_2OH} \text{2-R-oxazoline} \quad 39\text{-}90\%$$

R = Et, Ar,
  heterocyclic

$$\xrightarrow{\text{2-aminophenol (NH}_2\text{, OH)}} \text{2-R-benzoxazole} \quad 71\text{-}93\%$$

IV.I.2-8 H. Sliwa and A. Tartar, Tetrahedron Lett., 311 (1977).

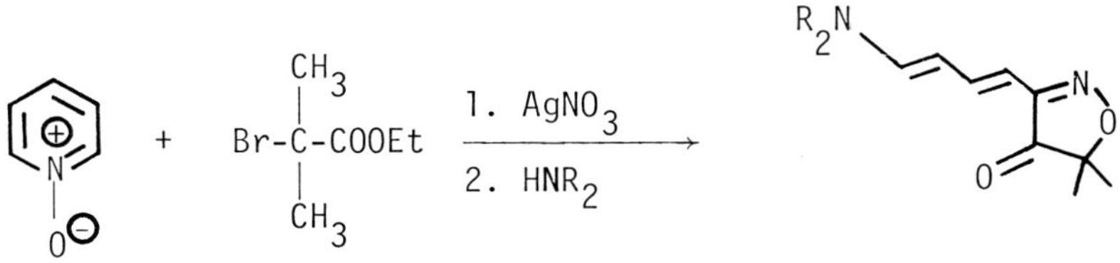

R_2 = Et_2, $(CH_2)_4$, $(CH_2)_5$

30-56%

IV.I.3-1 H. Boshagen and W. Geiger, Liebigs Ann. Chem., 20 (1977).

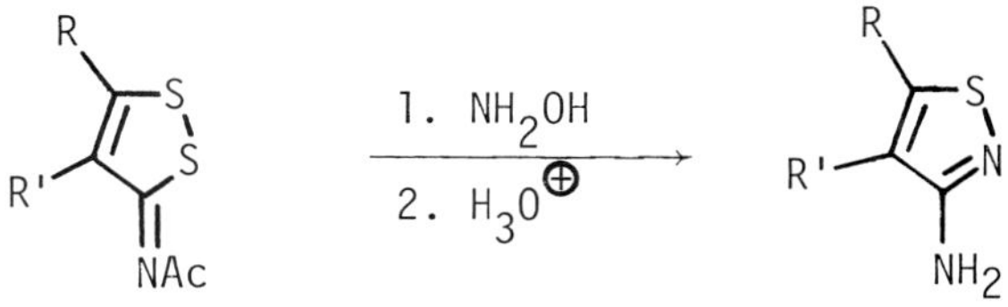

R,R' = H, Me, cyclic, benzo   ∼50-80%

IV.I.3-2 S. Kambe et al., Synthesis, 839 (1977).

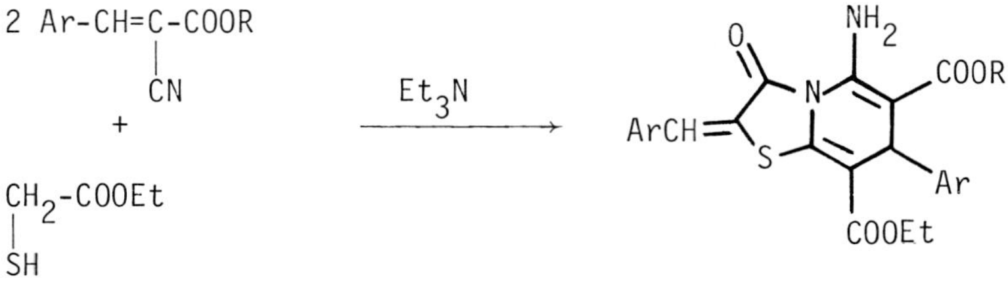

2 Ar-CH=C-COOR
       |
       CN
   +
   CH_2-COOEt
   |
   SH

Et_3N →

17-35%

IV.I.3-3   B. George and E. P. Papadopoulos, J. Org. Chem., 42, 441 (1977).

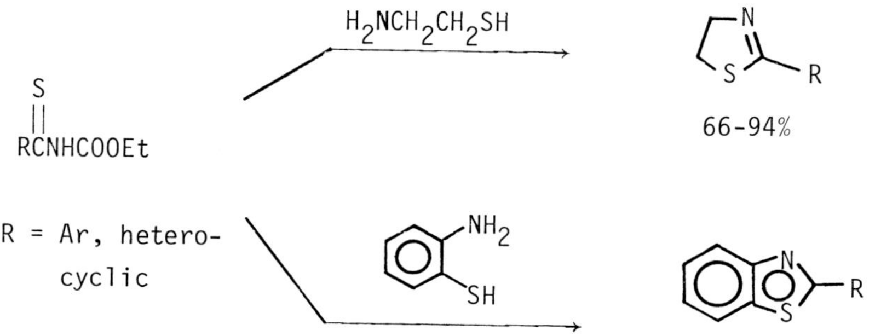

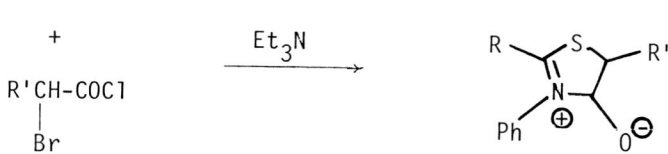

66-94%

R = Ar, hetero-
    cyclic

67-95%

IV.I.3-4   K. T. Potts et al., J. Org. Chem., 42, 1633 (1977).

RCSNHPh
   +   Et₃N
R'CH-COCl   ——————→
   |
   Br

generally ~50-80%

R = Ph, CN, SR, NR₂   Full paper, many examples.
R'= Ph, COOEt

IV.I.3-5   G. Rosini and A. Medici, Synthesis, 892 (1977).

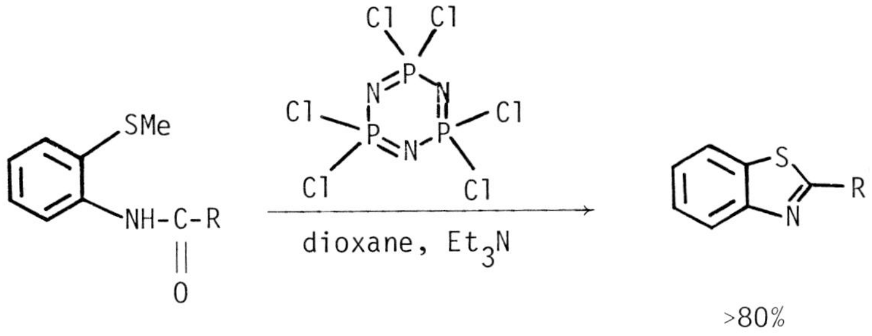

Ar-SMe, NH-C(=O)-R + (NPCl₂)₃ —dioxane, Et₃N→ benzothiazole-2-R   >80%

R = Me, n-octyl, Ph, adamantyl, Cl(CH$_2$)$_3$

IV.I.3-6   A. M. van Leusen and J. Wildeman, Synthesis, 501 (1977).

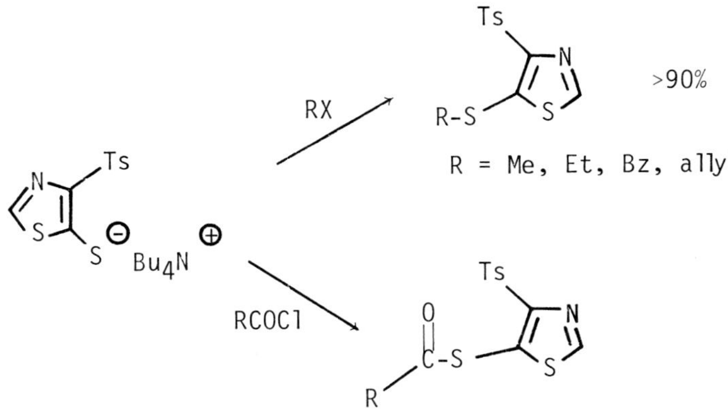

4-Ts-5-S⁻-thiazole Bu$_4$N⁺

RX → R-S-(5)-4-Ts-thiazole   >90%
R = Me, Et, Bz, allyl

RCOCl → R-C(=O)-S-(5)-4-Ts-thiazole
R = Me (86%), Ph (91%)

IV.I.3-7   G. Prota et al., Synthesis, 876 (1977).

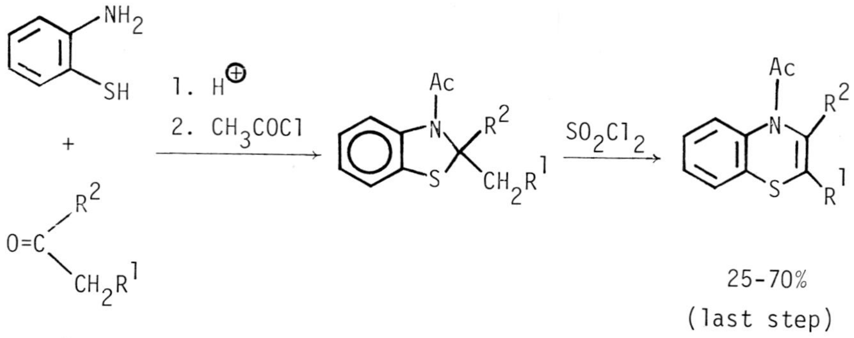

[Reaction scheme: 2-aminothiophenol + ketone $O=C(R^2)CH_2R^1$ → (1. $H^\oplus$, 2. CH_3COCl) → N-Ac benzothiazoline intermediate → (SO_2Cl_2) → 4-Ac-benzothiazine product, 25–70% (last step)]

R^1 = H, Ph, $-(CH_2)_4-$
R^2 = Me, Ar, $-(CH_2)_4-$

IV.I.5-1   D. Seebach et al., Chem. Ber., 110, 1879 (1977).

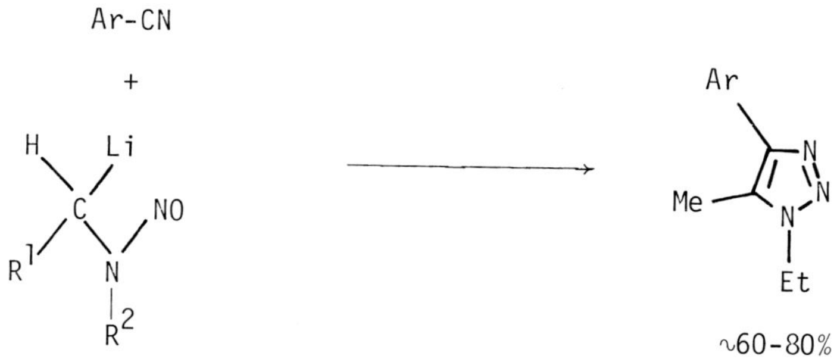

Ar-CN
+
[α-lithio nitrosamine: R^1-CH(Li)-N(NO)-R^2]

→ [1,2,3-triazole: 4-Ar, 5-Me, 1-Et]

~60–80%

R^1 = H, Me
R^2 = Me, i-Pr, t-Bu

IV.I.5-2   M. Bertrand, J. P. Dulcere and M. Santelli, Tetrahedron Lett., 1783 (1977).

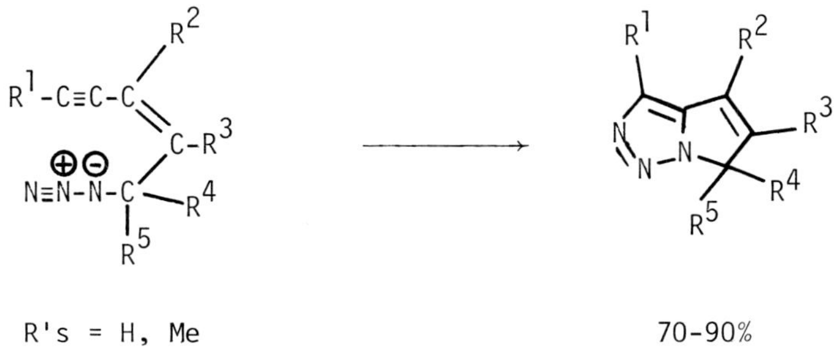

R's = H, Me   70-90%

IV.I.5-3   A. Elgavi and H. G. Viehe, Angew. Chem. Int. Ed., 16, 181 (1977).

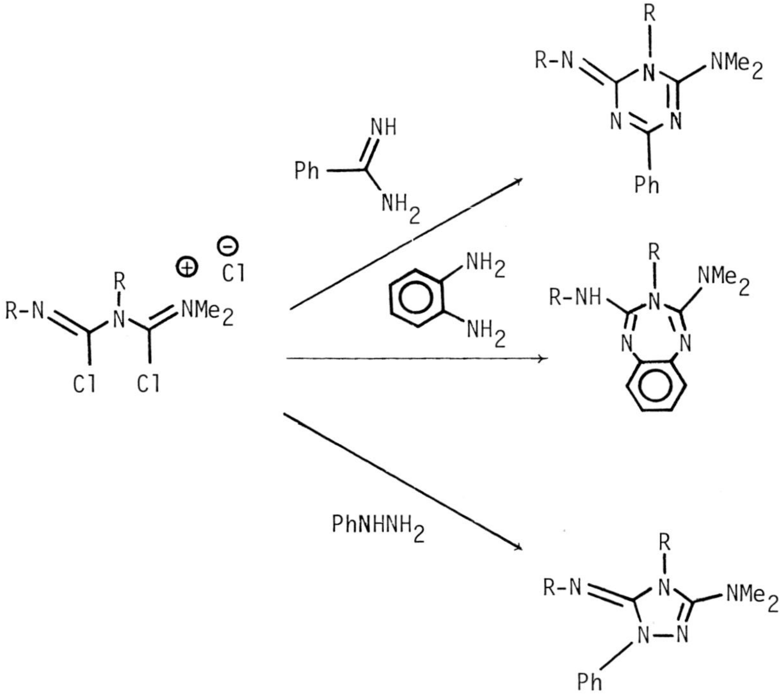

IV.I.5-4 W. Stadlbauer, O. Schmut, and T. Kappe, Monatsh. Chem., 108, 367 (1977).

[Scheme: 2-(ethoxycarbonylamino)pyridine or 3-R-2H-pyrido[1,2-a]pyrimidine-2,4(3H)-dione + Ph-N=C=O, 140-200° → pyrido-triazine-dione with N-Ph]

Widely varying yields.

R = Et, Ph, Bz

IV.I.5-5 A. V. Zeiger and M. M. Joullié, J. Org. Chem., 42, 542 (1977).

[Scheme: 1-amino-2-aminobenzimidazole with Pb(OAc)$_4$ / CH$_2$Cl$_2$ → benzotriazine-amine]

R = H, Me

R' = H, Me, Cl, CF$_3$

48-95%

IV.I.5-6 G. Ege, K. Gilbert, and H. Franz, Synthesis, 556 (1977).

[Scheme: 5-aminopyrazole 1. HNO$_2$, 2. Base, 3. R-C≡C-NR'$_2$ → pyrazolo-triazine]

∼60-80%

IV.I.5-7 B. Slanovnik _et al._, _Synthesis_, 491 (1977).

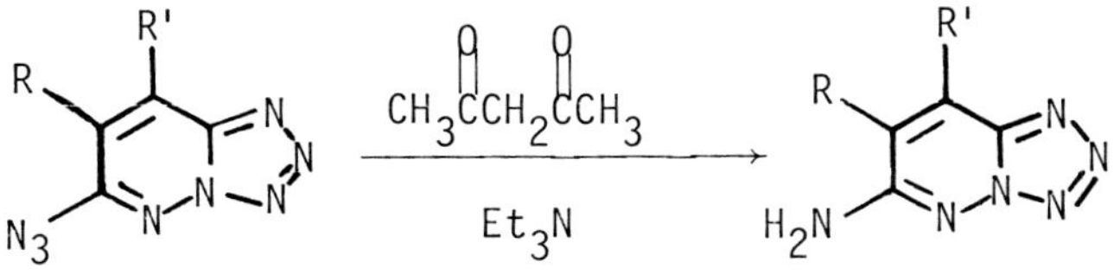
R'
R
N_3
$CH_3CCH_2CCH_3$
Et_3N
H_2N

R,R' = H, Me

77-100%

IV.I.6-1 W. Schroth and W. Kaufmann, _Z. Chem._, _17_, 331 (1977).

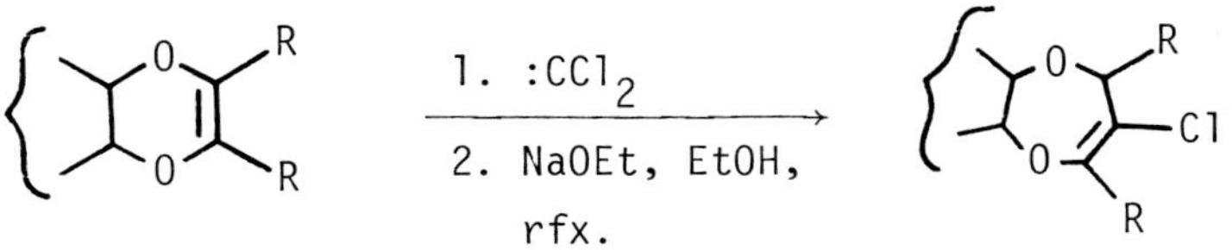
1. $:CCl_2$
2. NaOEt, EtOH,
rfx.
Cl

R = H, Ph

~60% (second step)

IV.I.6-2 K. Akiba, K. Ishikawa, and N. Inamoto, _Synthesis_, 861 (1977).

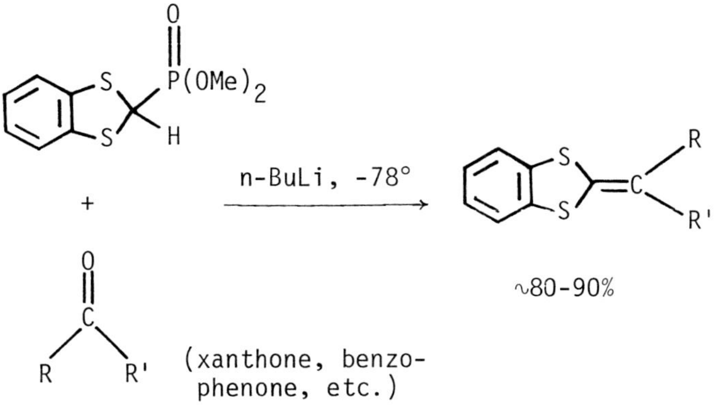
$P(OMe)_2$
n-BuLi, -78°
~80-90%
(xanthone, benzo-
phenone, etc.)

IV.I.6-3 C. Giordano and A. Belli, Synthesis, 193 (1977).

$$R^1-\overset{\overset{S}{\|}}{C}-NH-CH_2OH \quad \xrightarrow[BF_3\cdot Et_2O]{R^2-\overset{\overset{S}{\|}}{C}-R^3} \quad \text{(dithiazine product with } R^1, R^2, R^3\text{)}$$

R's = H, alkyl, aryl   Widely varying yields.

IV.J-1   E. I. Levkoeva and L. N. Yakhontov, Russ. Chem. Rev., 46, 565 (1977).

   Review: "The Synthesis of Heterocyclic Analogues of Prostaglandins"

IV.J-2   C. Morin and R. Beugelmans, Tetrahedron, 33, 3183 (1977).

   Review: "The Reactions of Hydroxylamine and Hydrazine with γ-Pyrones"

IV.J-3   G. R. Newkome et al., Chem. Rev., 77, 513 (1977).

   Review: "Construction of Synthetic Macrocyclic Compounds Possessing Subheterocyclic Rings, Specifically Pyridine, Furan, and Thiophene"

IV.J-4   K. D. Berlin et al., Chem. Rev., 77, 121 (1977).

   Review: "Polycyclic Carbon-Phosphorus Heterocycles"

V.A-1   R. A. Olofson and R. C. Schnur, Tetrahedron Lett., 1571 (1977).

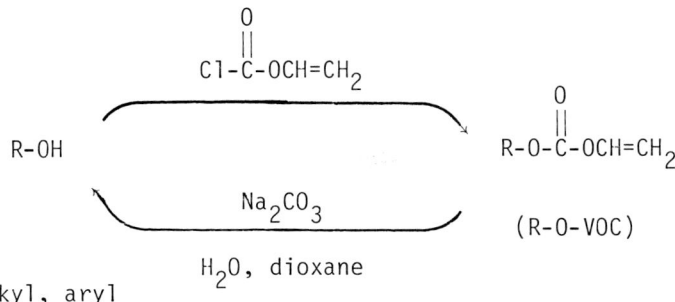

R = alkyl, aryl

Deprotection can be effected in the presence of N-VOC groups.

V.A-2   N. Miyashita, A. Yoshikoshi, and P. A. Grieco, J. Org. Chem., 42, 3772 (1977).

$$ROH \xrightarrow[\text{TsOH·pyr}]{\text{DHP}} ROTHP$$

>95%

This catalyst is much milder than most others used for tetrahydropyranylation, and is better for use with acid-sensitive alcohols.

V.A-3   M. E. Jung and M. A. Lyster, J. Org. Chem., 42, 3761 (1977).

$$R-O-R' \xrightarrow[\text{2. } H_2O]{\text{1. } Me_3SiI} ROH + R'I$$

R = C_6H_{11}, Ar, cyclic, etc.   100% 100%
R'= Me, Et, t-Bu, etc.

V.A-4   K. Abe et al., Chem. Lett., 817 (1977).

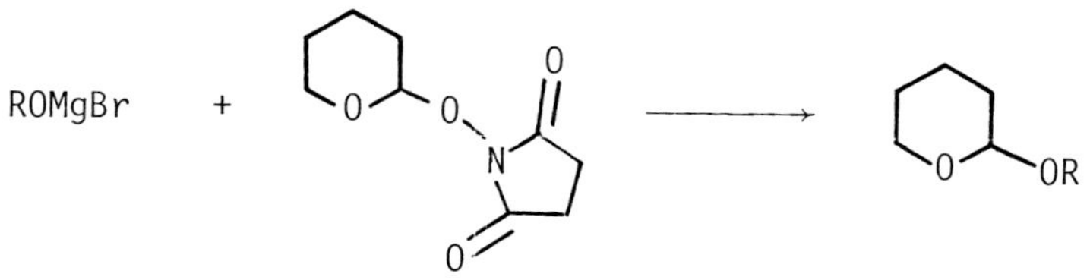

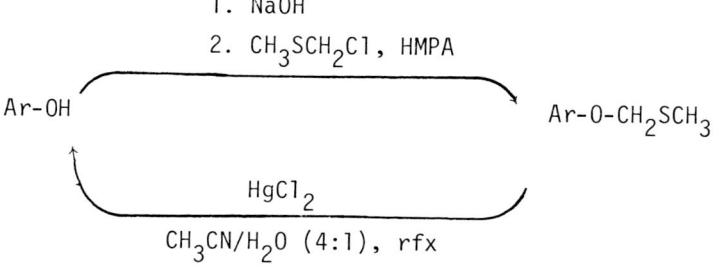

R =	Et	Bu	Chx	HC≡CCH$_2$-
% yield =	37	72	72	98

V.A-5   R. A. Holton and R. G. Davis, Tetrahedron Lett., 533 (1977).

$$\text{Ar-OH} \xrightarrow[\substack{\text{HgCl}_2 \\ \text{CH}_3\text{CN/H}_2\text{O (4:1), rfx}}]{\substack{\text{1. NaOH} \\ \text{2. CH}_3\text{SCH}_2\text{Cl, HMPA}}} \text{Ar-O-CH}_2\text{SCH}_3$$

V.A-6   F. Dardoize, M. Gaudemar, and N. Goasdoue, Synthesis, 567 (1977).

$$\text{ArOH} \xrightarrow[\text{2. H}_2\text{C(OMe)}_2]{\text{1. BrZnCH}_2\text{COOEt}} \text{ArOCH}_2\text{OCH}_3$$

45-72%

V.A-7   T.-L. Ho and G. A. Olah, Synthesis, 417 (1977).

$$Ar-O-R \xrightarrow{PhSiMe_3,\ I_2} ArOH$$

Ar = Ph, naphthyl   50-90%

R = Me, Et, Bz

V.A-8   W. H. Pirkle and J. R. Hauske, J. Org. Chem., 42, 2781 (1977).

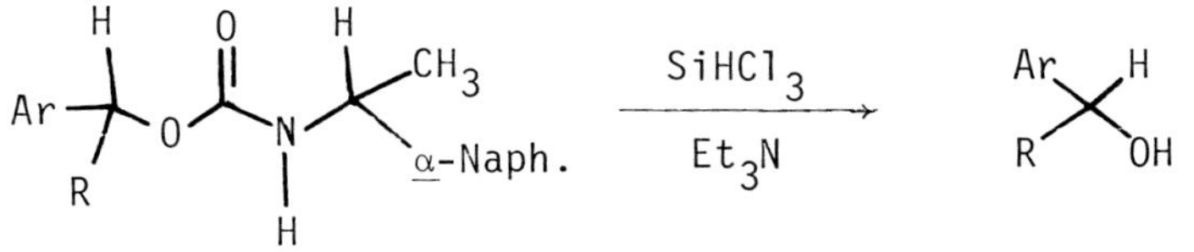

$$Ar\underset{R}{\overset{H}{-}}C-O-C(=O)-N(H)-CH(CH_3)(\underline{\alpha}\text{-Naph}) \xrightarrow[Et_3N]{SiHCl_3} \underset{R}{\overset{Ar}{>}}CH(OH)$$

R = alkyl, fluoroalkyl   ~80-90%

V.A-9   C. C. Leznoff and D. M. Dixit, Can. J. Chem., 55, 3351 (1977).

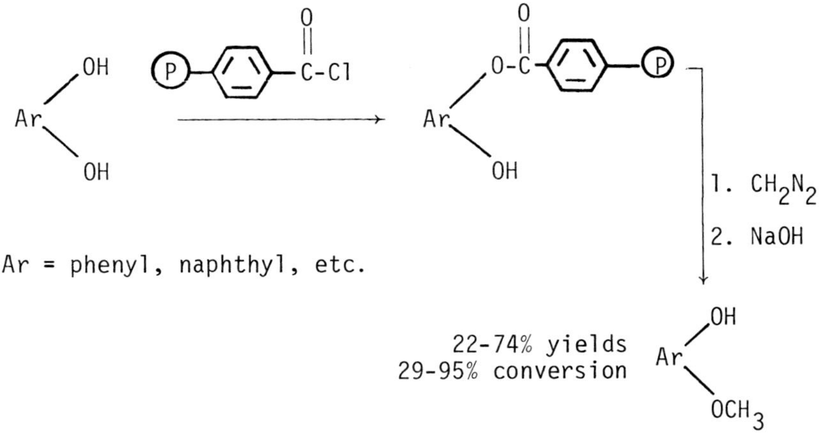

$$Ar(OH)_2 + \text{(P)}-C_6H_4-C(=O)Cl \longrightarrow Ar(OH)(O-C(=O)-C_6H_4-\text{(P)})$$

Ar = phenyl, naphthyl, etc.

1. CH_2N_2
2. NaOH

$$\rightarrow Ar(OH)(OCH_3)$$

22-74% yields
29-95% conversion

V.A-10 R. Arentzen and C. B. Reese, J. Chem. Soc., Chem. Commun., 270 (1977).

The following 5'-esters of thymidine are readily deacylated by treatment with dilute hydrazine in methanol/pyridine: acrylyl, crotonyl, 4-methoxycrotonyl, and 4-phenoxycrotonyl.

V.B-1 G. Barany and R. B. Merrifield, J. Am. Chem. Soc., 99, 7363 (1977).

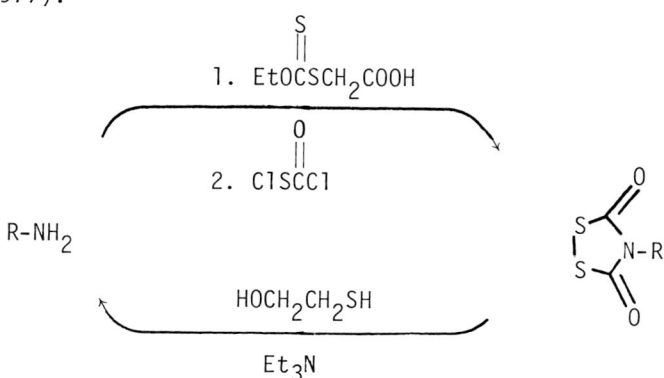

Does not racemize optically active amino acids.

V.B-2 M. Waki and J. Meienhofer, J. Org. Chem., 42, 2019 (1977).

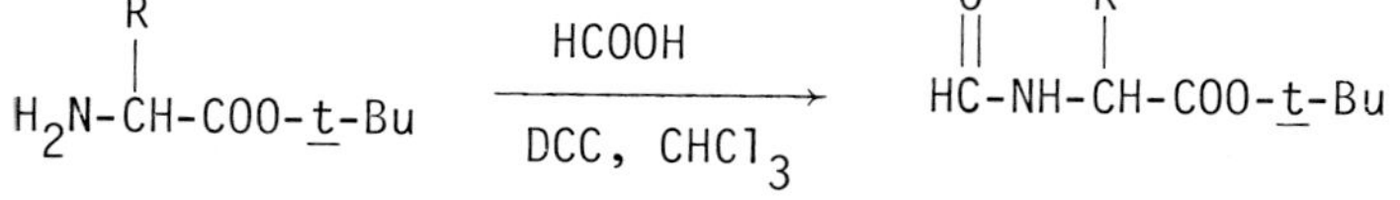

$$H_2N-CH(R)-COO-\underline{t}-Bu \xrightarrow[DCC, CHCl_3]{HCOOH} HC(=O)-NH-CH(R)-COO-\underline{t}-Bu$$

87% for leucine

PROTECTING GROUPS

V.B-3   U. Zehavi, J. Org. Chem., 42, 2819 (1977).

Use of 1,2-diphenylmaleyl (DPM) as a protecting group for amino functions. These derivatives are yellow and fluorescent, and can easily be followed chromatographically and determined quantitatively.

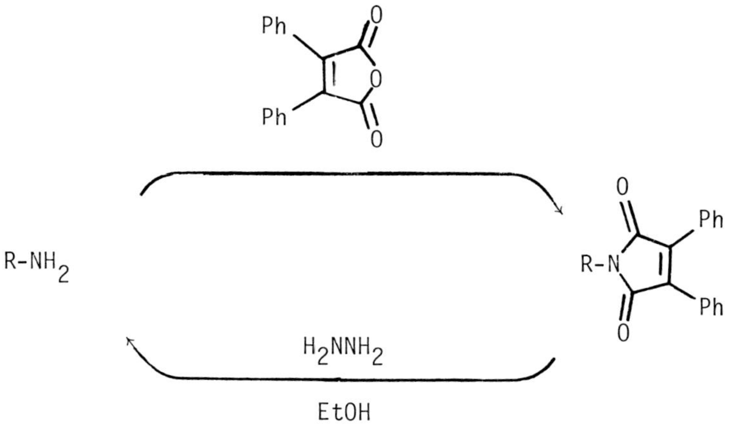

V.B-4   S. F. Brady, R. Hirschmann, and D. F. Veber, J. Org. Chem., 42, 143 (1977).

An investigation of the use of several new acid-labile amine protecting groups in peptide synthesis. The most promising appears to be 1-methylcyclobutyloxycarbonyl (1), which is more stable to 50% acetic acid than is the t-Boc group.

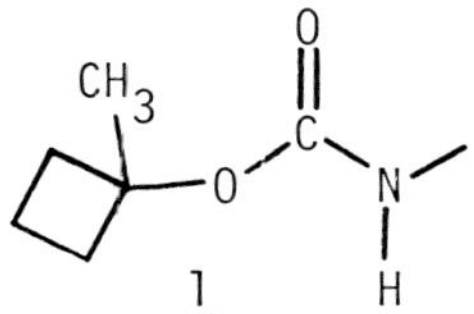

V.B-5 N. Kornblum and A. Scott, J. Org. Chem., 42, 399 (1977).

R-NH$_2$ + (9-anthrylmethyl p-nitrophenyl carbonate) $\xrightarrow{\text{CH}_3\text{S}^{\ominus}\text{ Na}^{\oplus} \text{ or CF}_3\text{COOH}}$ 9-anthrylmethyl carbamate (CH$_2$OCNHR)

R = alkyl, aryl; some 2° amines

77-97%

86% (R = p-ClPh)

V.B-6 H. Yajima et al., Chem. Pharm. Bull., 25, 740 (1977).

Ethanesulfonic acid in acetic acid or methylene chloride removes α-amino protecting groups such as Boc and Z(OMe). Z, benzyl ester, S-p-methoxybenzyl, and N^G-p-methoxybenzenesulfonyl groups are unaffected.

V.B-7 P. M. Hardy and D. J. Samworth, J. Chem. Soc., Perkin I, 1954 (1977).

Use of N,N'-isopropylidene dipeptides in peptide synthesis. Coupled by using DCC, and deprotected by hydrolysis under neutral conditions.

V.B-8   A. Abdipranoto, A. P. Hope, and B. Halpern, Aust. J. Chem., 30, 2711 (1977).

"The use of N^ε-2-hydroxyarylmethylene-protected ornithine and lysine derivatives in solid-phase peptide synthesis." The ketimine protecting group was found to be stable under anhydrous conditions and the urethane blocking group could be removed preferentially.

V.B-9   D. F. Veber et al., J. Org. Chem., 42, 3286 (1977).

Use of the isonicotinyloxycarbonyl protecting group for protection of the ε-amino group of lysine in peptide synthesis.

$$-\underset{H}{N}-\overset{O}{\underset{\|}{C}}-OCH_2-\text{pyridyl}$$

Put on using isonicotinyl p-nitrophenyl carbonate; removed by catalytic hydrogenation or Zn/HCl.

V.B-10   M. Itoh, D. Hagiwara, and T. Kamiya, Bull. Chem. Soc. Japan, 50, 718 (1977).

Use of oxime carbonates such as I as t-butoxycarbonylating reagents.

$$\underline{I} \qquad t\text{-BuO-}\overset{O}{\underset{\|}{C}}\text{-N=C}\underset{W}{\overset{Ph}{\diagup}} \qquad W = COOR, CN$$

V.B-11   R. A. Olofson, Y. S. Yamamoto, and D. J. Wancowicz, Tetrahedron Lett., 1563 (1977).

Use of the vinyloxycarbonyl (VOC) group for the protection of amine functions in peptide chemistry.

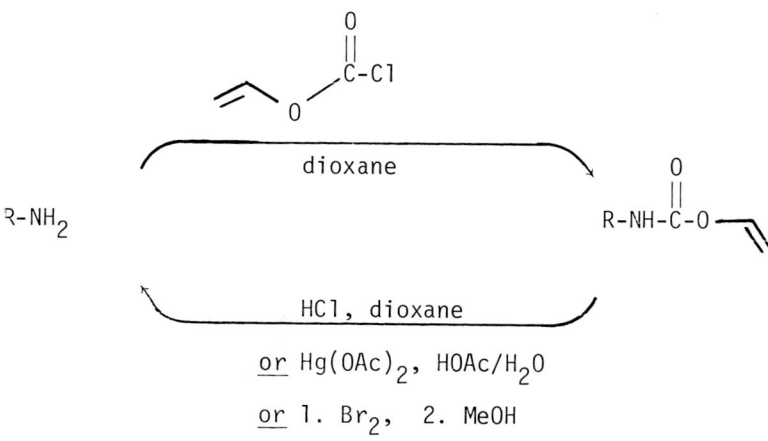

V.B-12   Y. S. Klausner and M. Chorev, J. Chem. Soc., Perkin I, 627 (1977).

Protection of the indole ring of tryptophan by benzyloxycarbonyl and 2,4-dichlorobenzyloxy-carbonyl groups using KF and 18-crown-6 as a catalyst for acylation with suitable carbonates (no racemization occurs). Removed by cat. H_2 or liquid HF or H_2NNH_2.

V.C-1 F. Dardoize, M. Gaudemar, and N. Goasdoue, Synthesis, 567 (1977).

$$\text{R-SH} \xrightarrow[\text{2. } H_2C(OMe)_2]{\text{1. BrZnCH}_2\text{COOEt}} \text{R-S-CH}_2\text{OCH}_3$$

R = aryl   ~80%

V.D-1 T. Ishimaru et al., Chem. Lett., 1313 (1977).

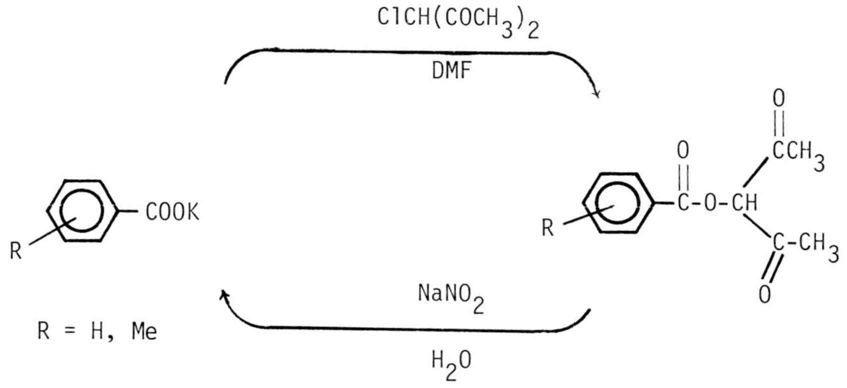

$$R\text{-}C_6H_4\text{-COOK} \xrightarrow[\substack{\text{NaNO}_2 \\ H_2O}]{\substack{\text{ClCH(COCH}_3)_2 \\ \text{DMF}}} R\text{-}C_6H_4\text{-C(O)-O-CH(COCH}_3)_2$$

R = H, Me

V.D-2 S.-S. Wang et al., J. Org. Chem., 42, 1286 (1977).

$$\underset{\text{(an amino acid)}}{\text{R-COOH}} \xrightarrow[\text{2. BzBr}]{\text{1. Cs}_2\text{CO}_3} \text{R-COOBz}$$

~70-90%

V.D-3   F. Dardoize, M. Gaudemar, and N. Goasdoue, Synthesis, 567 (1977).

$$\underset{R=\text{alkyl, aryl, benzyl}}{R-\overset{\overset{O}{\|}}{C}-OH} \xrightarrow[\text{2. } H_2C(OMe)_2]{\text{1. } BrZnCH_2COOEt} R-\overset{\overset{O}{\|}}{C}-OCH_2OCH_3$$

68-85%

V.D-4   M. E. Jung and M. A. Lyster, J. Am. Chem. Soc., **99**, 968 (1977).

$$R-\overset{\overset{O}{\|}}{C}-OR' + Me_3SiI \xrightarrow[\text{2. } H_2O]{\text{1. } CCl_4, 50°} RCOOH + R'I$$

R = alkyl, aryl, heterocyclic, etc.   >80%
R'= Me, Et, t-Bu

V.D-5   D. Liotta, W. Markiewicz, and H. Santiesteban, Tetrahedron Lett., 4365 (1977).

$$R-\overset{\overset{O}{\|}}{C}-O-R' \xrightarrow{NaSePh} R-\overset{\overset{O}{\|}}{C}-O^{\ominus}$$

R = 1°,2°,3° alkyl, Ph   >90%
R'= Me, Bz, n-alkyl, i-Pr

PROTECTING GROUPS

V.D-6 P. G. Gassman and W. N. Schenk, *J. Org. Chem.*, **42**, 918 (1977).

$$\text{R-C(=O)-OR'} \xrightarrow{2 \text{ KO-}\underline{t}\text{-Bu, } 1 \text{ H}_2\text{O}^{\oplus\ominus}} \text{R-C(=O)-O}^{\ominus} + \text{R'O}^{\ominus}$$

80-100%

Works well with benzoate, mesitylate, $\underline{t}$-butyl, and other hindered esters.

V.D-7 E. J. Corey and M. A. Tius, *Tetrahedron Lett.*, 2081 (1977).

$$\text{R-C(=O)-O-CH}_2\text{CH=CHPh} \xrightarrow[\text{2. KSCN, H}_2\text{O}]{\text{1. Hg(OAc)}_2\text{, MeOH}} \text{R-COOH}$$

R = Ph, Bz, $\underline{p}$-OMePh, cyclohexyl   ~90%

V.D-8 B. Castro et al., *Synthesis*, 413 (1977).

$$\text{Boc-NH-CH(R)-COOH} + \text{Ph-OH} \xrightarrow[\text{CH}_2\text{Cl}_2]{\text{BOP, Et}_3\text{N}} \text{Boc-NH-CH(R)-COOPh}$$

73-97%

BOP = benzotriazolyl-OP(NMe$_2$)$_3$ $\cdot$ PF$_6^{\ominus\oplus}$

V.D-9   T.-L. Ho and G. A. Olah, Synthesis, 417 (1977).

$$\underset{R-C-OR'}{\overset{O}{\|}} \xrightarrow{PhSiMe_3,\ I_2} R\text{-COOH}$$

R = Ph, Bz, n-alkyl   88-98%
R'= Me, Et, i-Pr, Bz

V.D-10   D. S. Kemp and J. Reczek, Tetrahedron Lett., 1031 (1977).

Use of Maq esters as carboxyl protecting groups in peptide synthesis. Prepared using Maq-OH and DCC with the carboxylic acid to be protected. The esters are readily soluble in CH_2Cl_2, EtOAc, and DMF, and their UV chromophore allows sensitive detection on TLC. Removed using sodium dithionite (dioxane-H_2O), photolysis (i-PrOH-morpholine), 9-hydroxyanthrone (DMF-Et_3N), or 9,10-dihydroxyanthracene on polystyrene resin.

Maq ester: [structure of 9,10-anthraquinone with $-CH_2-O-C(=O)-R$ substituent]

V.D-11   G. M. Anantharamaiah and K. M. Sivanandaiah, J. Chem. Soc., Perkin I, 490 (1977).

Use of transfer hydrogenation (cyclohexene and Pd/C) to remove N-benzyloxycarbonyl and benzyl ester groups in peptide synthesis.

PROTECTING GROUPS

V.E-1 E. C. Taylor and C.-S. Chiang, Synthesis, 467 (1977).

$$\underset{R}{\overset{O}{\underset{\|}{C}}}\underset{R'}{} \xrightarrow[\text{(adsorbed on montmorillonite)}]{HC(OMe)_3} \underset{R}{\overset{MeO \quad OMe}{\underset{}{C}}}\underset{R'}{}$$

R,R' = alkyl, Ph, Bz, cyclic, cinnamyl, H   >90%

V.E-2 J. P. Alazard, H. B. Kagan, and R. Setton, Bull. Soc. Chim. France, 499 (1977).

$$\underset{R}{\overset{O}{\underset{\|}{C}}}\underset{R'}{} \xrightarrow[\text{graphite bisulfate}]{HC(OEt)_3} \underset{R}{\overset{EtO \quad OEt}{\underset{}{C}}}\underset{R'}{}$$

R,R' = alkyl, aryl, cyclic   65-85%

Enol ethers may also be formed.

V.E-3 M. E. Jung, W. A. Andrus, and P. L. Ornstein, Tetrahedron Lett., 4175 (1977).

$$\underset{R'}{\overset{OR \quad OR}{\underset{}{C}}}\underset{R''}{} \xrightarrow[CHCl_3]{Me_3SiI, \text{ propene}} \underset{R'}{\overset{O}{\underset{\|}{C}}}\underset{R''}{}$$

R = Me, Et   ~80-90%

R',R" = H, Ph, n-alkyl, cyclic

V.E-4   F. Huet, M. Pellet, and J. M. Conia, Tetrahedron Lett., 3505 (1977).

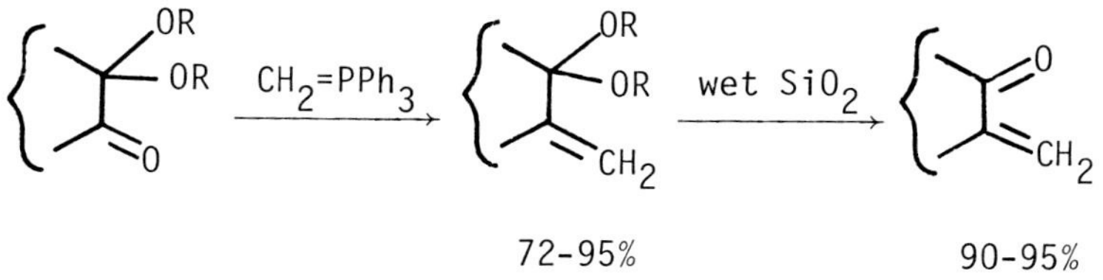

[cyclic ketone acetal] + $CH_2=PPh_3$ → [cyclic methylene acetal] 72-95% → wet SiO_2 → [cyclic methylene ketone] 90-95%

V.E-5   G. A. Olah, J. Welch, and M. Henninger, Synthesis, 308 (1977).

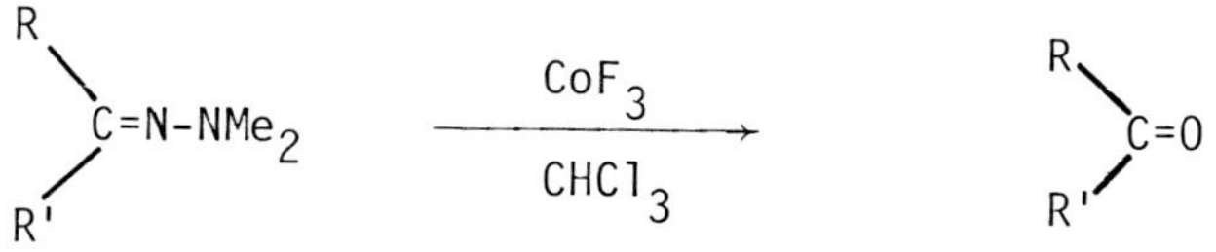

$$\underset{R'}{\overset{R}{>}}C=N-NMe_2 \xrightarrow[CHCl_3]{CoF_3} \underset{R'}{\overset{R}{>}}C=O$$

R = H, alkyl   46-94%
R' = alkyl, aryl

V.E-6   W. Lutz et al., Synthesis, 893 (1977).

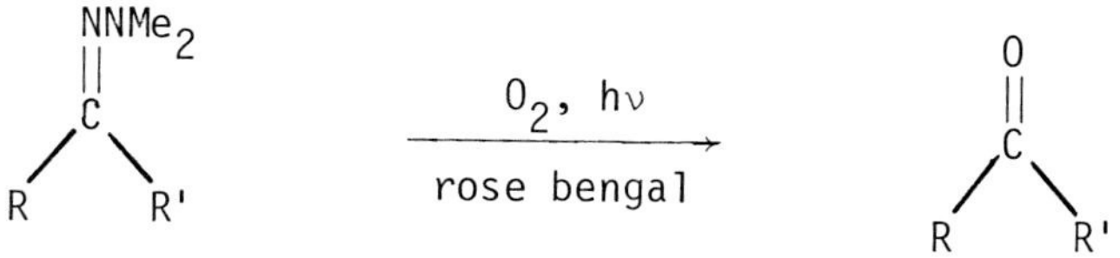

$$\underset{R \quad R'}{\overset{NNMe_2}{\underset{\|}{C}}} \xrightarrow[\text{rose bengal}]{O_2,\ h\nu} \underset{R \quad R'}{\overset{O}{\underset{\|}{C}}}$$

R,R' = H, alkyl   48-88%

PROTECTING GROUPS

V.E-7   V. V. Vakatkar et al., Chem. and Ind., 742 (1977).

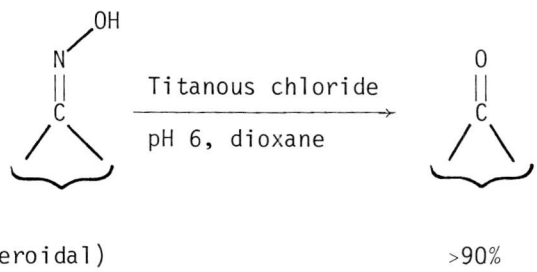

(steroidal)   >90%

Also works with semicarbazones.

V.E-8   C. R. Narayanan, P. S. Ramaswamy, and M. S. Wadia, Chem. and Ind., 454 (1977).

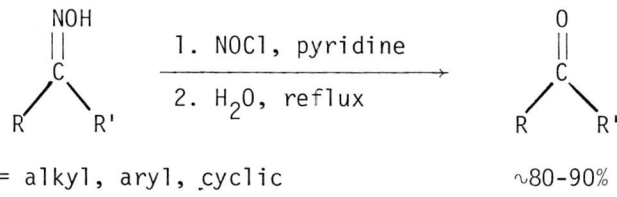

R,R' = alkyl, aryl, cyclic   ∿80-90%

V.E-9   C. R. Narayanan, P. S. Ramaswamy, and M. S. Wadia, Indian J. Chem., 15B, 578 (1977).

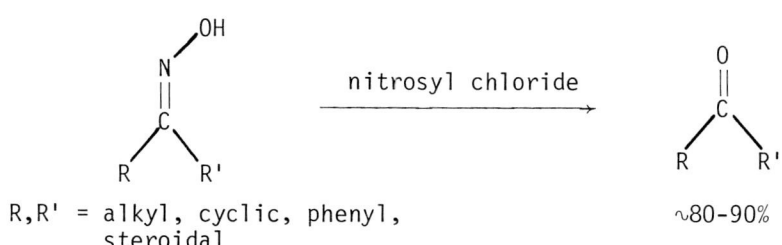

R,R' = alkyl, cyclic, phenyl,   ∿80-90%
       steroidal

V.E-10 D.H.R. Barton, D. J. Lester, and S. V. Ley, J. Chem. Soc., Chem. Commun., 445 (1977).

$$\left.\begin{array}{c}\text{Hydrazones}\\ \text{Oximes}\\ \text{Semicarbazones}\end{array}\right\} \xrightarrow{(PhSeO)_2O} \text{Ketones}\quad \sim 50\text{-}95\%$$

V.E-11 B. S. Ong and T. H. Chan, Synth. Commun., $\underline{7}$, 283 (1977).

$$\begin{array}{c}R\\ \diagdown\\ C=O\\ \diagup\\ R'\end{array} \xrightarrow[CHCl_3]{R''SH,\ ClTMS} \begin{array}{c}R\\ \diagdown\\ C(SR'')_2\\ \diagup\\ R'\end{array}$$

(R'' = Et, Ph)

R,R' = H, alkyl, aryl   generally >80%
(but not benzyl)

V.E-12 T. T. Takahashi, C. Y. Nakamura, and J. Y. Satoh, J. Chem. Soc., Chem. Commun., 680 (1977).

$$\left\{\!\!\!\begin{array}{c}S\\ \times\\ S\end{array}\!\!\!\right\} \xrightarrow{h\nu,\ O_2} \left\{\!\!\!\begin{array}{c}\cdot\\ \diagup\!=\!O\end{array}\!\!\!\right.$$

$\sim 60\text{-}80\%$

V.E-13   D.H.R. Barton, N. J. Cussans, and S. V. Ley, J. Chem. Soc., Chem. Commun., 751 (1977).

$$\underset{\text{C}}{\overset{\text{S}\frown\text{S}}{\diagdown}} \xrightarrow[\text{THF or CH}_2\text{Cl}_2]{\text{Ph}_2\text{Se}_2\text{O}_3} \underset{}{\overset{\text{O}}{\underset{\diagdown}{\overset{\|}{\text{C}}}}}$$

~60-90%
(ketones)

V.E-14   D. A. Evans et al., J. Am. Chem. Soc., 99, 5009 (1977).

$$\underset{R\quad R'}{\overset{O}{\underset{}{\overset{\|}{C}}}} \xrightarrow{\text{EtS-SiMe}_3} \underset{R'\quad\text{OSiMe}_3}{\overset{R\quad\text{SEt}}{C}}$$

R,R' = H, Ph, alkyl   ~80-90%

V.E-15   P. A. Grieco and Y. Yokoyama, J. Am. Chem. Soc., 99, 5210 (1977).

$$\text{RCH}_2\text{CHO} \xrightarrow[\text{Bu}_3\text{P, THF}]{\text{o-O}_2\text{N-C}_6\text{H}_4\text{-SeCN}} \text{RCH}_2\text{CH}(\text{CN})(\text{Se-C}_6\text{H}_4\text{-NO}_2)$$

R = n-alkyl, cyclohexenyl, etc.

V.E-16   C. C. Leznoff and S. Greenberg, Can. J. Chem., 54, 3824 (1976).

Use of divinylbenzene-styrene copolymer containing 1,3-diol groups as a monoblocking agent for dialdehydes:

$$\text{(P)}-\text{C}_6\text{H}_4-\text{CH}_2\text{OCH}_2-\underset{\underset{\text{CH}_3}{|}}{\overset{\overset{\text{CH}_2\text{OH}}{|}}{\text{C}}}-\text{CH}_2\text{OH}$$

$$R\begin{pmatrix}\text{CHO}\\\text{CHO}\end{pmatrix} \xrightarrow[\text{HCl, dioxane}]{\text{TsOH, mol. sieves}} \text{(P)}-\text{C}_6\text{H}_4-\text{CH}_2-\text{O}-\overset{\overset{\text{CH}_2-\text{C}-\text{CH}_3}{|}}{\underset{\underset{\text{R-CHO}}{|}}{\text{CH}_2-\text{O}-\text{CH}-\text{O}}}$$

R = alkyl, aryl

V.E-17   D.H.R. Barton et al., J. Chem. Soc., Perkin I, 1075 (1977).

"The selective formation of enolate anions provides protection for carbonyl groups against reduction by lithium aluminum hydride." Selective reductions of steroid diones and triones are used as examples.

V.E-18   M. Nazir, W. Kreiser, and H. H. Inhoffen, <u>Synthesis,</u> 466 (1977).

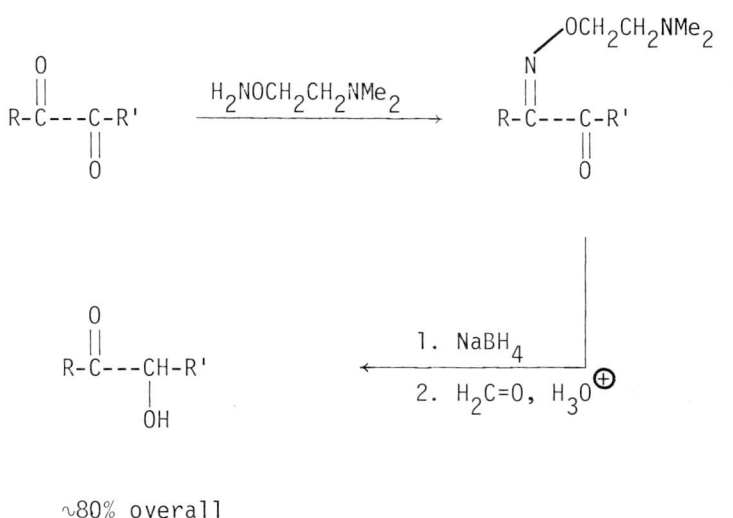

∿80% overall

V.F-1   G. W. Daub and E. E. van Tamelen, <u>J. Am. Chem. Soc.</u>, <u>99</u>, 3526 (1977).

Use of methyl phosphotriester intermediates in oligoribonucleotide synthesis. The methyl protecting group may be removed by LiSPh.

V.F-2   N. F. Sergeeva, Z. A. Shabarova, and M. A. Prokof'ev, <u>Doklady Chem.</u>, <u>234</u>, 280 (1977).

Blocking the Phosphodiester Groups of Oligodeoxyribonucleotides with Phosphoryltriimidazolide.

V.H-1 V. G. Granik, A. M. Zhidkova, and R. G. Glushkov,
Russ. Chem. Rev., 46, 361 (1977).

Review: "Advances in the Chemistry of the Acetals of Acid Amides and Lactams"

VI.A.1-1   N. Ishikawa and S. Sasaki, Chem. Lett., 483 (1977).

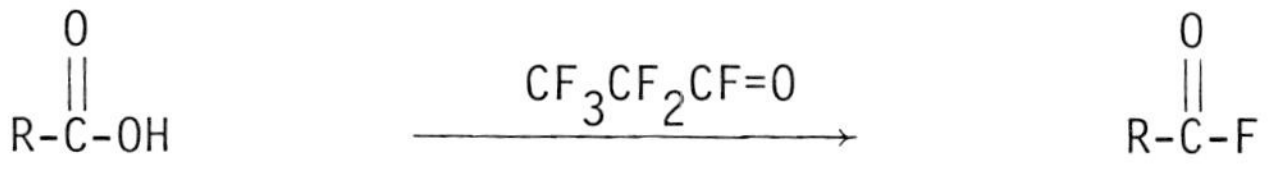

$$\underset{R-C-OH}{\overset{O}{\|}} \quad \xrightarrow{CF_3CF_2CF=O} \quad \underset{R-C-F}{\overset{O}{\|}}$$

R = Et, i-Pr, Ph   ~85%

VI.A.1-2   Y. Nagai et al., J. Chem. Soc., Chem. Commun., 808 (1977).

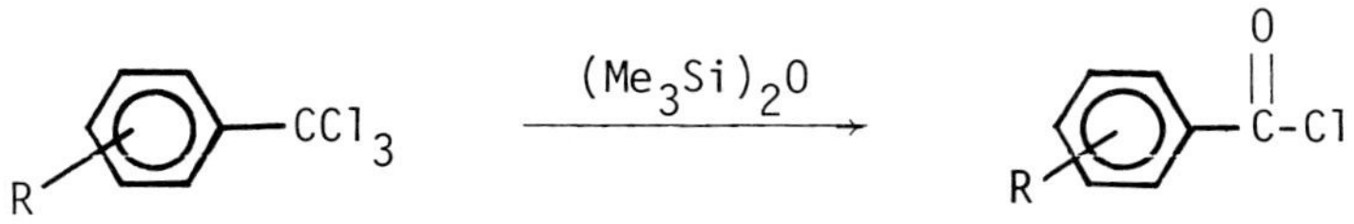

$$R-C_6H_4-CCl_3 \quad \xrightarrow{(Me_3Si)_2O} \quad R-C_6H_4-\underset{}{\overset{O}{\overset{\|}{C}}}-Cl$$

R = Cl, CCl$_3$ 74-84%

VI.A.1-3   H. Schick, S. Schwartz, and U. Eberhardt, J. Prakt. Chem., 319, 213 (1977).

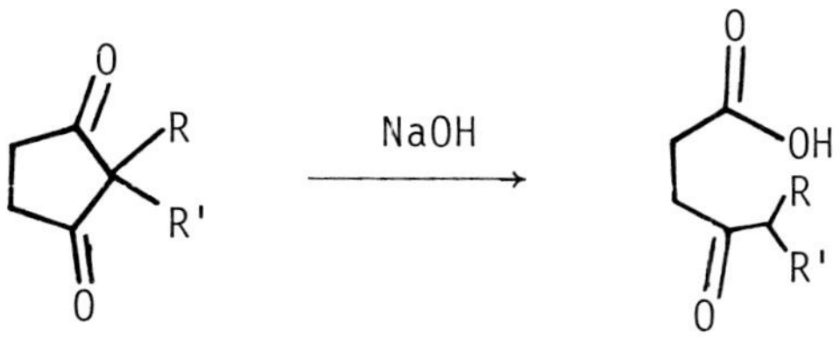

[cyclopentane-1,3-dione with R, R' substituents] $\xrightarrow{NaOH}$ [open-chain keto acid with R, R' substituents]

R = Me, Et   >90%
R'= allyl, CH$_2$COOH

VI.A.1-4 V. N. Odinokov, L. P. Zhemaiduk, and G. A. Tolstikov, Bull. Akad. USSR Chem., 25, 1790 (1977).

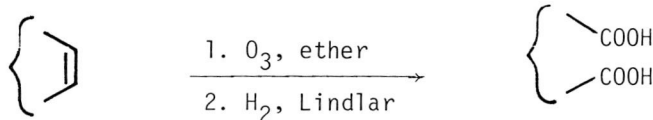

80-99%

VI.A.1-5 B. M. Trost and G. S. Massiot, J. Am. Chem. Soc., 99, 4405 (1977).

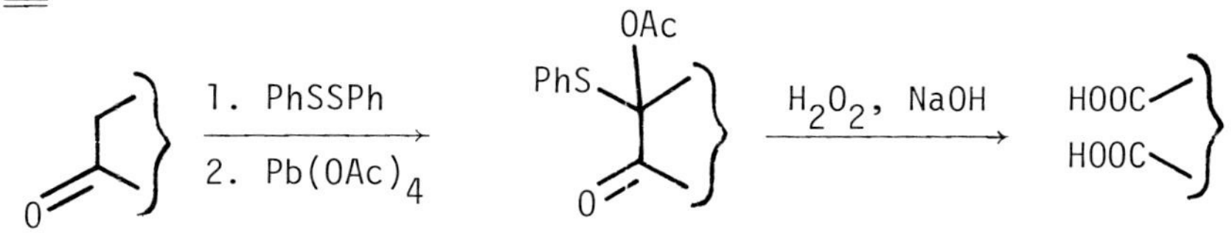

55-99%   65-99%

VI.A.1-6 F. M. Hauser and R. Rhee, Synthesis, 245 (1977).

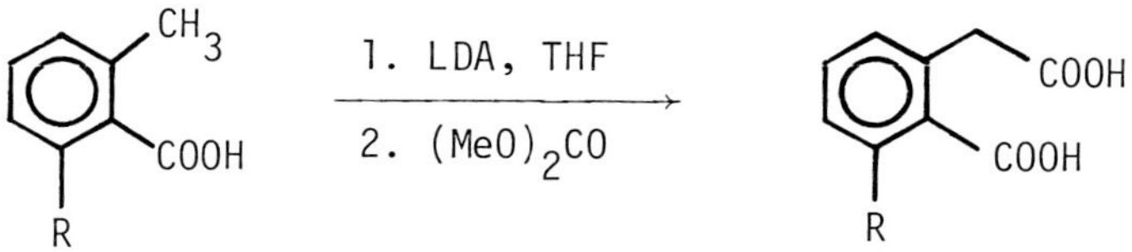

R = H, OMe   85-90%

USEFUL SYNTHETIC PREPARATIONS

VI.A.2-1 D.H.R. Barton and S. C. Narang, J. Chem. Soc., I, 1114 (1977).

$$R-NH_2 \xrightarrow[\text{2. Zn, AcOH}]{\text{1. } N_2O_4} R-OH$$

R = n-octyl, cycloalkyl ~30-80%

VI.A.2-2 K. Smith et al., J. Chem. Soc., Perkin I, 1172 (1977).

$$R_3B \; + \; \underset{\substack{}}{\text{Li}\overset{R'}{\underset{}{C}}\!\!\begin{pmatrix}S\\S\end{pmatrix}} \xrightarrow[\text{3. NaOH, } H_2O_2]{\text{2. HgCl}_2} R-\underset{R}{\overset{R'}{\underset{|}{\overset{|}{C}}}}-OH$$

R = n-alkyl, cycloalkyl ~80-90%

R' = H, n-Pr

VI.A.2-3 K. Isagawa, K. Tatsumi, and Y. Otsuji, Chem. Lett., 1117 (1977).

$$R-CH=CH_2 \xrightarrow[\text{2. Oxidation}]{\text{1. LiAlH}_4, \text{Cp}_2\text{Ti}(\text{AlH}_3)_2} RCH_2CH_2OH$$

R = Ph, n-alkyl, cyclohexenyl 34-86%

VI.A.2-4   H. C. Brown and N. Ravindriran, J. Org. Chem., 2533 (1977).

$$\underset{R}{CH_2=CH} \xrightarrow[\text{2. } H_2O_2, \text{ NaOH}]{\text{1. } H_2BCl \cdot SMe_2 \text{ or } H_2BCl \cdot OEt_2} \underset{R}{CH_2OH-CH_2-}$$

R = Ph, simple alkyl   generally >99%

VI.A.2-5   H. C. Brown and S. C. Kim, J. Org. Chem., 42, 1482 (1977).

$$PhCH=CH_2 \xrightarrow[\text{3. NaOH, } H_2O_2]{\text{1. } LiEt_3BH \quad \text{2. } CH_3SO_3H} \underset{OH}{PhCHCH_3}$$

90%

$$PhCH=CHCHO \xrightarrow{\text{same conditions}} \underset{OH \quad OH}{PhCHCH_2CH_2}$$

80%

VI.A.2-6   H. C. Brown and N. M. Yoon, Israel J. Chem., 15, 12 (1977).

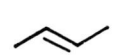 $\xrightarrow{\text{Diisopinocampheylborane}}$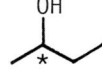

344

USEFUL SYNTHETIC PREPARATIONS

T. Cohen, A. G. Dietz, Jr., and J. R. Miser, Chem., **42**, 2053 (1977).

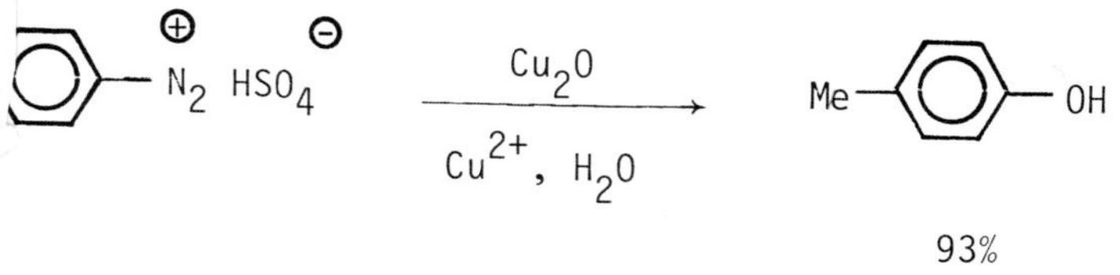

$$\text{Ph-N}_2^+ \text{HSO}_4^- \xrightarrow[\text{Cu}^{2+},\ H_2O]{Cu_2O} \text{Me-C}_6H_4\text{-OH}$$

93%

VI.A.2-8 N. J. Lewis, S. Y. Gabhe, and M. R. De La Mater, J. Org. Chem., **42**, 1479 (1977).

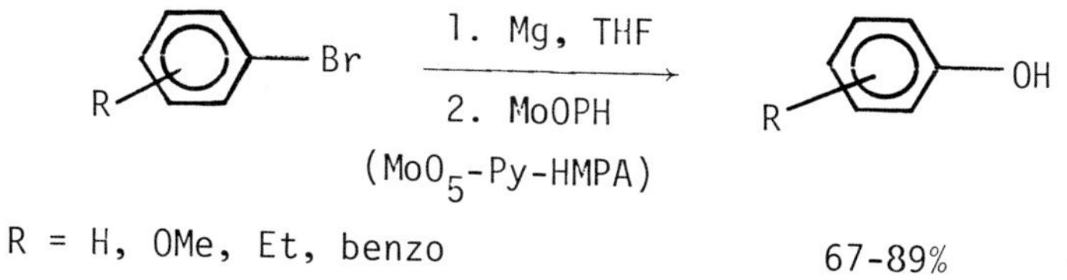

$$\text{R-C}_6H_4\text{-Br} \xrightarrow[\substack{2.\ \text{MoOPH} \\ (\text{MoO}_5\text{-Py-HMPA})}]{1.\ \text{Mg, THF}} \text{R-C}_6H_4\text{-OH}$$

R = H, OMe, Et, benzo   67-89%

VI.A.2-9 G. Zweifel, S. J. Backlund, and T. Leung, J. Am. Chem. Soc., **99**, 5194 (1977).

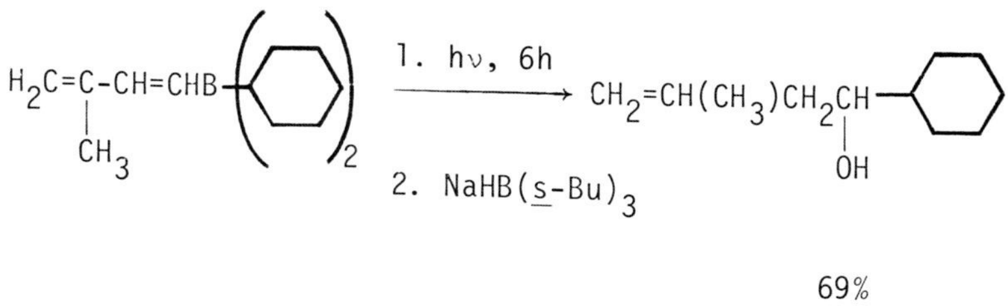

$$(H_2C=C(CH_3)\text{-CH=CHB})(\text{C}_6H_{11})_2 \xrightarrow[2.\ \text{NaHB}(\underline{s}\text{-Bu})_3]{1.\ h\nu,\ 6h} CH_2=CH(CH_3)CH_2CH(OH)\text{-C}_6H_{11}$$

69%

VI.A.3-1 V. P. Savel'yanov et al., J. Org. Chem. (USSR), 604 (1977).

$$\text{R-OH} \xrightarrow[\text{(on a column)}]{\text{HCl, trinonyl borate} \atop 130°, \text{ZnCl}_2} \text{R-Cl}$$

R = 1° alkyl   90-95%

VI.A.3-2 M. E. Jung and P. L. Ornstein, Tetrahedron Lett., 2659 (1977).

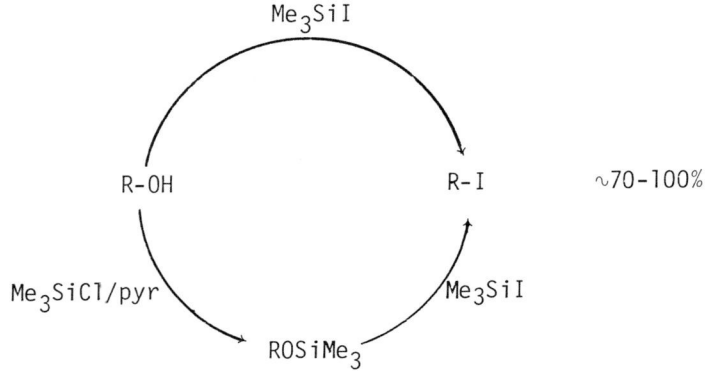

R = 1°,2°,3° alkyl, cycloalkyl

~70-100%

VI.A.3-3 K. Isagawa, K. Tatsumi, and Y. Otsuyi, Chem. Lett., 1117 (1977).

$$\text{RCH=CH}_2 \xrightarrow[\text{2. Br}_2]{\text{1. LiAlH}_4, \text{Cp}_2\text{Ti(AlH}_3)_2} \text{RCH}_2\text{CH}_2\text{Br}$$

R = n-alkyl, Ph   50-87%

USEFUL SYNTHETIC PREPARATIONS

VI.A.2-1   D.H.R. Barton and S. C. Narang, J. Chem. Soc., Perkin I, 1114 (1977).

$$R-NH_2 \xrightarrow[\text{2. Zn, AcOH}]{\text{1. }N_2O_4} R-OH$$

R = n-octyl, cycloalkyl   ~30-80%

VI.A.2-2   K. Smith et al., J. Chem. Soc., Perkin I, 1172 (1977).

$$R_3B \;+\; \underset{S}{\overset{R'}{Li\overset{|}{C}}}\!\!\diagdown\!\!\diagup \xrightarrow[\text{3. NaOH, }H_2O_2]{\text{2. HgCl}_2} \begin{array}{c} R' \\ | \\ R-C-OH \\ | \\ R \end{array}$$

R = n-alkyl, cycloalkyl   ~80-90%

R'= H, n-Pr

VI.A.2-3   K. Isagawa, K. Tatsumi, and Y. Otsuji, Chem. Lett., 1117 (1977).

$$R-CH=CH_2 \xrightarrow[\text{2. Oxidation}]{\text{1. LiAlH}_4, \text{Cp}_2\text{Ti(AlH}_3)_2} RCH_2CH_2OH$$

R = Ph, n-alkyl, cyclohexenyl   34-86%

VI.A.2-4   H. C. Brown and N. Ravindriran, J. Org. Chem., 42, 2533 (1977).

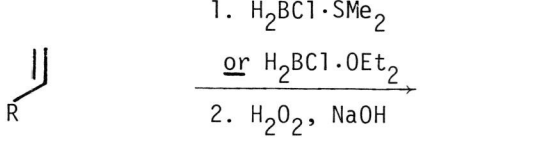

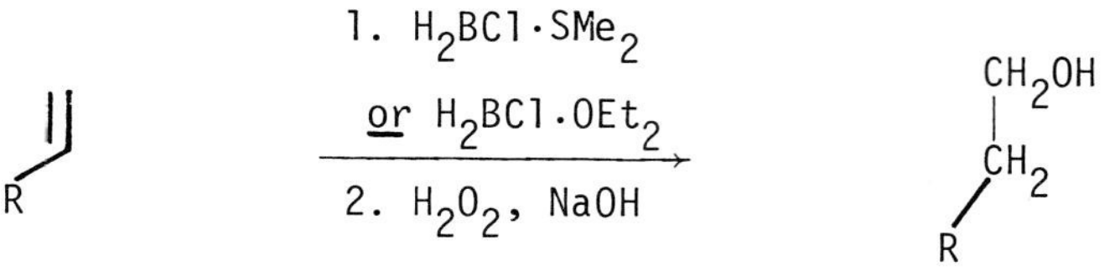

R = Ph, simple alkyl   generally >99%

VI.A.2-5   H. C. Brown and S. C. Kim, J. Org. Chem., 42, 1482 (1977).

$$PhCH=CH_2 \xrightarrow[\text{3. NaOH, }H_2O_2]{\text{1. LiEt}_3BH \quad \text{2. }CH_3SO_3H} \begin{array}{c} PhCHCH_3 \\ | \\ OH \end{array}$$

90%

$$PhCH=CHCHO \xrightarrow{\text{same conditions}} \begin{array}{c} PhCHCH_2CH_2 \\ |\quad\quad | \\ OH \quad OH \end{array}$$

80%

VI.A.2-6   H. C. Brown and N. M. Yoon, Israel J. Chem., 15, 12 (1977).

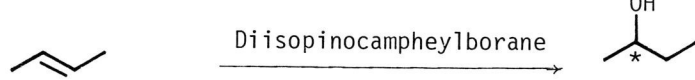

USEFUL SYNTHETIC PREPARATIONS

VI.A.2-7   T. Cohen, A. G. Dietz, Jr., and J. R. Miser, J. Org. Chem., 42, 2053 (1977).

$$\text{Me-C}_6\text{H}_4\text{-N}_2^{\oplus}\;\text{HSO}_4^{\ominus} \xrightarrow[\text{Cu}^{2+},\;\text{H}_2\text{O}]{\text{Cu}_2\text{O}} \text{Me-C}_6\text{H}_4\text{-OH}$$

93%

VI.A.2-8   N. J. Lewis, S. Y. Gabhe, and M. R. De La Mater, J. Org. Chem., 42, 1479 (1977).

$$\text{R-C}_6\text{H}_4\text{-Br} \xrightarrow[\substack{2.\;\text{MoOPH} \\ (\text{MoO}_5\text{-Py-HMPA})}]{1.\;\text{Mg, THF}} \text{R-C}_6\text{H}_4\text{-OH}$$

R = H, OMe, Et, benzo   67-89%

VI.A.2-9   G. Zweifel, S. J. Backlund, and T. Leung, J. Am. Chem. Soc., 99, 5194 (1977).

$$(\text{H}_2\text{C=C(CH}_3\text{)-CH=CH-B})_2\text{-C}_6\text{H}_{11} \xrightarrow[2.\;\text{NaHB(s-Bu)}_3]{1.\;h\nu,\;6h} \text{CH}_2\text{=CH(CH}_3\text{)CH}_2\text{CH(OH)-C}_6\text{H}_{11}$$

69%

VI.A.3-1   V. P. Savel'yanov et al., J. Org. Chem. (USSR), 13, 604 (1977).

$$\text{R-OH} \xrightarrow[\text{130°, ZnCl}_2]{\text{HCl, trinonyl borate}} \text{R-Cl}$$

R = 1° alkyl   (on a column)   90-95%

VI.A.3-2   M. E. Jung and P. L. Ornstein, Tetrahedron Lett., 2659 (1977).

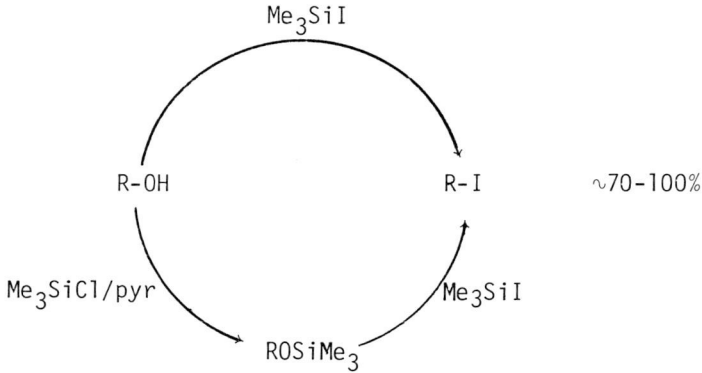

R = 1°,2°,3° alkyl, cycloalkyl

VI.A.3-3   K. Isagawa, K. Tatsumi, and Y. Otsuyi, Chem. Lett., 1117 (1977).

$$\text{RCH=CH}_2 \xrightarrow[\text{2. Br}_2]{\text{1. LiAlH}_4,\ \text{Cp}_2\text{Ti(AlH}_3)_2} \text{RCH}_2\text{CH}_2\text{Br}$$

R = n-alkyl, Ph   50-87%

USEFUL SYNTHETIC PREPARATIONS

VI.A.3-4   R. M. Magid, O. S. Fruchey, and W. L. Johnson, Tetrahedron Lett., 2999 (1977).

$$\underset{CH_3\quad CH_2OH}{\diagup\!\!=\!\!\diagdown} \xrightarrow{PPh_3,\ Cl_3CCCCl_3\ (\text{C}=\text{O})} \underset{CH_3\quad CH_2Cl}{\diagup\!\!=\!\!\diagdown}$$

98%

$$\underset{\underset{OH}{|}}{CH_2=CH-CH-CH_3} \xrightarrow{\text{same conditions}} \underset{\underset{Cl}{|}}{CH_2=CH-CH-CH_3}$$

94%

VI.A.3-5   R. Mornet and L. Gouin, Synthesis, 786 (1977).

$$\underset{R^2\quad\ CH_2N\diagdown R}{\overset{R^1}{\diagdown}\!\!\!\diagup\!\!=\!\!\diagup\!\!\diagdown\!\!R} \xrightarrow{ClCOOEt} \underset{R^2\qquad CH_2Cl}{\overset{R^1}{\diagdown}\!\!\!\diagup\!\!=\!\!\diagup\!\!\diagdown}$$

R^1 = H, $-\overset{O}{\underset{\|}{C}}$-alkyl

R^2 = 1°,2° alkyl, Ph, H

∼80%

VI.A.3-6   T. Kauffmann, H. Fischer, and A. Woltermann, Angew. Chem. Int. Ed., 16, 53 (1977).

$$R-X \xrightarrow[\text{2.}\ Br_2]{\text{1.}\ Ph_2\overset{O}{\underset{\|}{As}}CH_2Li} R-CH_2Br$$

X = halogen
R = 1° alkyl

∼50-80% overall

VI.A.3-7 T. Mukaiyama, S. Shoda, and Y. Watanabe, Chem. Lett., 383 (1977).

$$R\underset{H}{\overset{R'}{\underset{|}{C}}}-OH \quad \xrightarrow[\text{2. Et}_4\text{NCl}]{\text{1. Cl}-\underset{O}{\overset{\overset{Et}{\underset{|}{\overset{\oplus}{N}}}}{\underset{}{\bigcirc}}}\text{BF}_4,\ Et_3N} \quad Cl-\underset{H}{\overset{R'}{\underset{|}{C}}}R$$

R = alkyl, protected sugar

76-97%

>90% inversion

VI.A.3-8 A. R. Katritzky et al., Synthesis, 634 (1977).

$$R-NH_2 \quad \xrightarrow[\Delta]{\underset{\oplus\ I^\ominus}{\underset{Ph}{\bigcirc}\underset{O}{\bigcirc}\underset{Ph}{\bigcirc}\ Ph}} \quad R-I$$

52-85%

R = alkyl, benzyl, pyridinyl

VI.A.3-9 P. Gölitz and A. de Meijere, Angew. Chem. Int. Ed., 16, 854 (1977).

$$R-NH_2 \xrightarrow{CF_3NO} R-N=N-CF_3 \xrightarrow{h\nu} R-CF_3$$

R = n-alkyl, cycloalkyl

5-52%

VI.A.3-10 L. M. Yagupol'skii, N. V. Kondratenko, and
V. I. Popov, J. Org. Chem. (USSR), 13, 561 (1977).

$$Ph\text{-}CCl_3 \xrightarrow[\underline{or}\ Ph_2SbF_3]{PhSbF_4} PhCF_3$$

up to 91%

VI.A.3-11 T.-L. Ho, B.G.B. Gupta, and G. A. Olah, Synthesis, 676 (1977).

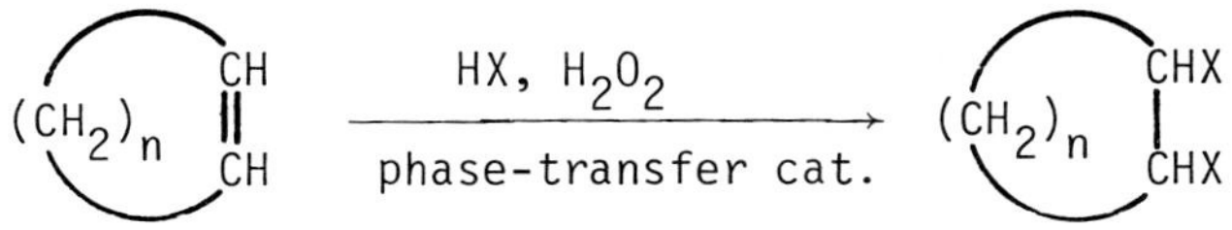

$$(CH_2)_n\begin{matrix}CH\\ \|\\ CH\end{matrix} \xrightarrow[\text{phase-transfer cat.}]{HX,\ H_2O_2} (CH_2)_n\begin{matrix}CHX\\ |\\ CHX\end{matrix}$$

n = 4,5,6

∼75% (X=Cl)
∼95% (X=Br)

VI.A.3-12 Y. Echigo, Y. Watanabe, and T. Mukaiyama, Chem. Lett., 1013 (1977).

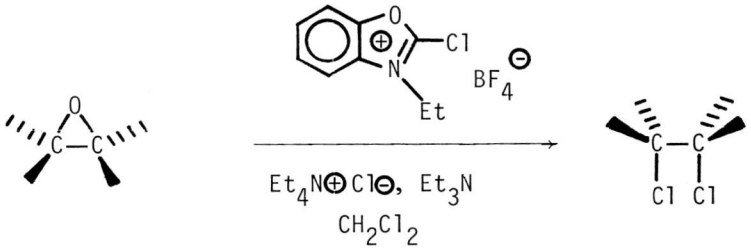

alkyl, aryl, cyclic substituents

∼60-90%

(syn functionalization)

VI.A.3-13   M. Zupan and A. Pollak, J. Org. Chem., 42, 1559 (1977).

$$\text{PhCH=CHR} \xrightarrow{\text{XeF}_2} \text{Ph-CH-CH-R} \atop \phantom{\text{Ph-CH}} \overset{|}{\text{F}}\ \overset{|}{\text{F}}$$

R = Me, Ph, t-Bu   53-64%

VI.A.3-14   S. A. Shackelford, R. R. McGuire, and J. L. Pflug, Tetrahedron Lett., 363 (1977).

Simplified benchtop procedures for the use of XeF_2 for alkene fluorination.

VI.A.3-15   D.H.R. Barton, Pure and Appl. Chem., 49, 1241 (1977).

Review: "The Invention of Reactions Useful for the Synthesis of Specifically Fluorinated Natural Products"

USEFUL SYNTHETIC PREPARATIONS

VI.A.3-16   D. Bethell, K. McDonald, and K. S. Rao, <u>Tetrahedron Lett.</u>, 1447 (1977).

$$Ph_2CN_2 \xrightarrow[Bu_4N^{\oplus} ClO_4^{\ominus},\ CH_2Cl_2]{H_2O,\ KHF_2} Ph_2CHF$$

~60%

VI.A.3-17   J. Kagan et al., <u>J. Org. Chem.</u>, <u>42</u>, 343 (1977).

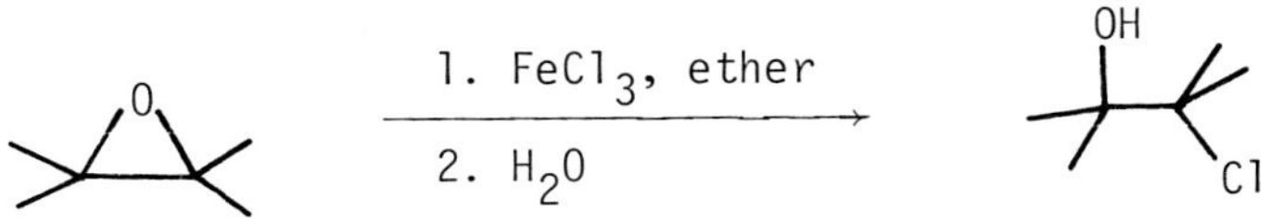

<img of trimethyl epoxide> → 1. FeCl$_3$, ether 2. H$_2$O → chlorohydrin product

55-78%

VI.A.3-18   F. M. Laskovics and E. M. Schulman, <u>Tetrahedron Lett.</u>, 759 (1977).

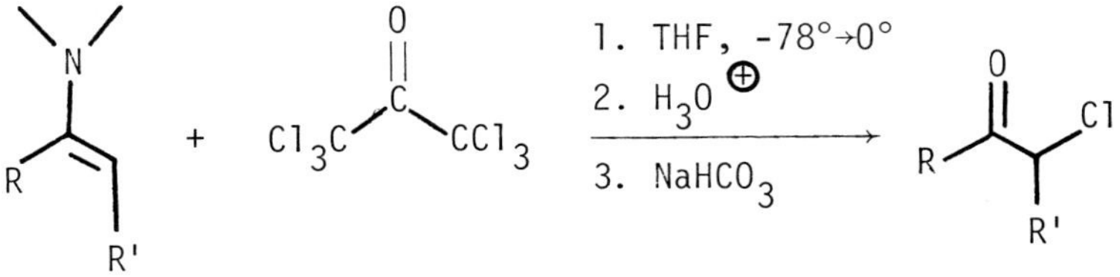

enamine + Cl$_3$C-C(O)-CCl$_3$ →
1. THF, -78°→0°
2. H$_3$O$^{\oplus}$
3. NaHCO$_3$

→ R-C(O)-CHCl-R'

61-65% for cyclopentanone, cyclohexanone, cycloheptanone

May give mixtures with compounds which can give two enamine isomers.

VI.A.3-19   H. Nakai and M. Kurono, Chem. Lett., 995 (1977).

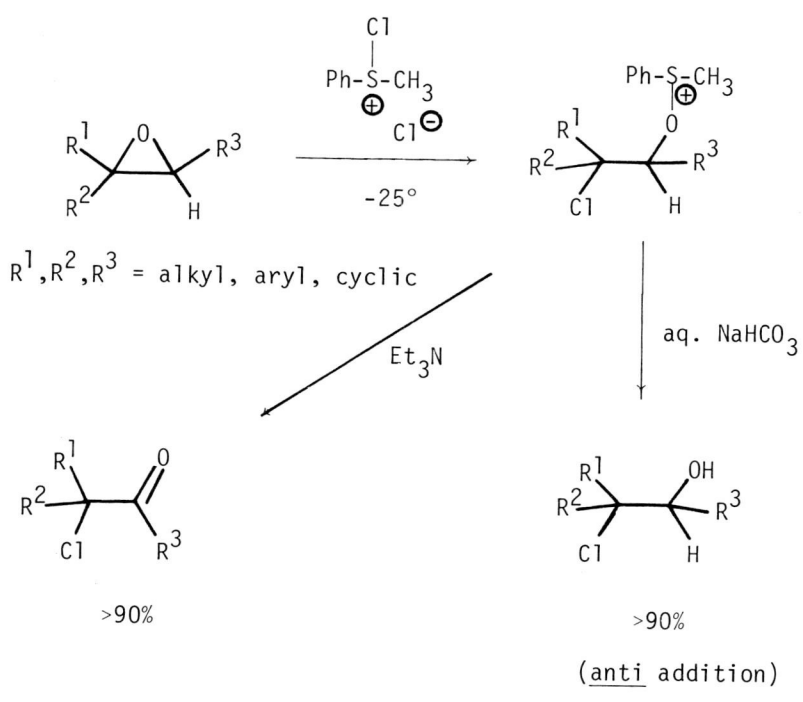

VI.A.3-20   V. Reutrakul and W. Kanghae, Tetrahedron Lett., 1225 (1977).

$$R\text{-}\underset{O}{\overset{\|}{C}}\text{-}H \xrightarrow[\text{2. } \Delta, \text{ xylene}]{\text{1. Ph-S-}\overset{\overset{O\;Li}{\|\;|}}{CH}\text{-Cl, THF, -78°}} R\text{-}\underset{O}{\overset{\|}{C}}\text{-}CH_2Cl$$

R = n-alkyl   76-95%

VI.A.3-21 S. H. Korzeniowski and G. W. Gokel, Tetrahedron Lett., 3519 (1977).

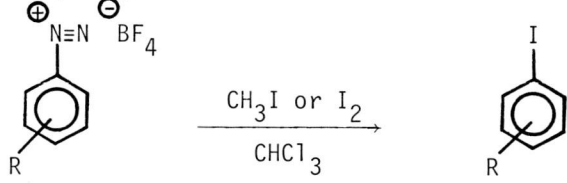

R = Me, OMe, Br, Cl, NO_2 ~50-90%

VI.A.3-22 M. P. Doyle, B. Siegfried, and J. F. Dellaria, Jr., J. Org. Chem., 42, 2426 (1977).

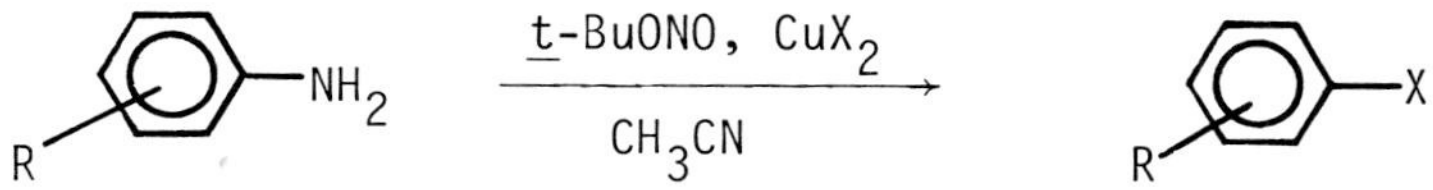

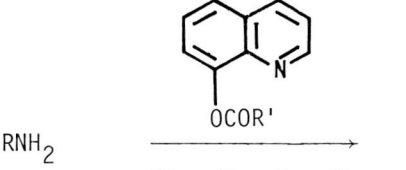

X = Cl, Br

R = NO_2, $COCH_3$, COOH, halogen, alkyl

~60-90%

VI.A.4-1 T.-L. Ho, Synth. Commun., 7, 393 (1977).

$$\text{quinoline-8-OCOR'}$$

RNH$_2$ $\xrightarrow{\text{R' = Me, Et, Ph}}$ RNHCOR'

R = Ph, Bz, c-Hx, etc.

73-98%

Also works with piperidine.

VI.A.4-2   M. K. Sahni et al., Synth. Commun., 7, 57 (1977).

$$R-NH_2 \xrightarrow[R' = Me, Ph]{R'-C(=O)-O-(salicylic\ acid/formaldehyde\ polymer)} R-NH-C(=O)-R'$$

R = H, Ph, CH_2COOEt ~80%

VI.A.4-3   A. Basha, M. Lipton, and S. M. Weinreb, Tetrahedron Lett., 4171 (1977).

$$R-C(=O)-O-R^1 \xrightarrow[CH_2Cl_2]{Me_2AlNR^2R^3} R-C(=O)-NR^2R^3$$

R = Ph, Bz, alkyl, aryl, conjugated, etc.   ~70-80%

R^1 = Me, Et

R^2, R^3 = H, alkyl, cyclic

VI.A.4-4   I. Butula et al., Synthesis, 704 (1977).

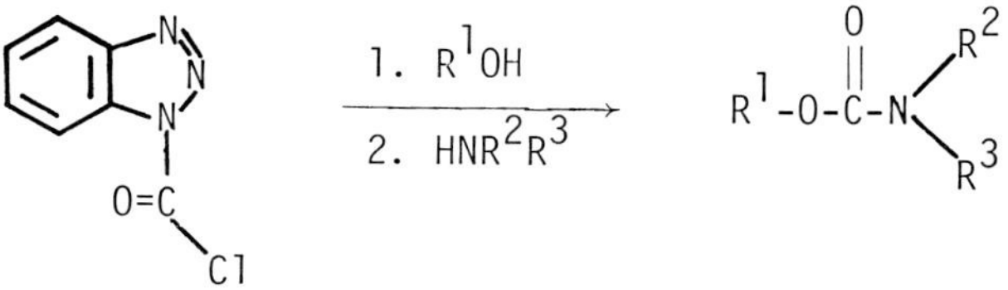

(benzotriazole-N-C(=O)-Cl) $\xrightarrow[2.\ HNR^2R^3]{1.\ R^1OH}$ $R^1-O-C(=O)-NR^2R^3$

R^1 = Et, Ph, Bz
R^2 = H, Et, n-alkyl, c-Hx

VI.A.4-5   R. Mukherjee, Indian J. Chem., 15B, 502 (1977).

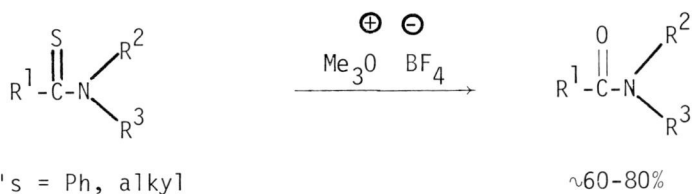

R's = Ph, alkyl   ~60-80%

VI.A.4-6   J. Palecek and J. Kuthan, Z. Chem., 17, 260 (1977).

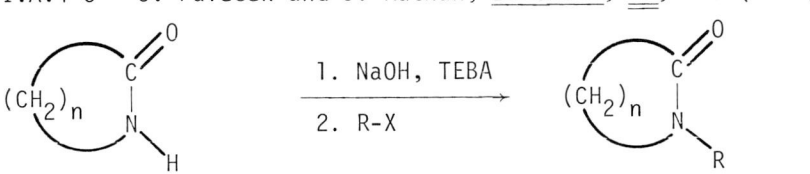

n = 3,5,7   Widely varying yields
RX = Me_2SO_4, BzBr, BuBr, etc.   (15-73%)

VI.A.4-7   A. Basha, J. Orlando, and S. M. Weinreb, Synth. Commun., 7, 549 (1977).

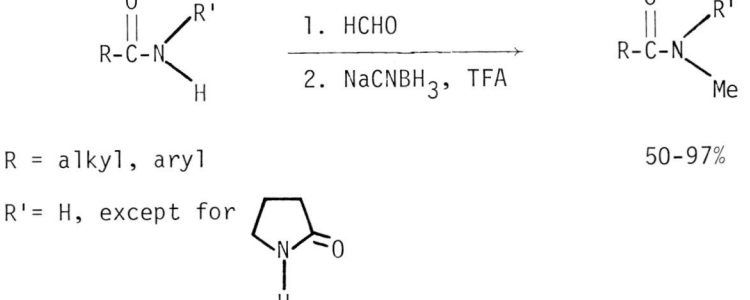

R = alkyl, aryl   50-97%

R'= H, except for

VI.A.4-8 J. D. Elliott and J. H. Jones, *J. Chem. Soc., Chem. Commun.*, 758 (1977).

$$\text{R-Br} \xrightarrow{K^{\oplus} \; {}^{\ominus}N(CO\text{-}O\text{-}t\text{-Bu})(CO\text{-}OMe)} \text{R-NH-Boc}$$

R = Bz, CH_2CO_2Et ∼70%

VI.A.4-9 G. Helmchen et al., *Angew. Chem. Int. Ed.*, **16**, 728 (1977).

$$\underset{\underset{NHNHTs}{|}}{R-C}\overset{O}{\diagup} \xrightarrow{Pb(OAc)_4} \underset{\underset{N=NTs}{|}}{R-C}\overset{O}{\diagup} \xrightarrow{BzNH_2} R-\overset{O}{\underset{}{C}}-NHBz$$

R = Me, t-Bu, Ph   Widely varying yields.

VI.A.4-10 D. C. Owsley and J. J. Bloomfield, *Synthesis*, 118 (1977).

$$\underset{R}{\text{C}_6\text{H}_4}\text{-NO}_2 \xrightarrow{\text{Fe, } CH_3COOH} \underset{R}{\text{C}_6\text{H}_4}\text{-NHCCH}_3\overset{O}{\|}$$

R = Me, OEt, Cl   ∼70-90%

USEFUL SYNTHETIC PREPARATIONS

VI.A.4-11   T.-L. Ho, J. Org. Chem., 42, 3755 (1977).

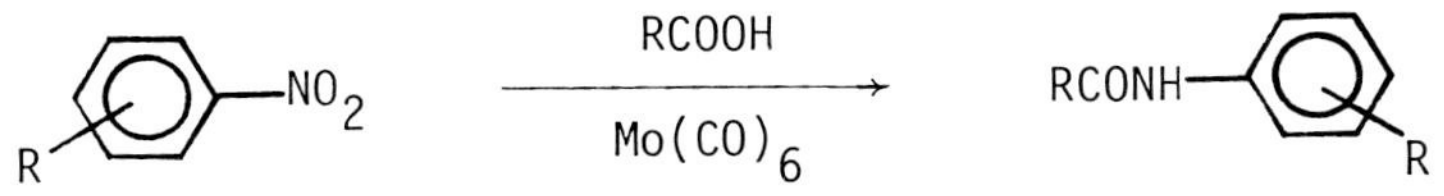

R–C₆H₄–NO₂ →[RCOOH / Mo(CO)₆] RCONH–C₆H₄–R

R = H, Me, OMe, COCH$_3$, CH=CHPh

46% (R = COCH$_3$)
85% (R = OMe)

VI.A.4-12   A. Basha and S. M. Weinreb, Tetrahedron Lett., 1465 (1977).

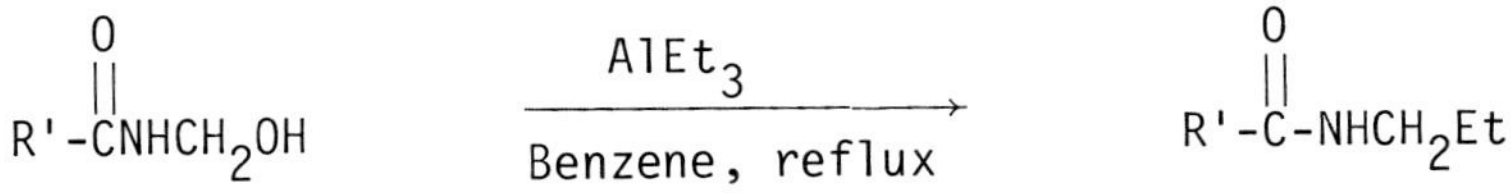

$$R'\text{-CONHCH}_2\text{OH} \xrightarrow[\text{Benzene, reflux}]{\text{AlEt}_3} R'\text{-CO-NHCH}_2\text{Et}$$

56-94%

VI.A.4-13   Y. Inamoto et al., Synthesis, 632 (1977).

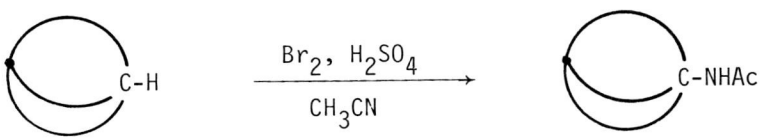

~90%

VI.A.4-14 R. W. Turner et al., Synthesis, 31 and 33 (1977).

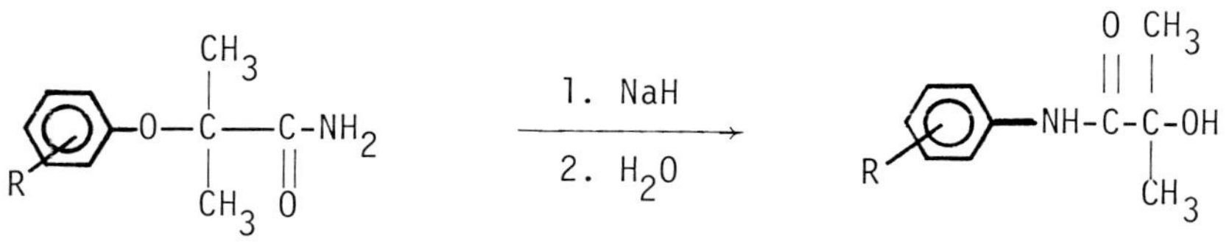

[Reaction: aryl ether with CH₃, C(CH₃)₂, C(=O)NH₂ group]
1. NaH
2. H₂O
→ Ar-NH-C(=O)-C(CH₃)₂-OH

R = NO_2, CF_3, COPh, etc.

~65%

VI.A.4-15 H. Driguez, J. M. Paton, and J. Lessard, Can. J. Chem., 55, 700 (1977).

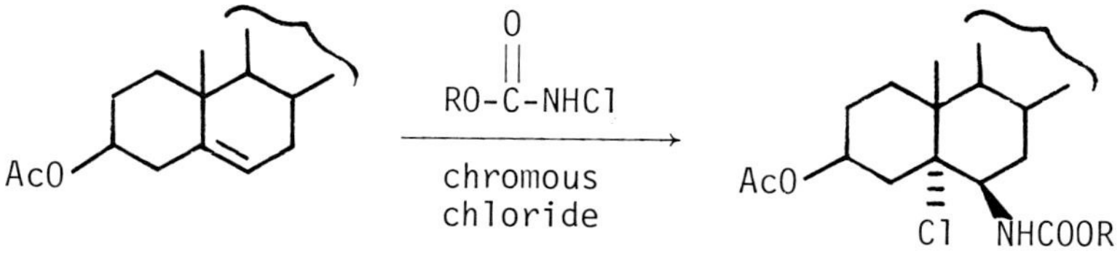

[Steroid with AcO and alkene]
RO-C(=O)-NHCl
chromous chloride
→ [Steroid product with Cl and NHCOOR]

79%

VI.A.4-16 S. Wadia et al., Synthesis, 35 (1977).

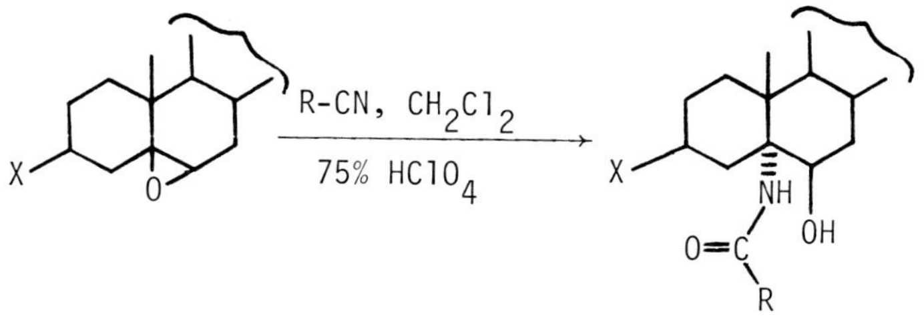

[Steroid epoxide]
R-CN, CH_2Cl_2
75% $HClO_4$
→ [Steroid with NH-C(=O)R and OH]

X = H, OH

R = Me, Ph

~30-40%

VI.A.4-17   K. A. Parker and J. J. Petratis, Tetrahedron Lett., 4561 (1977).

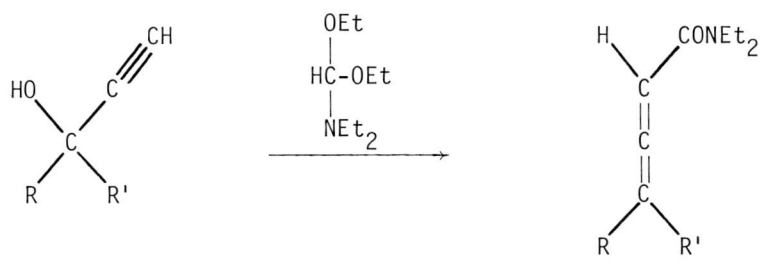

R,R' = cyclic, H, steroidal, alkyl   50-92%

VI.A.5-1   T. Mukaiyama et al., Chem. Lett., 635 (1977).

ROH $\xrightarrow[\text{2. LiN}_3]{\text{1. [pyridinium-F]} ^{\ominus}\text{OTs, Et}_3\text{N, Me}}$ R-N$_3$ $\xrightarrow[\text{or LiAlH}_4]{\text{H}_2/\text{Pd}}$ R-NH$_2$

R = 1°, 2°, allylic   75-94%

VI.A.5-2   A. Baiker and W. Richarz, Tetrahedron Lett., 1937 (1977).

R-OH $\xrightarrow[\text{CuO, Cr}_2\text{O}_3, \text{Na}_2\text{O}]{\text{HNMe}_2, 230°}$ R-NMe$_2$

R = long chain alkyl   96%

VI.A.5-3   N. L. Holy, Synth. Commun., 6, 539 (1976).

$$\text{R-MgBr} \xrightarrow{\overset{\oplus}{\text{Me}_2\text{N}=\text{CH}_2} \quad \overset{\ominus}{\text{OCOCF}_3}} \text{RCH}_2\text{NMe}_2$$

R = Ph, Tolyl, n-Bu   72-85%

VI.A.5-4   U. Schöllkopf et al., Liebigs Ann. Chem., 40 (1977).

$$\underset{\underset{H}{|}}{\overset{\overset{NH_2}{|}}{R-C-H}} \longrightarrow \longrightarrow \underset{\underset{H}{|}}{\overset{\overset{NC}{|}}{R-C-H}} \xrightarrow[\substack{2.\ R'Br \\ 3.\ H_3O^\oplus}]{1.\ BuLi} \underset{\underset{R'}{|}}{\overset{\overset{NH_2}{|}}{R-C-H}}$$

R = H, Ph, vinyl, 4-pyridyl   Up to 65% (last 3 steps)

Also works with aziridines.

VI.A.5-5   M. Botta, F. DeAngelis, and R. Nicoletti, Synthesis, 722 (1977).

$$\underset{R^2}{\overset{R^1}{>}}\!\!\text{NH} + R^3\text{-OH} \xrightarrow{(\underline{t}\text{-BuO})_3\text{Al, Ra-Ni}} \underset{R^2}{\overset{R^1}{>}}\!\!\text{N-R}^3$$

R^1 = H, Me;
R^2 = Ph, $\Big\}$ cyclic   >94%

R^3 = n-Pr, i-Pr

VI.A.5-6 A. Zwierzak and J. B. Piotrowicz, Angew. Chem. Int. Ed., 16, 107 (1977).

$$R-NH_2 \xrightarrow[\substack{2. \ R'X, \ NaOH, \\ Bu_4N^{\oplus} \ HSO_4^{\ominus}}]{1. \ (EtO)_2P(O)H} (EtO)_2P(O)NRR' \xrightarrow{HCl} HN{\overset{R}{\underset{R'}{\diagdown}}} \cdot HCl$$

R = Et, Bz, c-Hx   >80%

R' = 1° alkyl

VI.A.5-7 G. Charles et al., Tetrahedron Lett., 3469 (1977).

$$R-NH_2 \xrightarrow{CH_2O, \ CH_3OH} R-N(CH_2OCH_3)_2 \xrightarrow{NaCNBH_3} R-NMe_2$$

R = steroidal   ~90%

VI.A.5-8 B. S. Cowlagi, M. A. Dave, and A. B. Kulkarni, Indian J. Chem., 14B, 904 (1976).

$$Ar-NO_2 \xrightarrow[H_2, \ Ra-Ni]{MeOH, \ CH_2O} Ar-NMe_2$$

~60%

VI.A.5-9 Y. Tamura, J. Minamikawa, and M. Ikeda, Synthesis, 1 (1977).

Review: "O-Mesitylenesulfonylhydroxylamine and Related Compounds--Powerful Aminating Reagents"

VI.A.5-10 Y. Kikugawa and M. Kawase, Chem. Lett., 1279 (1977).

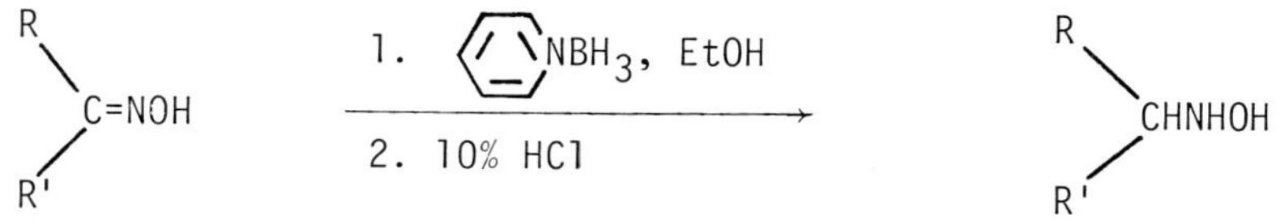

$$\underset{R'}{\overset{R}{>}}C=NOH \xrightarrow[\text{2. 10\% HCl}]{\text{1. } C_5H_5NBH_3, \text{ EtOH}} \underset{R'}{\overset{R}{>}}CHNHOH$$

R = 1° alkyl, aryl   74-92%
R'= H, 1° alkyl

VI.A.5-11 N. H. Khan et al., Synth. Commun., 7, 71 (1977).

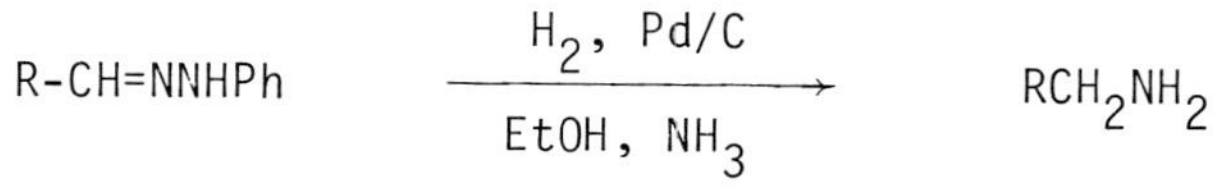

$$R-CH=NNHPh \xrightarrow[\text{EtOH, NH}_3]{H_2, \text{ Pd/C}} RCH_2NH_2$$

~50-80%

VI.A.5-12 E. P. Kyba and A. M. John, Tetrahedron Lett., 2737 (1977).

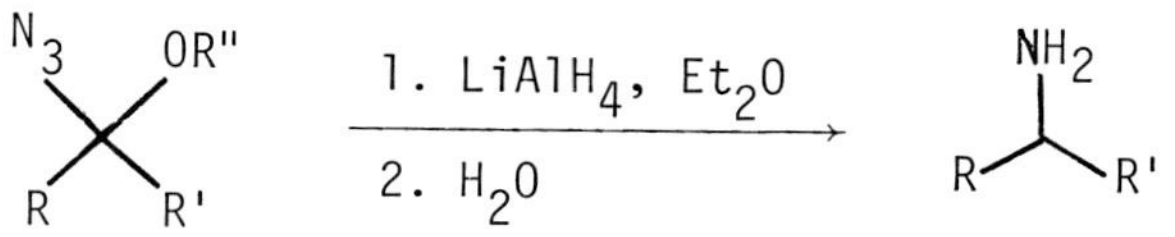

$$\underset{R \quad R'}{\overset{N_3 \quad OR''}{\diagdown\diagup}}\hspace{-1em}\diagup\diagdown \xrightarrow[\text{2. H}_2O]{\text{1. LiAlH}_4, \text{ Et}_2O} \underset{R \quad R'}{\overset{NH_2}{\diagdown\diagup}}$$

R = Ph, t-Bu, chx,   >83%
R'= H, Me,   -(CH$_2$)$_5$-, etc.

USEFUL SYNTHETIC PREPARATIONS

VI.A.5-13   M. Cinquini and F. Cozzi, J. Chem. Soc., Chem. Commun., 723 (1977).

$$\text{Me-}\underset{*}{\text{C}_6\text{H}_4}\text{-}\underset{\underset{O}{\|}}{S}\text{-N=C}\underset{Ph}{\overset{R}{\diagdown}} \xrightarrow[\text{2. H}^{\oplus}\text{, MeOH}]{\text{1. LiAlH}_4} \underset{Ph}{\overset{R}{\underset{|}{H-\overset{*}{C}-NH_2}}}$$

R = Me, Et, α-Naphthyl

∼40-60% overall
57-80% ee

VI.A.5-14   M. Chastrette, G. Axiotis, and R. Gauthier, Tetrahedron Lett., 23 (1977).

$$ROCH_2\text{-}C\equiv N \xrightarrow{R'M} \underset{R'}{\overset{R}{\diagdown}}C=NM \xrightarrow[\text{2. H}_2O]{\text{1. R''M'}} R\text{-}\underset{R''}{\overset{R'}{\underset{|}{\overset{|}{C}}}}\text{-NH}_2$$

R = Me, Et, cyclic   M = MgBr, Li   53-78%
R' = Me, n-Pr, n-Bu, allyl
R" = n-Bu, allyl

VI.A.5-15   A. Zwierzak and I. Podstawczynska, Angew. Chem. Int. Ed., 16, 702 (1977).

$$Ph_2\overset{O}{\overset{\|}{P}}NH_2 \xrightarrow[\text{PhH, phase-transfer cat.}]{R\text{-}X, \text{NaOH}} Ph_2\overset{O}{\overset{\|}{P}}NHR \xrightarrow[\text{PhH, phase-transfer cat.}]{R'X, \text{NaOH}} Ph_2\overset{O}{\overset{\|}{P}}NRR'$$

R,R' = 1°,2° alkyl, allyl, Bz

$$Ph_2P(O)NHR \xrightarrow[\text{THF}]{\text{HCl}} NH_2R \quad \sim 40\text{-}80\%$$

$$Ph_2P(O)NRR' \xrightarrow[\text{THF}]{\text{HCl}} NHRR' \quad \sim 42\text{-}80\%$$

VI.A.5-16   G. H. Posner and D. Z. Rogers, J. Am. Chem. Soc., 99, 8208 and 8214 (1977).

$$\text{cyclopentene oxide} \xrightarrow[\text{Al}_2\text{O}_3]{\underline{n}\text{-BuNH}_2} \text{trans-2-(butylamino)cyclopentanol}$$

~60-70%

Reaction fails with cyclooctene oxide and larger rings.

VI.A.5-17   A. I. Meyers and R. Gabel, J. Org. Chem., 42, 2653 (1977).

$$\text{Ar-oxazoline(OMe)} \xrightarrow[\text{THF}]{\text{LiNR}_2} \text{Ar-oxazoline(NR}_2\text{)} \xrightarrow[\text{MeOH}]{\text{H}^{\oplus}} \text{Ar-COOMe(NR}_2\text{)}$$

Y = H, OMe   41-98%

R = H, Et, $\underline{i}$-Pr, $\underline{t}$-Bu

VI.A.5-18   C. Jutz, A. F. Kirschner, and R.-M. Wagner, Chem. Ber., 110, 1259 (1977).

$$\underset{\text{ClO}_4^{\ominus}}{R^1R^2N^{\oplus}=CH-CH=CH-NR^3R^4} \xrightarrow{\text{NaBH}_4} R^1R^2N-CH_2CH_2CH_2-NR^3R^4$$

R's = Me, Bu, cyclic, Ph, Het   ~60-80%

USEFUL SYNTHETIC PREPARATIONS

VI.A.5-19   K. Niedenzu, Pure and Appl. Chem., **49**, 745 (1977).

Review: "The Aminoboronation Reaction"

VI.A.5-20   G. W. Gribble, R. W. Leiby, and M. N. Sheehan, Synthesis, 856 (1977).

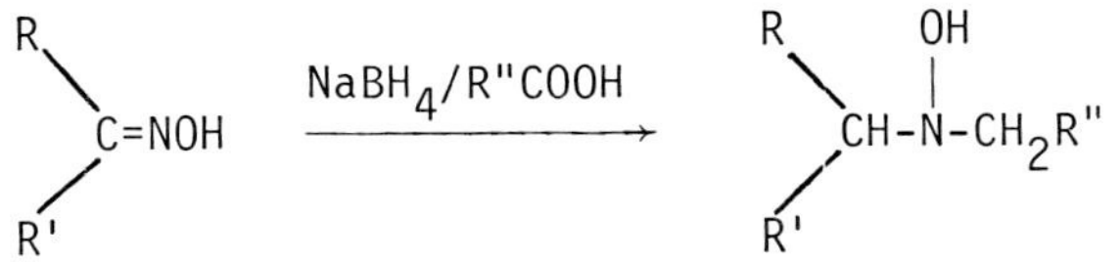

$$\underset{R'}{\overset{R}{>}}C=NOH \xrightarrow{NaBH_4/R''COOH} \underset{R'}{\overset{R}{>}}CH-\underset{\underset{}{|}}{\overset{OH}{N}}-CH_2R''$$

R,R' = H, Me, Et,   36-87%
       t-Bu, Ph, Bz, -(CH$_2$)$_5$-

R" = Me, Et, i-Pr, n-Pr

VI.A.6-1   E. L. Compere, Jr. and D. A. Weinstein, Synthesis, 852 (1977).

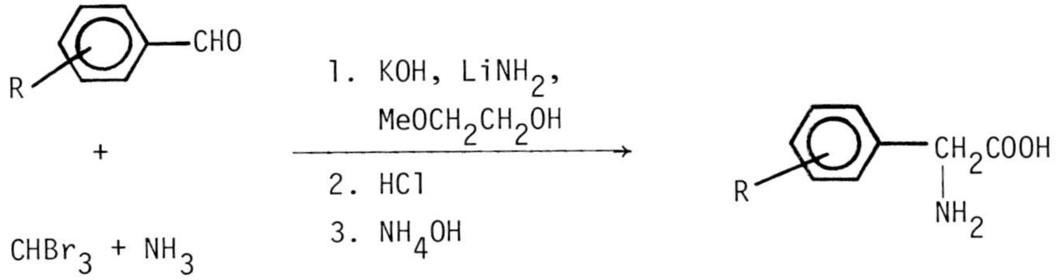

R-C$_6$H$_4$-CHO

+

CHBr$_3$ + NH$_3$

1. KOH, LiNH$_2$,
   MeOCH$_2$CH$_2$OH
2. HCl
3. NH$_4$OH

→ R-C$_6$H$_4$-CH(NH$_2$)CH$_2$COOH

R = H, Cl, F, Me, OMe   33-83%

VI.A.6-2 N. H. Khan, A. A. Siddiqui, and A. R. Kidwai, Indian J. Chem., 15B, 573 (1977).

$$\underset{R-C-COOH}{\overset{NNHPh}{\|}} \xrightarrow{H_2, \, Pd/C} \underset{R-CH-COOH}{\overset{NH_2}{|}}$$

R = alkyl, aryl, etc.   Widely varying yields.

VI.A.6-3 T. Iwasaki and K. Harada, J. Chem. Soc., Perkin I, 1730 (1977).

$$\underset{PhCH_2N=C-COOR^2}{\overset{R^1}{|}} \xrightarrow[\text{2. } H_2, \, Pd/C]{\text{1. } R^3X, \text{ electrolysis}} \underset{\underset{R^3}{|}}{\overset{R^1}{\underset{|}{H_2N-C-COOH}}}$$

R^1 = Me, H   36-86%

R^2 = Et, Bz

R^3 = Me, Et, Bz, CH_2CN, CH_2COOEt

VI.A.6-4 M. D. Fryzuk and B. Bosnich, J. Am. Chem. Soc., 99, 6262 (1977).

$$\underset{HOOC \quad\quad NHCOMe}{\overset{R}{\diagdown\!\!=\!\!\diagup}} \xrightarrow[H_2, \, THF]{\text{Chiral Rh(I)phosphine}} \underset{HOOC-\overset{*}{C}H-NHCOMe}{\overset{CH_2R}{|}}$$

R = H, Ph, alkyl, etc.   74-100% ee

VI.A.6-5 W. S. Knowles et al., J. Am. Chem. Soc., 99, 5946 (1977).

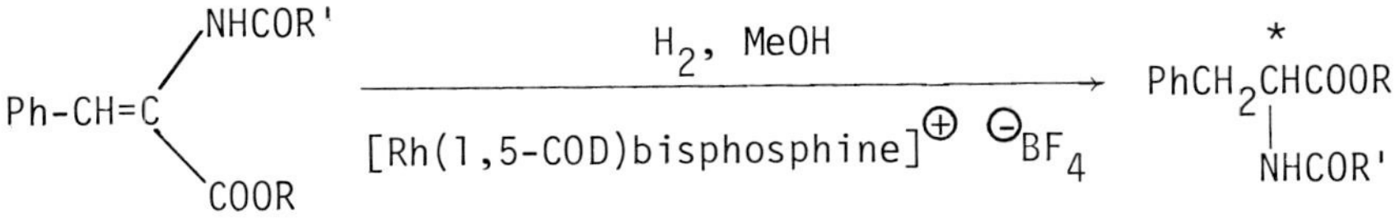

$$\text{Ph-CH=C}\begin{smallmatrix}\text{NHCOR'}\\ \\ \text{COOR}\end{smallmatrix} \xrightarrow[{[\text{Rh}(1,5\text{-COD})\text{bisphosphine}]^{\oplus}\ ^{\ominus}\text{BF}_4}]{\text{H}_2,\ \text{MeOH}} \text{PhCH}_2\overset{*}{\text{CH}}\text{COOR} \atop \text{NHCOR'}$$

R = H, Me
R'= Me, Ph

23-96% ee

VI.A.6-6 J. J. Fitt and H. W. Gschwend, J. Org. Chem., 42, 2639 (1977).

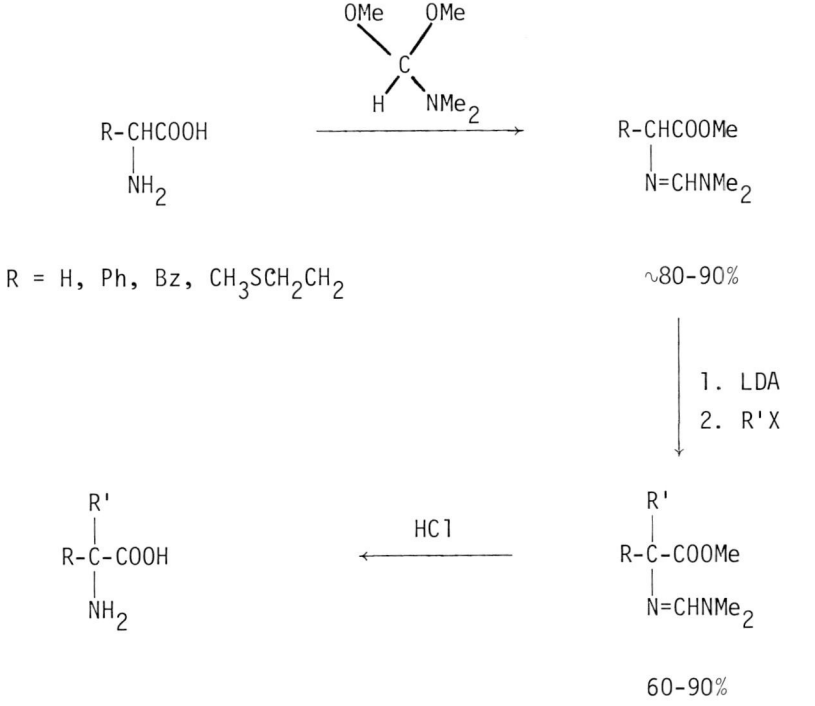

R = H, Ph, Bz, CH$_3$SCH$_2$CH$_2$

∿80-90%

1. LDA
2. R'X

60-90%

R'X = MeI, n-PrI, ethylene oxide, cinnamyl bromide, p-chlorobenzyl chloride, etc.

VI.A.6-7 S. G. Ramaswamy and E. Adams, J. Org. Chem., 42, 3440 (1977).

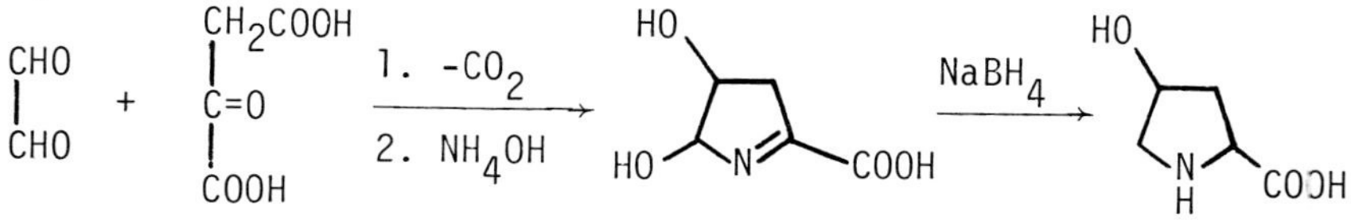

CHO + CH$_2$COOH, C=O, COOH → (1. -CO$_2$; 2. NH$_4$OH) → [HO-substituted pyrroline-COOH] → NaBH$_4$ → [HO-substituted pyrrolidine-COOH]

26-41%

A one-pot synthesis of 4-hydroxyproline.

VI.A.6-8 D. Ben-Ishai, R. Moshenberg, and J. Altman, Tetrahedron, 33, 1533 (1977).

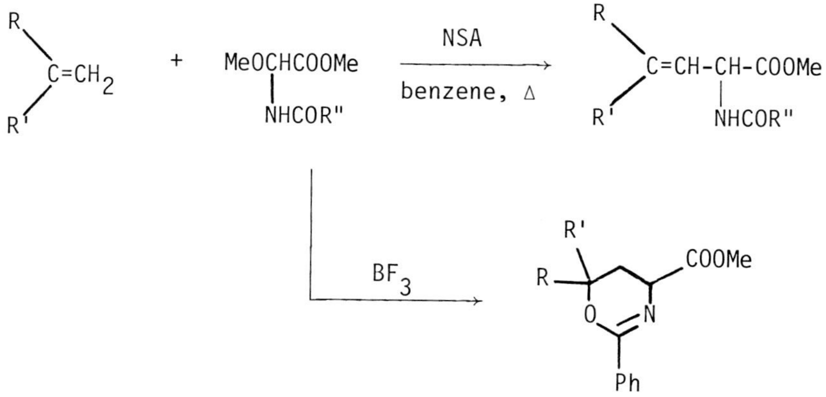

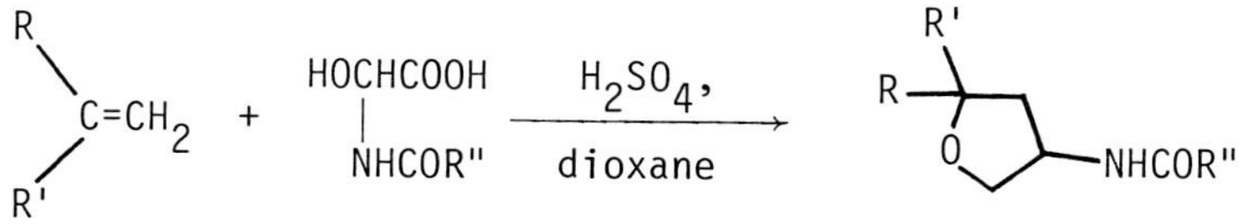

$RR'C=CH_2$ + MeOCHCOOMe(NHCOR'') → NSA, benzene, Δ → $RR'C=CH-CH(NHCOR'')-COOMe$

↓ BF$_3$

[6-membered ring: R, R', COOMe, N=C-Ph, O]

$RR'C=CH_2$ + HOCHCOOH(NHCOR'') → H$_2$SO$_4$, dioxane → [tetrahydrofuran ring with R, R', NHCOR'']

R,R' = H, Me, Ph, OMe, OAc
R'' = Ph, PhCH$_2$O

The above products are converted by standard methods to the corresponding amino acids.

USEFUL SYNTHETIC PREPARATIONS

VI.A.6-9   K. Achiwa, Chem. Lett., 777 (1977).

Use of APPM and PPPM as chiral biphosphine ligands in the asymmetric synthesis of (R)- and (S)-N-benzyloxycarbonyl-alanine.

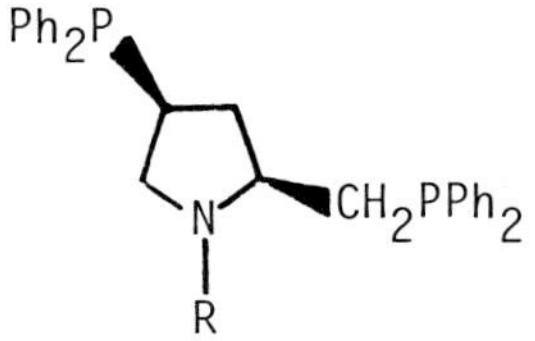

APPM, R = CH_3CO-

PPPM, R = $(CH_3)_3CCO-$

Yields are 85-95%, with 59% ee for (R) and 21% ee for (S).

VI.A.7-1   D. J. Burton and J. L. Hahnfeld, J. Org. Chem., 42, 828 (1977).

$CFCl_3$ + n-BuLi + olefin $\xrightarrow[\text{THF, hexane}]{-116°}$ [cyclopropane with F, Cl]

0-49%

$CFBr_3$ + n-BuLi + olefin $\xrightarrow[\text{THF, hexane}]{-116°}$ [cyclopropane with F, Br]

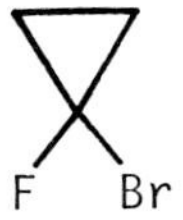

0-73%

VI.A.7-2   M. Fedorynski, Synthesis, 783 (1977).

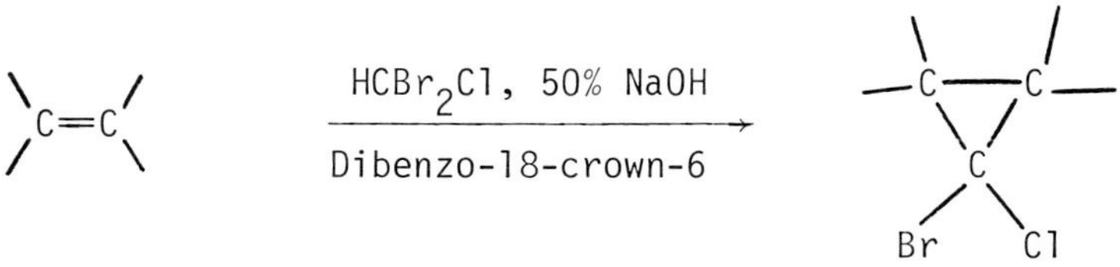

$$\text{>C=C<} \xrightarrow[\text{Dibenzo-18-crown-6}]{\text{HCBr}_2\text{Cl, 50\% NaOH}} \begin{array}{c} \text{-C---C-} \\ \text{\\C/} \\ \text{Br}\quad\text{Cl} \end{array}$$

∼50-70%

VI.A.7-3   R. A. Abramovitch, V. Alexanian, and J. Roy, J. Chem. Soc., Perkin I, 1928 (1977).

$$\text{PhSO}_2\text{CHN}_2 \xrightarrow{\text{heat or } h\nu} \text{PhSO}_2\text{CH:}$$

VI.A.7-4   L. T. Scott and M. A. Minton, J. Org. Chem., 42, 3757 (1977).

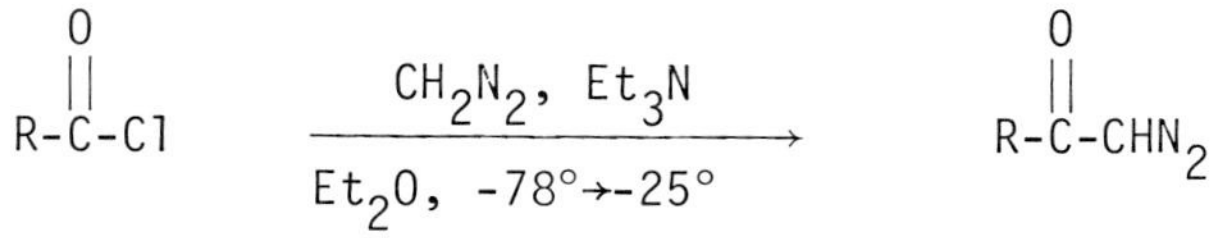

$$\underset{\text{R = alkyl, Bz}}{\text{R-C(=O)-Cl}} \xrightarrow[\text{Et}_2\text{O, } -78° \to -25°]{\text{CH}_2\text{N}_2,\ \text{Et}_3\text{N}} \underset{86\text{-}96\%}{\text{R-C(=O)-CHN}_2}$$

Requires only one equivalent of CH_2N_2.

USEFUL SYNTHETIC PREPARATIONS

VI.A.8-1 L. Duhamel and J.-M. Poirier, J. Am. Chem. Soc., 99, 8356 (1977).

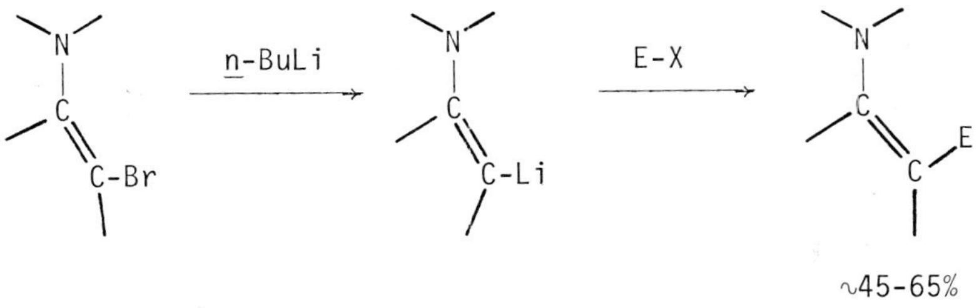

E-X = H_2O, MeI, EtI, BuI, I_2, MeCHOH-CHO

VI.A.8-2 F. M. Laskovics and E. M. Schulman, J. Am. Chem. Soc., 99, 6672 (1977).

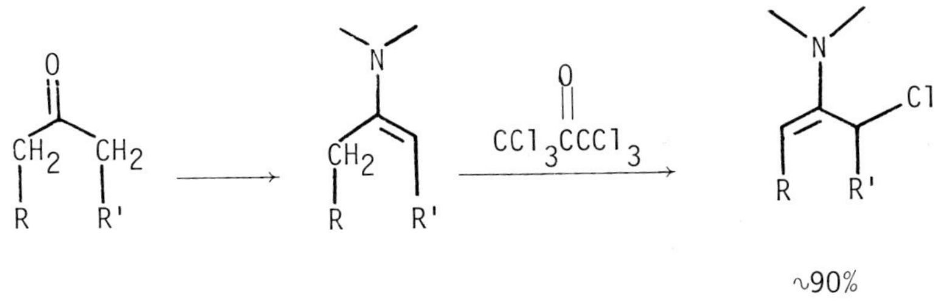

VI.A.8-3 H. G. Viehe, B. LeClef, and A. Elgavi, Angew. Chem. Int. Ed., 16, 182 (1977).

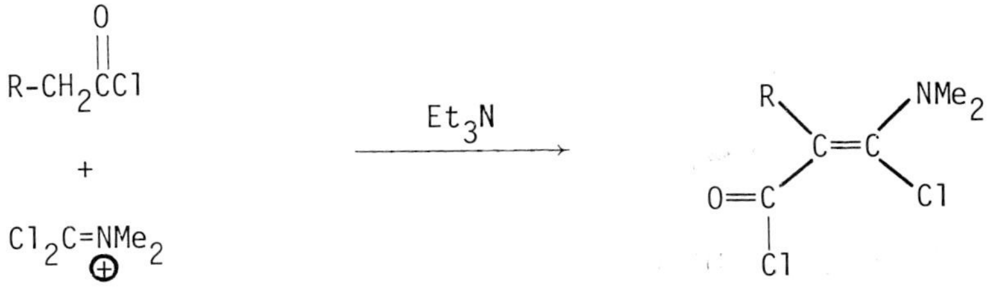

R = Me, Et, Cl, Ph

VI.A.8-4   H. L. Yale and E. R. Spitzmiller, J. Het. Chem., 14, 1419 (1977).

[Reaction scheme: 2-aminopyridine (R = Me, Br, Cl) + H$_2$N-C(R')=CHCOOEt → 2-(substituted amino)pyridine with =CHCOOEt group; R' = Me, Ph; 11-78%]

R = Me, Br, Cl
R'= Me, Ph

11-78%

VI.A.8-5   M. Pulst, S. Steingrüber, and E. Kleinpeter, Z. Chem., 17, 93 (1977).

[Reaction scheme: Ph,Ph-C(SH)=C(CHO) + HNRR' →(−H$_2$O) Ph,Ph-C(S)=C(CH-NRR')]

R,R' = Me, Et, Bz, -(CH$_2$)$_4$-, etc.   25-89%

VI.A.8-6   B. de Jeso and J.-C. Pommier, J. Chem. Soc., Chem. Commun., 565 (1977).

   Partial methanolysis of organotin or -magnesium salts of imines leads to secondary enamines which are stable in aprotic media.

VI.A.8-7   J. V. Greenhill, Chem. Soc. Rev., 6, 277 (1977).

   Review: "Enaminones"

USEFUL SYNTHETIC PREPARATIONS

VI.A.9-1   S. Holand and R. Epsztein, Synthesis, 706 (1977).

$$\underset{\underset{HO\ \ OH}{|\ \ \ |}}{\overset{\overset{R'}{|}}{R-C-CHR''}} \quad \xrightarrow[\text{glyme}]{\text{TsCl, NaOH}} \quad \overset{R'}{\underset{O}{R-C\!\!-\!\!-\!\!-\!\!CHR''}}$$

55-85%

R,R" = H, aryl, alkenyl, alkynyl
R' = H, Ph

VI.A.9-2   G. A. Kraus and M. J. Taschner, Tetrahedron Lett., 4575 (1977).

$$\underset{R}{\overset{O}{\|}}\underset{R'}{\|} \quad \xrightarrow{\text{LiCH}_2\text{COOEt}} \quad \underset{\underset{OH}{|}}{\overset{R}{\underset{R'}{|}}}\!\!\!\!\!\!\overset{O}{\underset{}{\|}}\!\!\!\!\!\!\text{OEt} \quad \xrightarrow[\text{2. I}_2,\ -78°]{\text{1. 2 LDA}} \quad \underset{R'}{\overset{R}{\rtimes}}\!\!\overset{O}{\underset{H}{\rtimes}}\!\!\text{COOEt}$$

R = Ph, n-alkyl, olefinic
R'= H, Me

31-48%

VI.A.9-3   Y. Tamura et al., Synthesis, 693 (1977).

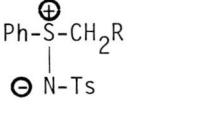

1. NaH, THF
2. PhCHO

45-80%

∼90% trans

VI.A.9-4   P. Sundararaman and W. Herz, J. Org. Chem., 42, 813 (1977).

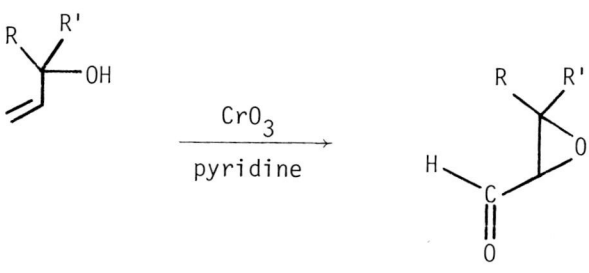

R,R' = alkyl, cyclic

50-81%

VI.A.9-5   I. G. Tishchenko, V. N. Sytin, and I. F. Revinskii, J. Org. Chem. (USSR), 13, 1062 (1977).

$$Hx-\overset{O}{\underset{\|}{C}}-CH=CHNMe_2 \quad \xrightarrow{\begin{array}{c}1.\ RMgX\\ 2.\ H_2O_2,\ ^{\ominus}OH\end{array}} \quad Hx-\overset{O}{\underset{\|}{C}}-\underset{O}{\triangle}-R$$

R = 1°,2°,3° alkyl

∼80%

VI.A.10-1   A. R. Katritzky et al., J. Chem. Soc., Chem. Commun., 701 (1977).

$$RCH_2NH_2 \quad \xrightarrow{\begin{array}{c}1.\ \text{[2,4,6-triphenylpyrylium BF}_4\text{]}\\ 2.\ R'COONa\end{array}} \quad R-O-\overset{O}{\underset{\|}{C}}-R'$$

R = alkyl, Bz, picolyl
R'= Me, Ph

∼60-70%

USEFUL SYNTHETIC PREPARATIONS

VI.A.10-2   M. Yamashita et al., Chem. Lett., 1355 (1977).

$$RO^{\ominus} + R'X \xrightarrow{Fe(CO)_5} R'-\overset{O}{\underset{\|}{C}}-OR$$

R = 1°,2° alkyl, Bz   20-80%
R' = 1° alkyl, CH_2COOEt
X = I, Br

VI.A.10-3   M. K. Sahni et al., Synth. Commun., 7, 57 (1977).

$$R-OH \xrightarrow{R'-\overset{O}{\underset{\|}{C}}-O-\text{(salicylic acid/formaldehyde polymer)}} R-O-\overset{O}{\underset{\|}{C}}-R'$$

R = Me, Et, Bu   R' = Me, Ph   ∿80%

VI.A.10-4   S.-S. Wang et al., J. Org. Chem., 42, 1286 (1977).

$$R-COOH \xrightarrow[\text{2. BzBr}]{\text{1. } Cs_2CO_3} R-COOBz$$

(an amino acid)

∿70-90%

VI.A.10-5   A. R. Banks, R. F. Fibiger, and T. Jones, J. Org. Chem., 42, 3965 (1977).

$$CH_2=C(CH_3)COCl + ROH \xrightarrow{\text{3 Å molecular sieves}} CH_2=C(CH_3)COOR$$

R = cyclopentyl, 3°alkyl, Ph   ∿80-90%

VI.A.10-6 J. C. Chottard, E. Mulliez, and D. Mansuy, J. Am. Chem. Soc., 99, 3531 (1977).

$$R-CH_2CH_2CH_2OH \xrightarrow{\underline{trans}-[PtCl_2(C_2H_4)AcIm]} R-CH_2CH_2CH_2OAc$$

AcIm = (N-acetylimidazole)

R = Ph, 2,4-dimethyl-6-pyridyl

~70%

VI.A.10-7 O. Achmatowicz, Jr. and G. Grynkiewicz, Tetrahedron Lett., 3179 (1977).

$$R-OH + \underset{R'}{\text{(Ar-COOH)}} \xrightarrow[\text{benzene}]{PPh_3, \; HOOC-\overset{O}{\underset{\|}{C}}-COOH} \underset{R'}{\text{(Ar-C(=O)-OR)}}$$

R = benzyl, menthyl, octyl
R' = H, NO_2

~50-80%
(inversion on R)

VI.A.10-8 S. Inaba and I. Ojima, Tetrahedron Lett., 2009 (1977).

$$2RR'C=C\begin{matrix}OMe\\OSiMe_3\end{matrix} \xrightarrow[2. \; H_2O]{1. \; TiCl_4, \; CH_2Cl_2} \begin{matrix}RR'C-COOMe\\|\\RR'C-COOMe\end{matrix}$$

73-80%

R,R' = H, Me, Ph, $-(CH_2)_5-$

VI.A.10-9   W. L. Mock and M. E. Hartman, J. Org. Chem., 42, 459 (1977).

$$\underset{R}{\overset{O}{\underset{\|}{C}}}\underset{R'}{} + N_2CHCOOEt \xrightarrow{Et_3O^{\oplus} BF_4^{\ominus}} \underset{R}{\overset{O}{\underset{\|}{C}}}\underset{CHR'}{\overset{COOEt}{|}}$$

A study of the synthetic scope of this reaction.

VI.A.10-10   S. C. Auteri, D. W. Cameron, and C. B. Drake, Aust. J. Chem., 30, 2479 and 2487 (1977).

[cyclic diol with OH, OH] $\xrightarrow[R'SO_3H]{RCOOH}$ [cyclic product with O-C(=O)-R and OSO$_2$R']

R = long-chain alkyl
R' = Ph, tolyl, mesityl, etc.

Widely varying yields.

VI.A.10-11   J. E. Herz et al., Synth. Commun., 7, 383 (1977).

[bicyclic structure with OH and C≡CH] $\xrightarrow[\text{2. RCOCl}]{\text{1. TlOEt}}$ [bicyclic structure with OCR(=O) and C≡CH]

R = 1-adamantyl,
    CMe$_2$n-Bu

70-80%

VI.A.10-12   J. H. Clark and J. M. Miller, Tetrahedron Lett., 599 (1977).

$$R\text{-COOH} \xrightarrow[\text{KF, DMF}]{\text{Br-CH}_2\text{-C(O)-Ph}} R\text{-C(O)-O-CH}_2\text{C(O)Ph}$$

R = Me, Et, t-Bu, Ph, etc.   >90%

VI.A.10-13   S. L. Beaucage and K. K. Ogilvie, Tetrahedron Lett., 1691 (1977).

MTrO—[furanose with Thymine, OSiMe$_2$-t-Bu] $\xrightarrow[\text{pyridine or THF}]{\text{R-C(O)-O-C(O)-R}}$ MTrO—[furanose with Th, O-C(O)-R]

MTr = p-monomethoxyltrityl

>95%, R = Me, Ph, t-Bu

60%, R = CH$_3$(CH$_2$)$_{12}$-

VI.A.11-1   M. Yamashita and Y. Takegami, Synthesis, 803 (1977).

$$R\text{-OH} + R'\text{-X} \xrightarrow{\text{Ni(acac)}_2} R\text{-O-R}'$$

R,R' = Ph, Bz, n-alkyl   65-90%

X = Cl, Br

USEFUL SYNTHETIC PREPARATIONS

VI.A.11-2   M. J. Diem, D. F. Burow, and J. L. Fry, *J. Org. Chem.*, **42**, 1801 (1977).

$$R-OH + R'_3O^{\oplus} BF_4^{\ominus} \longrightarrow ROR'$$

R = 1°,2° alkyl   ~70%

R' = Me, Et

VI.A.11-3   G. Gelbard and S. Colonna, *Synthesis*, 113 (1977).

$$\text{Resin } (Ar-O^{\ominus}) + R-X \longrightarrow Ar-O-R$$

>90%

Ar = Ph, naphthyl, etc.

R = Me, allyl, Bu, i-Pr

VI.A.11-4   G. H. Posner and D. Z. Rogers, *J. Am. Chem. Soc.*, **99**, 8208 and 8214 (1977).

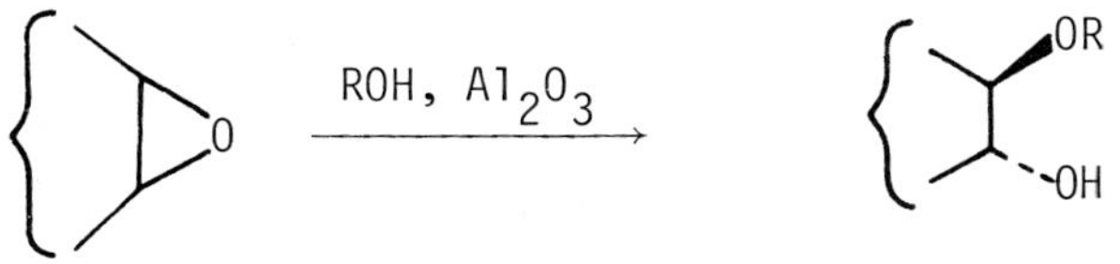

$$\text{epoxide} \xrightarrow{ROH, Al_2O_3} \text{OR, OH product}$$

R = Me, Bz, allyl   47-98%

Full paper, many examples.

VI.A.11-5   M. Cavazza and R. Cabrino, Synthesis, 298 (1977).

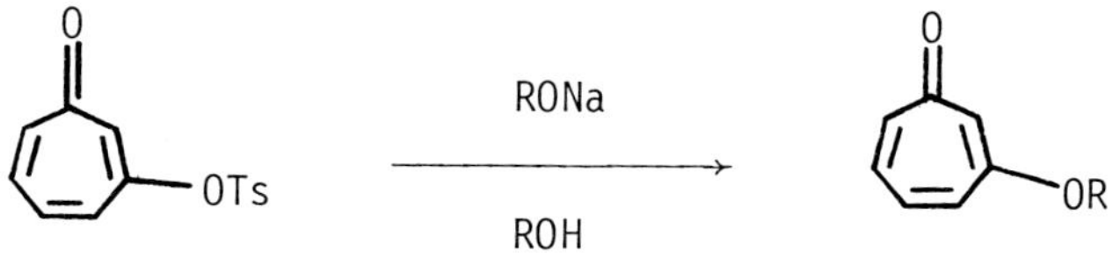

R = Et, Ph   30-80%

VI.A.11-6   F. J. Williams et al., J. Org. Chem., 42, 3425 (1977).

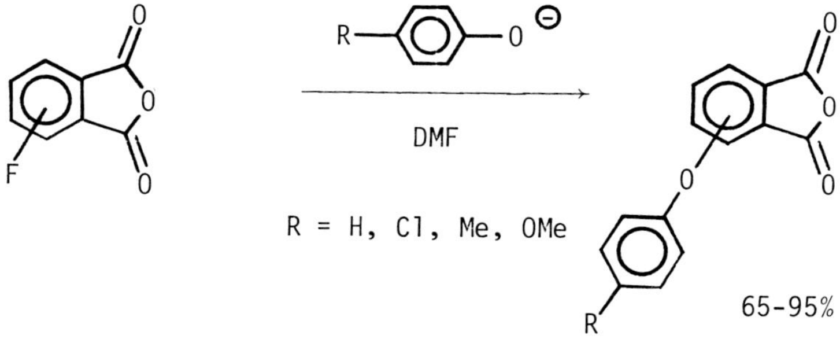

R = H, Cl, Me, OMe

65-95%

VI.A.11-7   J. Emert et al., J. Org. Chem., 42, 2012 (1977).

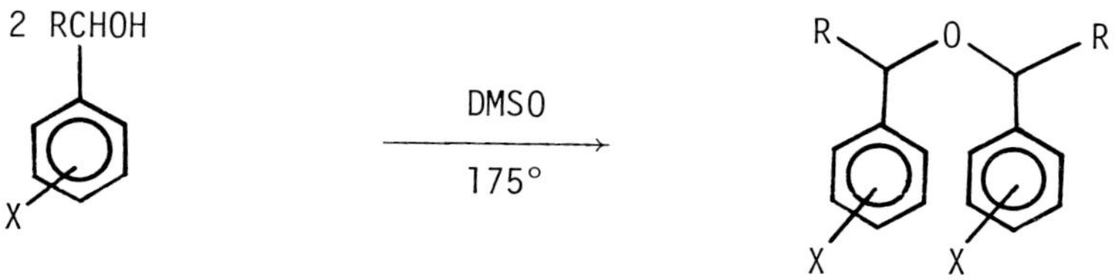

2 RCHOH

DMSO
175°

R = H, Me
X = H, F

∼90%

VI.A.11-8   K. C. Nicolaou and Z. Lysenko, Tetrahedron Lett., 1257 (1977).

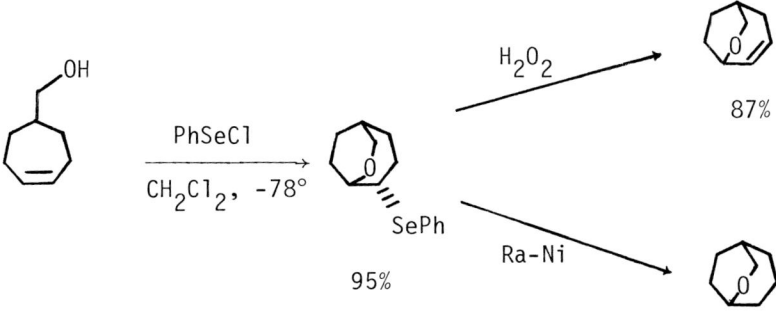

A general technique for ether formation.

VI.A.12-1   G. A. Olah et al., Synthesis, 677 (1977)

$$R_4Si + R'-\overset{O}{\underset{\|}{C}}-Cl \xrightarrow[CH_2Cl_2]{AlCl_3} R-\overset{O}{\underset{\|}{C}}-R'$$

R = Et, n-Bu   41-82%

R' = Et, i-Pr, c-Hx

VI.A.12-2   E. Keinan and Y. Mazur, J. Am. Chem. Soc., 99, 3861 (1977).

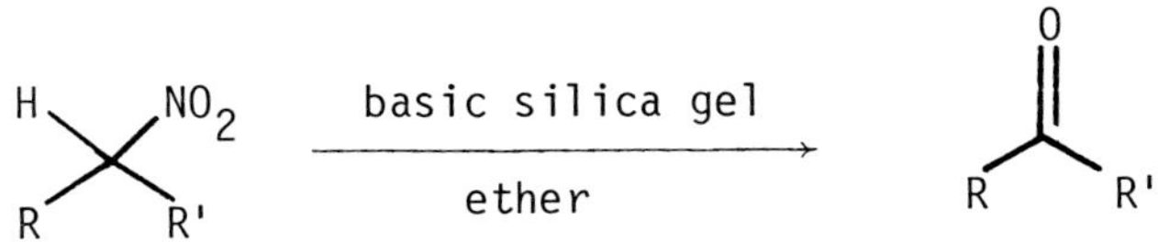

$$\underset{RR'}{\overset{HNO_2}{\diagdown\!\!\!\diagup}} \xrightarrow[\text{ether}]{\text{basic silica gel}} \underset{RR'}{\overset{O}{\|}}$$

R,R' = H, n-alkyl, cyclic   ~80-90%

VI.A.12-3   O. Possel and A. M. van Leusen, Tetrahedron Lett., 4229 (1977).

$$TosCH_2N=C \xrightarrow[\substack{1.\ RX \\ 2.\ R'X}]{\substack{NaH \\ Me_2SO/Et_2O}} Tos-\underset{\underset{R'}{|}}{\overset{\overset{R}{|}}{C}}-N=C \xrightarrow{H_3O^{\oplus}} \underset{R'}{\overset{R}{\diagdown}}C=O$$

R,R' = 1°,2° alkyl, Bz, -(CH$_2$)$_6$-   ∼40-80%
X = Cl, Br, I

VI.A.12-4   P. A. Bartlett et al., Tetrahedron Lett., 331 (1977).

$$\underset{R\ \ \ \ R'}{\overset{H\ \ \ \ NO_2}{\diagdown C \diagup}} \xrightarrow[\substack{2.\ t\text{-BuOOH}, \\ VO(acac)_2, \\ benzene}]{1.\ \underline{t}\text{-BuOK}} \underset{R\ \ \ \ R'}{\overset{O}{\diagdown C \diagup}}$$

R,R' = H, alkyl   ∼50-80%

VI.A.12-5   T. Shono et al., Tetrahedron Lett., 3625 (1977).

$$CH_3OCH_2COOMe \xrightarrow{2\ RMgX} CH_3OCH_2\underset{\underset{R}{|}}{\overset{\overset{R}{|}}{C}}-OH \xrightarrow{\substack{anodic \\ oxidation}} \underset{R}{\overset{R}{\diagdown}}C=O$$

R	Hydroxy ether, %	Ketone, %
n-Pr	85	85
i-Pr	70	88
n-Bu	82	80
i-Bu	83	80
cyclohexyl	48	80

VI.A.12-6   B. M. Trost and Y. Tamaru, *J. Am. Chem. Soc.*, **99**, 3101 (1977).

$$\text{R}_2\text{C}(\text{H})\text{COOH} \xrightarrow[\text{2. MeSSMe}]{\text{1. 2 LDA}} \text{R}_2\text{C}(\text{SMe})\text{COOH} \xrightarrow{\text{NCS, EtOH}} \text{R}_2\text{C=O}$$

∼50-60% overall

VI.A.12-7   T. Mukaiyama, Y. Echigo, and M. Shiono, *Chem. Lett.*, 179 (1977).

Ph-C(H)(OH)-C(R)(OH)-R' $\xrightarrow{\text{2-chloro-N-methylpyridinium } FSO_3^{\ominus}}$ Ph-C(OH)(R)-C(=O)-R'

R = Ph, Bz

R' = H, Me, Et, Ph, Bz

54-88%

VI.A.12-8   D. van Leusen and A. M. van Leusen, *Tetrahedron Lett.*, 4233 (1977).

$$\text{TosCHRN=C} \xrightarrow[\text{2. R'COCl}]{\text{1. n-BuLi}} \underset{\underset{\text{Tos}}{|}}{\overset{\overset{\text{N=C}}{|}}{\text{R-C}}}\text{—COR'} \xrightarrow{H_3O^{\oplus}} \text{R-C(=O)-C(=O)-R'}$$

R = Ph, p-OMePh

R' = Ph, Me, Et, Bz

∼50-70%

VI.A.12-9   J. B. Hendrickson, K. W. Bair, and P. M. Keehn,
J. Org. Chem., 42, 2935 (1977).

$$R\text{-}CH_2\text{-}SO_2CF_3 \xrightarrow[\text{2. R'CH}_2X]{\text{1. NaH}} R\text{-}\underset{\underset{\displaystyle SO_2CF_3}{|}}{\overset{\overset{\displaystyle CH_2R'}{|}}{CH}} \xrightarrow[\text{2. TosN}_3]{\text{1. 2 NaH}} \underset{R}{\overset{R'}{\diagdown}}C=C\underset{N_3}{\diagdown}$$

R = Ph, alkyl

R' = alkyl

$$\underset{R}{\overset{R'}{\diagdown}}CH\text{-}C(=O)\text{-}R \xleftarrow[\text{2. } H_3O^{\oplus}]{\text{1. P(OEt)}_3 \text{ cyclohexane}}$$

∼80%

VI.A.12-10   M. Asaoka, N. Sugimura, and H. Takei, Chem. Lett.,
171 (1977).

[butenolide with R^1] $\xrightarrow[R^2C(OEt)_3]{Ac_2O}$ [3-(alkoxyalkylidene)butenolide with R^1, R^2, OEt]

R^1 = Ph, p-CH$_3$OPh
R^2 = Me, Et

$\downarrow$ HOAc, H$_2$O

$R^1\text{-}C(=O)\text{-}CH_2\text{-}CH_2\text{-}C(=O)\text{-}R^2$

51-83%

USEFUL SYNTHETIC PREPARATIONS

VI.A.12-11   T. E. Cole and R. Petit, <u>Tetrahedron Lett.</u>, 781 (1977).

$$R\text{-}\underset{\underset{O}{\|}}{C}\text{-}Cl \xrightarrow{3\ NMe_4 \cdot HFe(CO)_4} R\text{-}CHO$$

R = 1°, 2°, 3° alkyl, Ar, cycloalkyl, heterocyclic   75-100%

VI.A.12-12   A. Ohta <u>et al.</u>, Synthesis, 792 (1977).

$$R\text{-}CH_2\text{-}NO_2 \xrightarrow[CH_3OH]{CrCl_2} R\text{-}CHO$$

R = 1° alkyl   32-66%

VI.A.12-13   L. A. Paquette, W. D. Klobucar, and R. A. Snow, <u>Synth. Commun.</u>, <u>6</u>, 575 (1976).

$$R\text{-}CH_2X \xrightarrow[\text{2. NCS}]{\text{1. NaSPH}} R\text{-}\underset{Cl}{\overset{SPh}{CH}} \xrightarrow{Na_2CO_3} R\text{-}CHO$$

R = aryl, cyclohexyl
X = Cl, Br   ~80%

$$\underset{R'}{\overset{R}{>}}C{=}CH_2 \xrightarrow[\text{2. HgCl}_2,\ CdCO_3]{\text{1. PhSH, AIBN, h}\nu} \underset{R'}{\overset{R}{>}}CH\text{-}CHO$$

R, R' = H, alkyl, cyclic   24-43%

VI.A.12-14   W. Ertel and K. Friedrich, *Chem. Ber.*, __110__, 86 (1977).

$$Ar-H + \underset{Cl}{\overset{H}{>}}C=C\underset{Y}{\overset{COOEt}{<}} \xrightarrow[\text{2. }H_2O]{\text{1. }AlCl_3} Ar-CHO$$

~40-60%

Y = COOEt, CN

Ar = subst. Ph, naphthyl, etc.

VI.A.12-15   P. Magnus *et al.*, *J. Am. Chem. Soc.*, __99__, 4536 (1977).

$$\underset{R\quad R'}{\overset{O}{\|}}C \xrightarrow{\underset{ClCHSiMe_3}{\overset{Li}{|}}} \underset{R\quad\quad H}{\overset{O}{\triangle}}\substack{R'\quad\quad SiMe_3} \xrightarrow[MeOH]{BF_3 \cdot Et_2O} \underset{R\quad R'}{\overset{H\quad CHO}{>C<}}$$

R,R' = H, Ph, cycloalkyl   ~70-90%   ~80-90%

VI.A.12-16   G. Doleschall, *Tetrahedron Lett.*, 381 (1977).

$$R-CH_2-COOH \longrightarrow \longrightarrow RCH_2-\underset{Ph}{\overset{Ph}{\underset{N-N}{N\bigoplus}}}SMe \longrightarrow R-\underset{OAc}{\overset{|}{CH}}-\underset{Ph}{\overset{Ph}{\underset{N-N}{N\bigoplus}}}SMe$$

$$\downarrow$$

R = alkyl, MeOOC-(CH$_2$)$_3$-, PhCO(CH$_2$)$_2$-

R-CHO

31-56% overall

VI.A.12-17   P. C. Traas, H. J. Takken, and H. Boelens, Tetrahedron Lett., 2027 (1977).

$$\underset{R}{\overset{R'}{>}}\!\!=\!\!O \xrightarrow[\text{DMF}]{POCl_3} \underset{R}{\overset{R'}{>}}\!\!=\!\!\underset{Cl}{\overset{CHO}{<}} \xrightarrow{H_2,\ Pd/C} \underset{R}{\overset{R'}{>}}\!\!=\!\!\underset{H}{\overset{CHO}{<}}$$

R,R' = cyclic, Bz, cyclopropyl, n-alkyl   60-95%

VI.A.12-18   S. Murai et al., Angew. Chem. Int. Ed., 16, 789 (1977).

$$\text{oxetane-(CH}_2)_n \xrightarrow[CO,\ Co_2(CO)_8]{HSiEt_2Me} \begin{array}{c} CH_2\text{-OSiEt}_2Me \\ (CH_2)_n \\ CH_2\text{-CHO} \end{array}$$

n = 0, 1, 2   ~40-50%

VI.A.12-19   J. M. Reuter and R. G. Salomon, J. Org. Chem., 42, 3360 (1977).

$$\xrightarrow[\Delta]{(PPh_3)_3RuCl_2}$$

84%

Several additional examples, yields 56-92%.

VI.A.12-20 V. Reutrakul and W. Kanghae, Tetrahedron Lett., 1377 (1977).

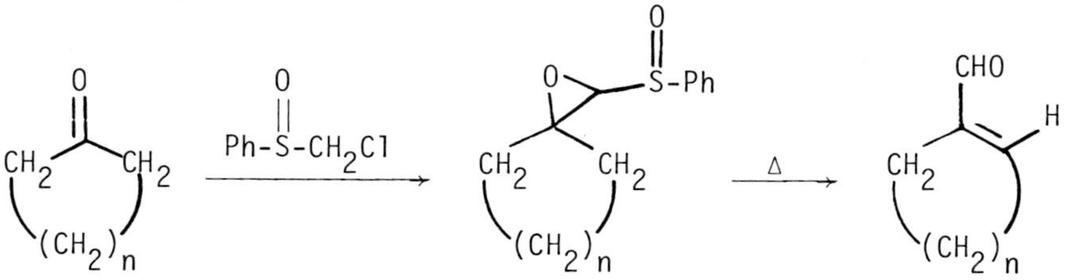

[cyclic ketone with $(CH_2)_n$ ring] + Ph-S-CH$_2$Cl → [spiro epoxide with S(O)-Ph] →Δ→ [cycloalkylidene acetaldehyde, CHO]

n = 1-5   40-90%

VI.A.12-21 V. E. Gunn and J.-P. Anselme, J. Org. Chem., 42, 754 (1977).

$$ArC(O)-CH_2Br + Et_2NOH \longrightarrow ArC(O)-C(O)-H + Et_2NH \cdot HBr$$

Ar = Ph, β-naphthyl,   55-78%
    p-BrPh, m-MeOPh

VI.A.12-22 S. Ncube, A. Pelter, and K. Smith, Tetrahedron Lett., 255 (1977).

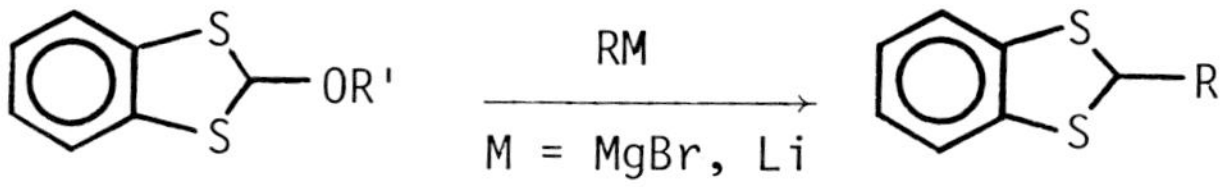

[benzo-1,3-dithiole-2-OR'] $\xrightarrow{RM}$ [benzo-1,3-dithiole-2-R]

M = MgBr, Li
R = n-Pr, i-Pr,
    chx, Bu, Ph

USEFUL SYNTHETIC PREPARATIONS

VI.A.13-1   F. Campagna, A. Carotti, and G. Casini, Tetrahedron Lett., 1813 (1977).

$$\text{R-}\overset{\overset{\text{O}}{\|}}{\text{C}}\text{-NH}_2 \xrightarrow[\text{pyridine, rt}]{(\text{CF}_3\text{CO})_2\text{O}} \text{R-C} \equiv \text{N}$$

R = alkyl, aryl   ~70-90%

VI.A.13-2   C. R. Harrison, P. Hodge, and W. J. Rogers, Synthesis, 41 (1977).

$$\text{R-}\overset{\overset{\text{O}}{\|}}{\text{C}}\text{NH}_2$$

or

R-CH=NOH

$$\xrightarrow{\text{CCl}_4 \cdot \text{PPh}_2-\text{\textcircled{P}}} \text{R-C} \equiv \text{N}$$

76-100%

R = 1° alkyl, steroidal, substituted Ph

VI.A.13-3   M. D. Dowle, J. Chem. Soc., Chem. Commun., 220 (1977).

$$\text{R-}\overset{\overset{\text{S}}{\|}}{\text{C}}\text{NH}_2 \xrightarrow{\text{EtOOC-N=N-COOEt}} \text{R-C} \equiv \text{N}$$

~60-70%

VI.A.13-4 O. H. Oldenziel, D. van Leusen, and A. M. van Leusen, J. Org. Chem., **42**, 3114 (1977).

$$\underset{R\quad R'}{R-\overset{O}{\underset{\|}{C}}-R'} \xrightarrow[\underline{t}\text{-BuOK}]{C=N-CH_2Ts} \underset{R\quad R'}{\overset{H\quad C\equiv N}{\diagdown\diagup}}$$

R,R' = cyclic, Ph, 1°,2°,3° alkyl   ~70-80%

VI.A.13-5 D. M. Orere and C. B. Reese, J. Chem. Soc., Chem. Commun., 280 (1977).

$$\underset{R\quad R'}{\overset{O}{\underset{\|}{C}}} \xrightarrow[\text{2. KCN, MeOH, reflux}]{\text{1. }(\underline{i}\text{-Pr})_3PhSO_2NHNH_2(TPSH)} \underset{R\quad R'}{\overset{H\quad CN}{\diagdown C \diagup}}$$

R,R' = H, alkyl, cyclic   47-74%

VI.A.13-6 C. Bernhart and C.-G. Wermuth, Synthesis, 338 (1977).

$$X\text{-}C_6H_4\text{-}\underset{\underset{OH}{\overset{\|}{N}}}{\overset{H}{\underset{\|}{C}}} \xrightarrow[CH_3CN]{H_3C-C\equiv C-NEt_2} X\text{-}C_6H_4\text{-}CN$$

X = H, Cl, NO$_2$, OMe   69-80%

USEFUL SYNTHETIC PREPARATIONS

VI.A.13-7   T. Kametani et al., Synthesis, 245 (1977).

$$\text{R-C}_6\text{H}_4\text{-CH}_2\text{NH}_2 \xrightarrow[\text{pyridine}]{\text{CuCl, O}_2} \text{R-C}_6\text{H}_4\text{-C}\equiv\text{N}$$

R = H, OMe, OBz, etc.   35-41%

VI.A.13-8   J. K. Rasmussen, Chem. Lett., 1295 (1977).

$$\text{PhCH=NOH} \xrightarrow[\text{Me}_3\text{SiCl}]{\text{KCN, 18-crown-6}} \text{Ph-CN}$$

∼40-50%

VI.A.13-9   K. Yamamura and S.-I. Murahashi, Tetrahedron Lett., 4429 (1977).

$$\underset{R'}{\overset{R}{>}}C=C\underset{Br}{\overset{H}{<}} \xrightarrow[\text{crown ether, benzene}]{\text{KCN, Pd(PPh)}_4} \underset{R'}{\overset{R}{>}}C=C\underset{CN}{\overset{H}{<}}$$

R,R' = H, Ph, Me   ∼80-90%

VI.A.13-10 A. Loupy, K. Sogadji, and J. S. Penne, Synthesis, 126 (1977).

R-CHO $\xrightarrow[\text{THF or DMF}]{X_2\overset{O}{\overset{\|}{P}}-CH_2CN}$ R-CH=CHCN

R = aryl, i-Pr   X = OEt or Ph   ~90%

VI.A.13-11 C. L. Liotta, A. M. Dabdoub, and L. H. Zalkow, Tetrahedron Lett., 1117 (1977).

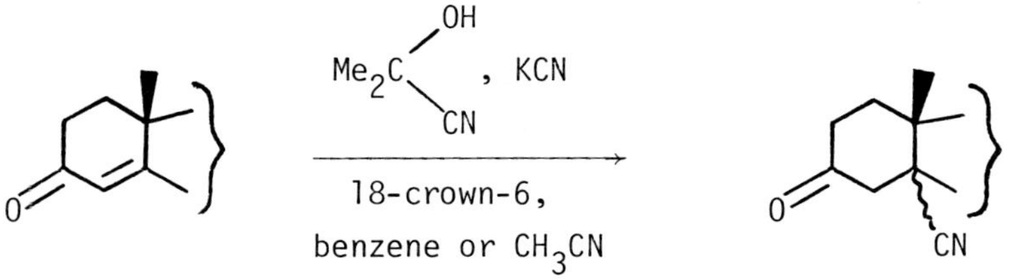

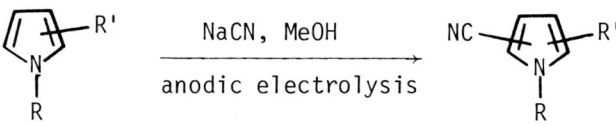

83-86%

VI.A.13-12 K. Yoshida, J. Am. Chem. Soc., 99, 6111 (1977).

[pyrrole]-R' $\xrightarrow[\text{anodic electrolysis}]{NaCN, MeOH}$ NC-[pyrrole]-R'

R = Me, Ph
R' = one or more Me, benzo (indoles)

(CN may be on one of the methyl groups)

Widely varying yields.

VI.A.13-13   F. E. Ziegler and P. A. Wender, *J. Org. Chem.*, **42**, 2001 (1977).

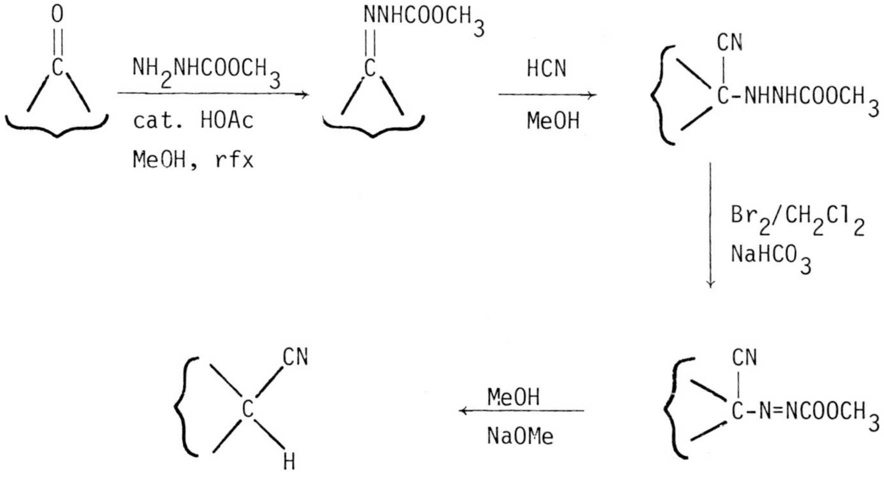

VI.A.13-14   F. Pochat, *Tetrahedron Lett.*, 3813 (1977).

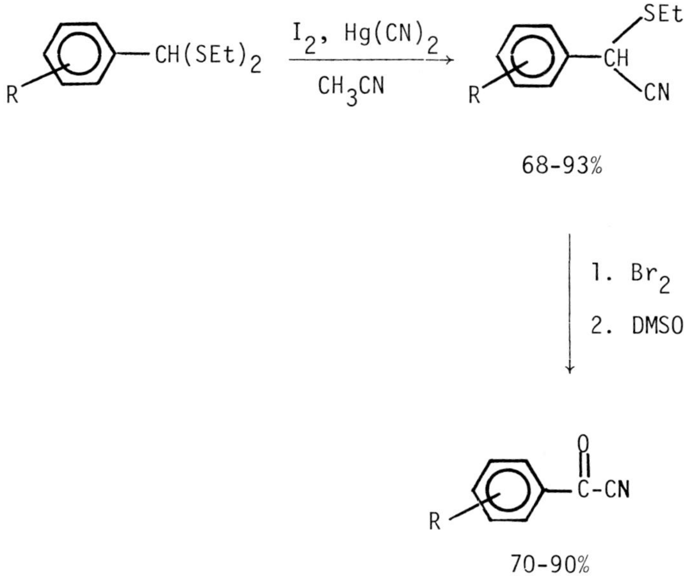

VI.A.13-15   A. N. Volkov and A. N. Nikol'skaya, Russ. Chem. Rev., 46, 374 (1977).

   Review: "α-Cyanoacetylenes"

VI.A.13-16   G. Skorna and I. Ugi, Angew. Chem. Int. Ed., 16, 259 (1977).

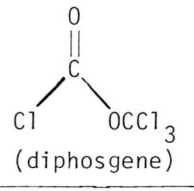

R-NH-CHO $\xrightarrow{\text{(diphosgene)}}$ R-NC

R = Me, t-Bu, c-Hx, Bz, Ar   47-98%

VI.A.13-17   G. Domschke, R. Beckert, and R. Mayer, Synthesis, 275 (1977).

R-N=S=O $\xrightarrow[\text{Benzene}]{\text{CHCl}_3, \text{KOH}}$ R-N≡C

R = aryl, cyclohexyl   >75%

VI.A.13-18   Y. Echigo, Y. Watanabe, and T. Mukaiyama, <u>Chem. Lett.</u>, 697 (1977).

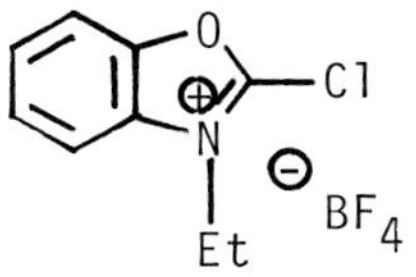

RNHCHO $\xrightarrow[\text{Et}_3\text{N}]{\text{[benzoxazolium Cl, Et, BF}_4^-\text{]}}$ RNC

R = 1°,2° alkyl, aryl, Bz   69-81%

VI.A.14-1   G. Gelbard and S. Colonna, <u>Synthesis</u>, 113 (1977).

Resin-(NO_2^-)
+ $\longrightarrow$ R-NO_2
R-X

R-X	Yield, %
MeCHCOOEt | Br	95
Me_2C-COOEt | Br	60
<u>n</u>-BuBr	47
<u>i</u>-PrBr	29
Bz-Br	87

VI.A.14-2   I. Sh. Shvarts et al., Bull. Akad. USSR Chem., 25, 1589 (1977).

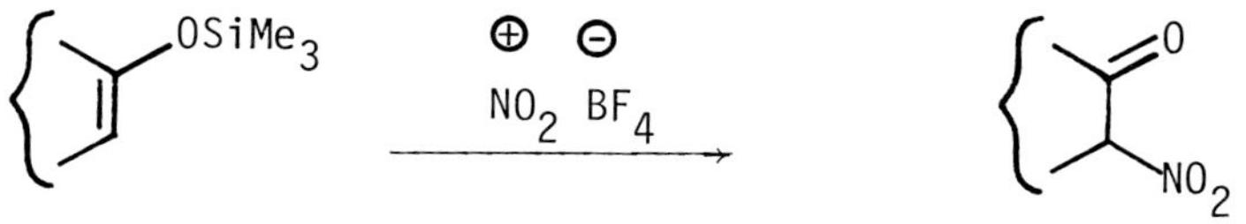

[cyclopentene-OSiMe$_3$] + NO$_2^{\oplus}$ BF$_4^{\ominus}$ → [cyclopentanone-NO$_2$]

37-90%

VI.A.14-3   V. Jager and J. Gunther, Angew. Chem. Int. Ed., 16, 246 (1977).

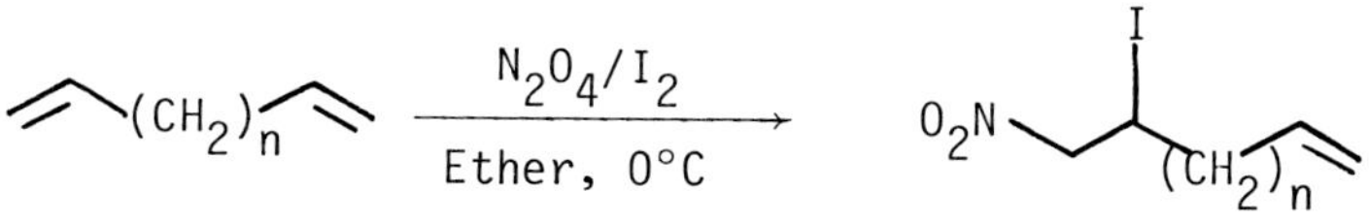

CH$_2$=CH-(CH$_2$)$_n$-CH=CH$_2$ $\xrightarrow{N_2O_4/I_2, \text{ Ether, } 0°C}$ O$_2$N-CH(I)-(CH$_2$)$_n$-CH=CH$_2$

n = 2,3,4   >80%

VI.A.14-4   H. Suzuki, Synthesis, 217 (1977).

Review: "Side-Reactions in Aromatic Nitration: Some Synthetic Aspects"

VI.A.15-1   D. Cech and A. Holy, Coll. Czech. Chem. Commun., 42, 2246 (1977).

Review: "Preparation of 2-Pyrimidone Nucleosides from Uracil Nucleosides"

VI.A.15-2 R. L. Shone, Tetrahedron Lett., 993 (1977).

[Reaction scheme: BzO-protected furanose oxocarbenium with Ph group reacts with 1. bis(trimethylsilyloxy)pyrimidine, 2. CH₃OH, Amberlite IR-45 to give the nucleoside product.]

50-75%
for
different bases

VI.A.15-3 F. Ramirez et al., J. Org. Chem., 42, 3144 (1977).

Synthesis of deoxyribooligonucleotides by means of cyclic enediol pyrophosphates formed using di(1,2-dimethylethenylene)pyrophosphate.

VI.A.15-4 R. Youssefyeh et al., Tetrahedron Lett., 435 (1977).

"Synthetic Routes to 4'-Hydroxymethylnucleosides" based on the aldol coupling of furanose-5'-aldehydes with formaldehyde accompanied by Cannizzaro reduction by excess formaldehyde.

VI.A.15-5 J.-L. Fourrey, G. Henry, and P. Jouin, J. Am. Chem. Soc., 99, 6753 (1977).

Use of a pyrimidine S-nucleoside photorearrangement in a multistep synthesis of pseudonucleosides.

VI.A.15-6   F. Ramirez et al., Synthesis 451 (1977).

Application of Cyclic Enediol Pyrophosphates to the Synthesis of Deoxyribonucleotides.

VI.A.15-7   S.Y.-K. Tan and B. F.-Reid, Can. J. Chem., 55, 3996 (1977).

"Synthetic Routes to 2',3'-Cyclopropanated and 2',3'-Unsaturated Nucleosides"

VI.A.15-8   R. R. Schmidt, J. Karg, and W. Guillard, Chem. Ber., 110, 2433 (1977).

"Synthesis of Pyrazole Nucleosides via Ribosylhydrazines"

VI.A.15-9   T. K. Bradshaw and D. W. Hutchinson, Chem. Soc. Rev., 6, 43 (1977).

Review: "5-Substituted Pyrimidine Nucleosides and Nucleotides"

VI.A.15-10   V. Amarnath and A. D. Broom, Chem. Rev., 77, 183 (1977).

Review: "Chemical Synthesis of Oligonucleotides"

USEFUL SYNTHETIC PREPARATIONS

VI.A.16-1   M. P. Doyle, S. B. Williams, and C. C. McOsker, Synthesis, 717 (1977).

$$\left[R-CH_2-CH(R')-O \right]_3 B \xrightarrow[100°]{BF_3 \cdot Et_2O} R-CH=CH-R'$$

R's = alkyl, cyclic   63-89%

VI.A.16-2   G. W. Francis and J. F. Berg, Acta Chem. Scand. B, 31, 721 (1977).

$$-\underset{HO}{\overset{|}{C}}-\underset{H}{\overset{|}{C}}- \xrightarrow[180°]{CuSO_4} \;\; \underset{}{>}C=C\underset{}{<}$$

∿45-90%

for secondary alcohols

VI.A.16-3   G. H. Posner, G. M. Gurria, and K. A. Babiak, J. Org. Chem., 42, 3173 (1977).

$$X-\text{cyclopentyl} \xrightarrow[24 \text{ hr}, Et_2O]{Al_2O_3,\; 25°} \text{cyclopentene}$$

X = OMs, OTs   ∿60-80%

Substrates are cycloalkanes, steroids, and open-chain alkanes

VI.A.16-4 R. Tanikaga et al., Synthesis, 299 (1977).

$$R-H + H_2C=CH-\overset{\overset{O}{\|}}{S}-Ph \xrightarrow[\substack{2.\ H_2O \\ 3.\ \Delta}]{1.\ \text{Base, THF}} R-CH=CH_2$$

R = SPh, OBz, pyrimidine, ~70% overall
alkyl containing cyano, keto,
nitro, and carboethoxy groups

VI.A.16-5 M. Kim and J. D. White, J. Am. Chem. Soc., 99, 1172 (1977).

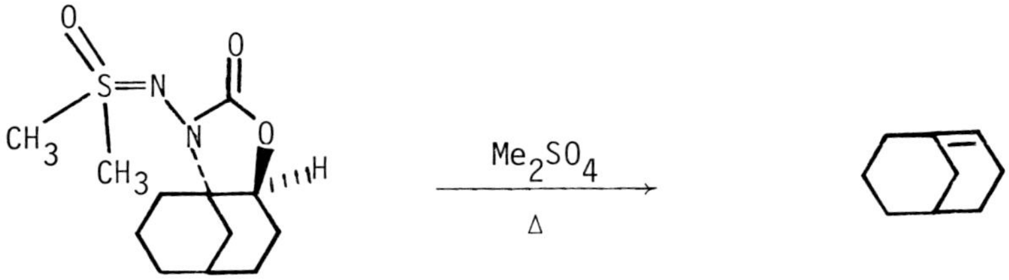

$\xrightarrow[\Delta]{Me_2SO_4}$

53% isolated

VI.A.16-6 T.-H. Chan, Accounts Chem. Res., 10, 442 (1977).

Review: "Alkene Synthesis via β-Functionalized Organosilicon Compounds"

USEFUL SYNTHETIC PREPARATIONS

VI.A.16-7 E. W. Colvin and B. J. Hamill, J. Chem. Soc., Perkin I, 869 (1977).

$$\underset{R \quad R'}{R_2C=O} \xrightarrow[\text{BuLi}]{\text{Me}_3\text{SiCHN}_2} R-C\equiv C-R'$$

R,R' = Ph, Me, H, benzoyl   22-80%

VI.A.16-8 H. Hommes and L. Brandsma, Rec. Trav. Chim. Pays-Bas, 96, 160 (1977).

$$C_4H_9C\equiv C-(CH_2)_nCH_2R \xrightarrow[\text{1,3-diaminopropane}]{\text{NaNH}_2} HC\equiv C-(CH_2)_{n+4}CH_2R$$

R = H, OH   ~80-90%
n = 0-5

VI.A.16-9 J. L. Coke, H. J. Williams, and S. Natarajan, J. Org. Chem., 42, 2380 (1977).

$$\text{dimedone-dione} \xrightarrow[\text{X = Cl, Br}]{\text{PX}_3} \text{vinyl halide enone} \xrightarrow[\substack{\text{1. CH}_3\text{Li} \\ \text{2. }\Delta}]{}$$

$$CH_3\overset{O}{\underset{\|}{C}}CH_2C(CH_3)_2CH_2C\equiv CH$$

70%

VI.A.16-10   M. M. Midland, J. Org. Chem., 42, 2650 (1977).

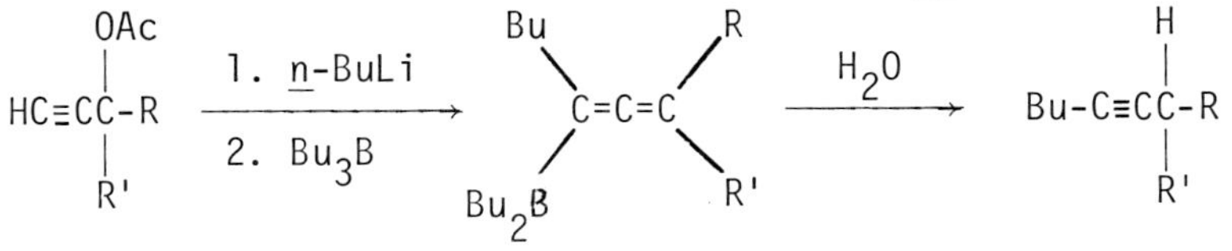

HC≡CC(OAc)(R')-R → 1. n-BuLi 2. Bu₃B → Bu\C=C=C/R / Bu₂B \R' → H₂O → Bu-C≡CC(H)(R')-R

R,R' = H, Ph, alkyl, cycloalkyl   60-91%

VI.A.16-11   J. S. Kiely, P. Boudjouk, and L. L. Nelson, J. Org. Chem., 42, 2626 (1977).

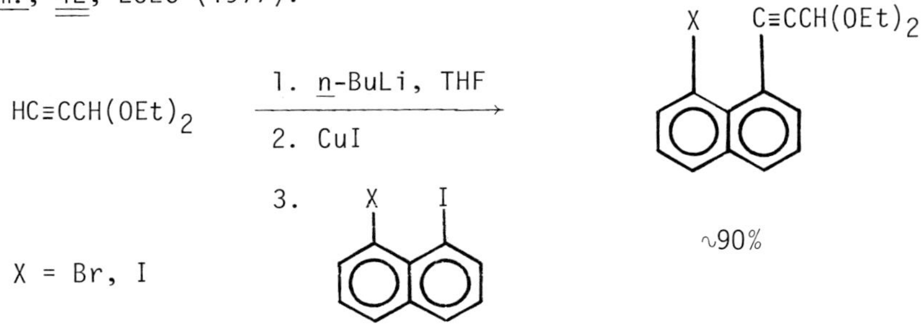

HC≡CCH(OEt)₂ → 1. n-BuLi, THF   2. CuI   3. 1,X-I-naphthalene (X = Br, I) → 1-X-8-[C≡CCH(OEt)₂]-naphthalene

X = Br, I   ~90%

VI.A.17-1   M. Waki and J. Meienhofer, J. Am. Chem. Soc., 99, 6075 (1977).

A study of peptide synthesis using the four-component condensation (Ugi reaction).

$$P^1\text{-COOH} + H_2NR^1 + R^2CHO + CN\text{-}P^2 \longrightarrow \longrightarrow P^1CO\text{-}NHCHR^2CO\text{-}NHP^2$$

USEFUL SYNTHETIC PREPARATIONS

VI.A.17-2   S. Nozaki, A. Kimura, and I. Muramatsu, <u>Chem. Lett.</u>, 1057 (1977).

$$\text{Boc-Ala-OH} + \text{H-Gly-OEt·HCl} \xrightarrow[\substack{\text{CH}_2\text{-CH}_2, \text{H}_2\text{O, HOBt} \\ |\ \ \ | \\ \text{Cl}\ \ \text{Cl}}]{\text{WSCD·HCl, Et}_3\text{N}} \text{Boc-Ala-Gly-OEt} \quad 79\%$$

VI.A.17-3   M. K. Sahni <u>et al.</u>, <u>Indian J. Chem.</u>, <u>15B</u>, 481 (1977).

"Sulphonates of Copoly(ethylene-N-hydroxymaleimide) as Coupling Reagents for Synthesis of Amides and Peptides". The advantages of the polymeric reagent over solution reactions are discussed.

VI.A.17-4   H. Ito, N. Takamatsu, and I. Ichikizaki, <u>Chem. Lett.</u>, 539 (1977).

Use of Bz-N=C=N-Et in peptide synthesis. Formation of N-acylurea and racemization are suppressed relative to the case using DCC.

VI.A.17-5   T. Teramoto, T. Kurosaki, and M. Okawara, <u>Tetrahedron Lett.</u>, 1523 (1977).

Use of 3-hydroxyhydantoin, (hydantoin ring with R substituent and N-OH) as an acyl activating group in synthesis. Yields 87-91% for simple reactions where R=CH_3 or $\underline{i}$-C_4H_9.

VI.A.17-6   C. DiBello et al., Tetrahedron Lett., 1135 (1977).

Use of DCC/HOBt for solid-phase coupling of arginyl peptides. Yields up to 96%.

VI.A.17-7   G. A. Zheltukhina et al., J. Gen. Chem. (USSR), 47, 1112 (1977).

"Modification of the Azide Method for the Condensation of Peptide Fragments on a Polymer". (Addition of N-hydroxysuccinimide and a gradual increase in temperature -10→20° over six days.)

VI.A.17-8   G. Jung et al., Angew. Chem. Int. Ed., 16, 642 (1977).

Use of insoluble polymer-bound reagents and soluble peptide carriers for chain elongation of peptides. No side products are formed in the actual coupling step, and excess reagents do not have to be removed.

VI.A.17-9   M. A. Tilak and J. A. Hoffman, J. Org. Chem., 42, 2098 (1977).

Use of a large excess of protected amino acid azide in peptide synthesis. The excess azide subsequently hydrolyzed and removed, giving "analytically pure peptides in high yields".

USEFUL SYNTHETIC PREPARATIONS

VI.A.17-10   M. Ueki and S. Ikeda, Chem. Lett., 869 (1977).

Use of diphenylphosphinothioyl (Ppt)-amino acids in solid phase peptide synthesis to produce tryptophan-containing peptides.

VI.A.17-11   D. Yamashiro, J. Org. Chem., 42, 523 (1977).

Use of the 4-bromobenzyl group as a side-chain protecting group for aspartic acid and serine and the 4-chlorobenzyl group for threonine in solid-phase peptide synthesis. Stable to 50% CF_3COOH in CH_2Cl_2, and removed by HF for 10 min at 0°.

VI.A.17-12   R. B. Merrifield et al., Tetrahedron Lett., 4001 (1977).

$$\text{Peptide-C(=O)-O-CH(R)-C(=O)-C}_6\text{H}_4\text{-P} \xrightarrow[\text{crown ether}]{\text{KCN}} \text{Peptide-COO}^{\ominus} + \text{N}\equiv\text{C-CH(R)-C(=O)-C}_6\text{H}_4\text{-P}$$

~90%

VI.A.17-13   V. J. Hornby, D. A. Upson, and N. S. Agarwal, J. Org. Chem., 42, 3552 (1977).

Comparative use of benzhydrylamine and chloromethylated resins in solid-phase synthesis of carboxamide terminal peptides.

VI.A.17-14 J. M. Schlatter, R. H. Mazur, and O. Goodmonson, Tetrahedron Lett., 2851 (1977); D. A. Jones, Jr., Tetrahedron Lett., 2853 (1977).

Peptide—(P) $\xrightarrow[\text{DMF}]{H_2, Pd(OAc)_2}$ Peptide-OAc

---

VI.A.17-15 A. E. Vasil'ev et al., J. Gen. Chem. (USSR), 47, 1500 and 1505 (1977).

Synthesis of amino acid derivatives of dextran using:
a. the cyanogen bromide method
b. activated esters of carboxydextran and their aminolysis with salts of amino acids

---

VI.A.17-16 Yu V. Mitin and M. P. Zapevalova, Russ. Chem. Rev., 46, 449 (1977).

Review: "Methods of Synthesis of Peptides"

VI.A.18-1 P. F. Hudrlik, J. Am. Chem. Soc., 99, 1993 (1977).

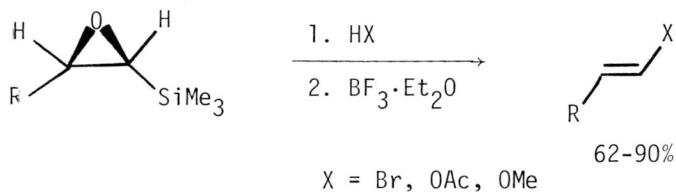

R = n-Pr, n-Hex

X = Br, OAc, OMe

62-90%

Reversed stereochemistry can be obtained in some cases.

VI.A.18-2 J. Cousseau and L. Gouin, *J. Chem. Soc., Perkin I*, 1797 (1977).

$$R-C\equiv C-R' \xrightarrow{Et_3\overset{\oplus}{N}H \;\; \overset{\ominus}{HCl_2}} \underset{Cl}{\overset{R}{>}}C=C\underset{R'}{\overset{H}{<}}$$

R = Ph, CH_2Cl, COOMe
R' = H, Me, $\underline{t}$-Bu, COOMe

∼50-90%

VI.A.18-3 S. Uemura et al., *J. Chem. Soc., Perkin I*, 676 (1977).

$$R-C\equiv C-R' \xrightarrow[MeCN]{CuCl_2-LiCl} \underset{Cl}{\overset{R}{>}}C=C\underset{R'}{\overset{Cl}{<}}$$

R = alkyl, Ph
R' = H, alkyl, Ph

∼40-90%

$$\xrightarrow[MeCN]{CuCl_2-I_2} \underset{Cl}{\overset{R}{>}}C=C\underset{R'}{\overset{I}{<}} \;\; + \;\; \underset{Cl}{\overset{R}{>}}C=C\underset{I}{\overset{R'}{<}}$$

∼70-100%

VI.A.18-4 H. Westmijze, J. Meijer, and P. Vermeer, *Rec. Trav. Chim. Pays-Bas*, **96**, 168 (1977).

$$R'-C\equiv CH \xrightarrow[\text{2. NBS, NCS, or } I_2]{\text{1. [RCuBr]MgX, THF}} \underset{R}{\overset{R'}{>}}C=C\underset{X}{\overset{H}{<}}$$

R' = H, alkyl, Ph
R = alkyl

>90%

X = Cl, Br, I

VI.A.18-5   A. B. Levy, P. Talley, and J. A. Dunford, *Tetrahedron Lett.*, 3545 (1977).

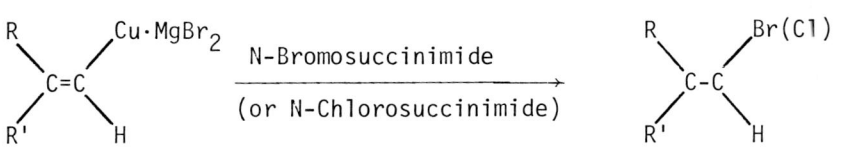

R = n-alkyl
R' = H, Ph, n-alkyl   52-98%

VI.A.18-6   W. G. Salmond, *Tetrahedron Lett.*, 1239 (1977).

$$R\text{-CHO} + (Me_2N)_3P=CCl_2 \longrightarrow RCH=CCl_2$$

R = alkyl, Ph, steroidal   85-94%

VI.A.18-7   P. Johncock, *Synthesis*, 551 (1977).

$$F_2CCl\text{-}CFCl\text{-}R \xrightarrow[EtOH]{Ph_3P} F_2C=CF\text{-}R$$

                                                                    98%

VI.A.18-8   R. B. Miller and G. McGarvey, *Synth. Commun.*, **7**, 475 (1977).

$$\underline{n}\text{-BuCH=CHSiMe}_3 \xrightarrow{\begin{array}{c}1.\ X_2,\ CH_2Cl_2\\ 2.\ \text{Alumina, pentane}\end{array}} \underline{n}\text{-BuCH=CHX}$$

X = Cl, Br   62-75% isolated

USEFUL SYNTHETIC PREPARATIONS

VI.A.18-9 S. Raucher, Tetrahedron Lett., 3909 (1977).

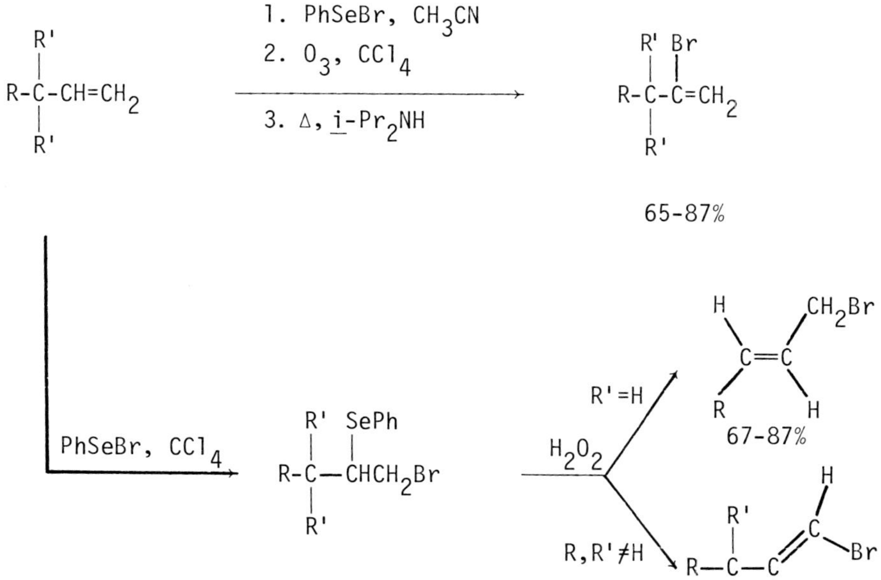

R = H, n-alkyl, Ph, i-Pr

R' = H, Me, Et

VI.A.18-10 C. Earnshaw, C. J. Wallis, and S. Warren, J. Chem. Soc., Chem. Commun., 314 (1977).

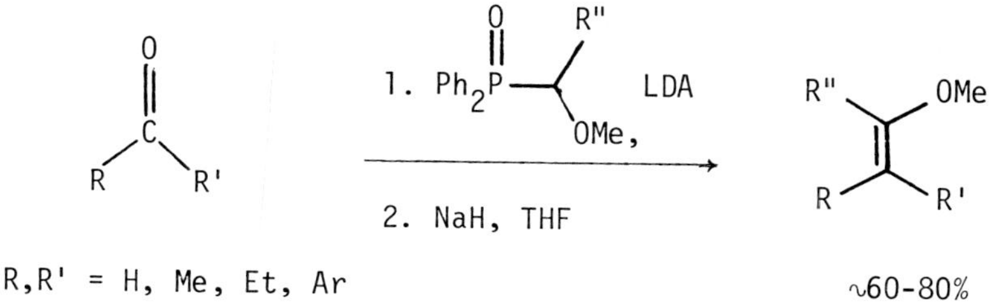

R,R' = H, Me, Et, Ar

R'' = H, Me

VI.A.18-11   G. A. Gareev and Yu. M. Belousov, J. Org. Chem. (USSR), 13, 606 (1977).

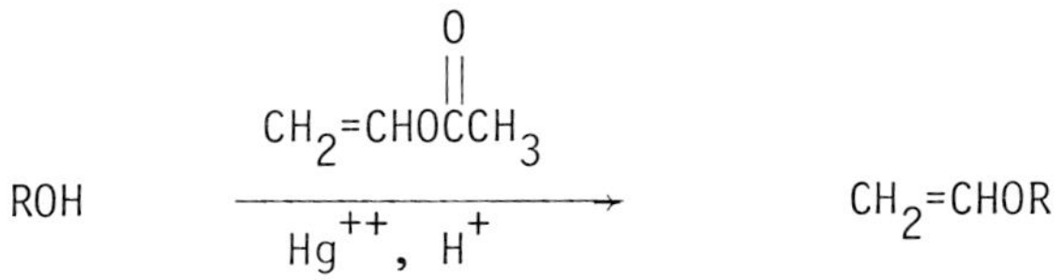

ROH $\xrightarrow[\text{Hg}^{++}, \text{H}^+]{\text{CH}_2=\text{CHOCCH}_3 \atop \|\atop \text{O}}$ CH$_2$=CHOR

R = haloalkyl   "up to 45-50%"

VI.A.18-12   G. M. Rubottom, R. C. Mott, and D. S. Krueger, Synth. Commun., 7, 327 (1977).

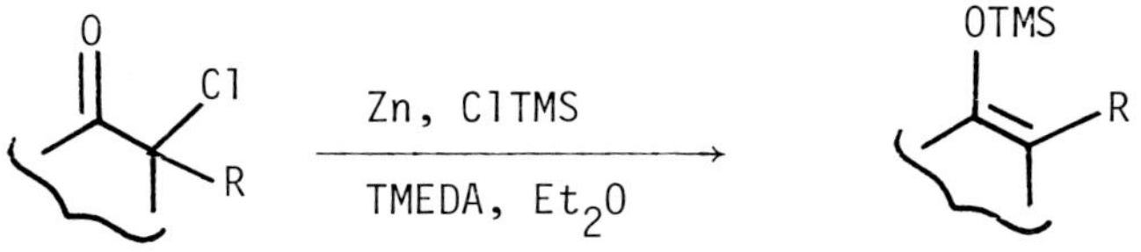

$\xrightarrow[\text{TMEDA, Et}_2\text{O}]{\text{Zn, ClTMS}}$

R = Me, Ph, cycloalkyl   51-85%

VI.A.18-13   H. Sakurai, K. Myoshi, and Y. Nakadaira, Tetrahedron Lett., 2671 (1977).

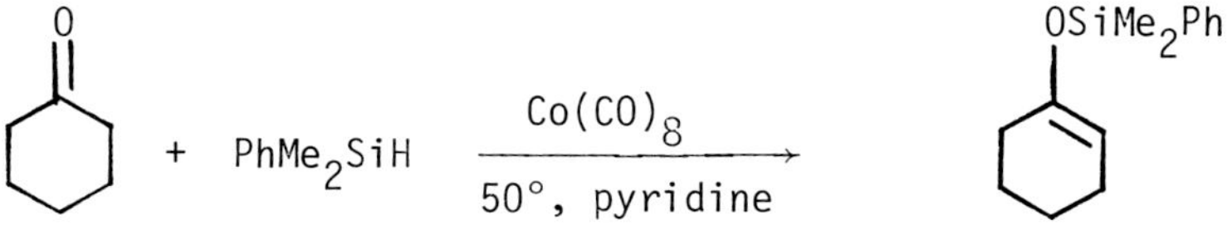

cyclohexanone + PhMe$_2$SiH $\xrightarrow[50°, \text{pyridine}]{\text{Co(CO)}_8}$ 1-(OSiMe$_2$Ph)-cyclohexene

                                                85%

VI.A.18-14/VI.A.19-1   F. Akiyama, Bull. Chem. Soc. Japan, 50, 936 (1977).

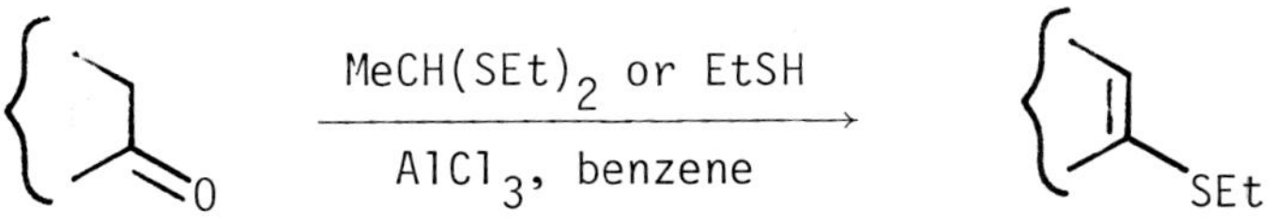

cyclopentanone $\xrightarrow{\text{MeCH(SEt)}_2 \text{ or EtSH}}_{\text{AlCl}_3, \text{ benzene}}$ cyclopentenyl-SEt

42-81%

VI.A.19-2   K.A.M. Walker, Tetrahedron Lett., 4475 (1977).

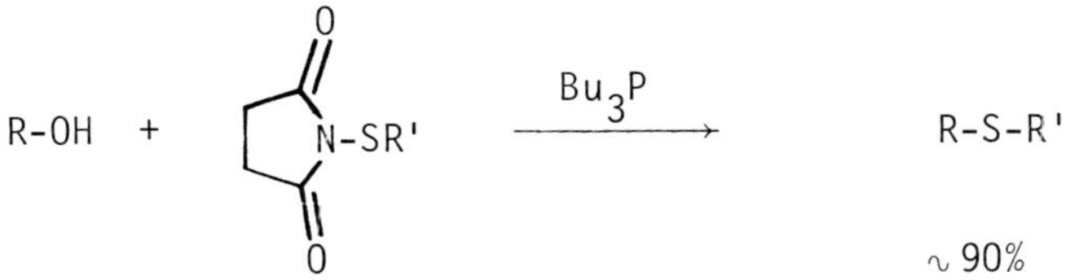

R-OH + (N-thiosuccinimide N-SR') $\xrightarrow{\text{Bu}_3\text{P}}$ R-S-R'

∼90%

R = benzyl, cholesteryl, etc.

R' = Ph, p-ClPh

VI.A.19-3   M. M. Screttas and C. G. Screttas, J. Org. Chem., 42, 1462 (1977).

R-O-R' + PhSH $\xrightarrow[\text{or TFA}]{\text{HClO}_4}$ RSPh

R' = H, Me

R = alkyl, benzyl   Widely varying yields.

VI.A.19-4   E. H. Gold, V. Piotrowski, and B. Z. Weiner, *J. Org. Chem.*, 42, 554 (1977).

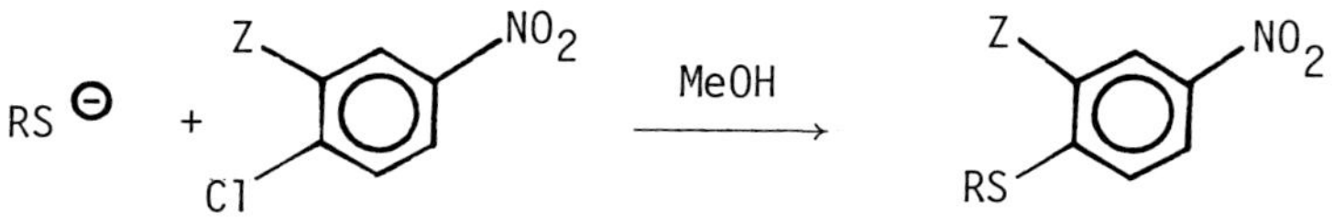

RS$^\ominus$ + [2-Z-4-NO$_2$-chlorobenzene] $\xrightarrow{\text{MeOH}}$ [2-Z-4-NO$_2$-SR-benzene]

~90%

R = 1°, 2° alkyl, Ph   Z = H, CF$_3$

VI.A.19-5   N. M. Kolbina et al., *J. Org. Chem. (USSR)*, 12, 1678 (1977).

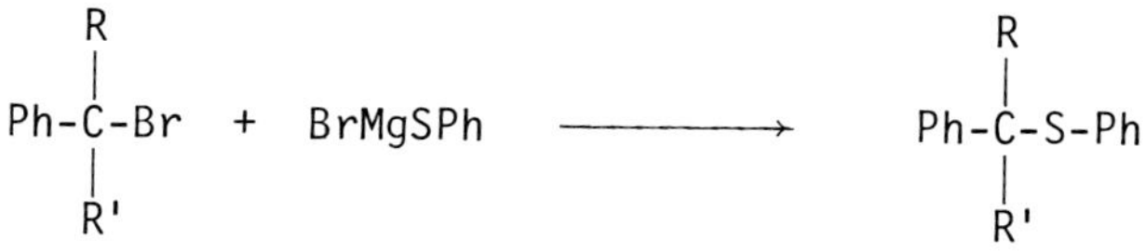

Ph-C(R)(R')-Br + BrMgSPh $\longrightarrow$ Ph-C(R)(R')-S-Ph

R, R' = H, Me, Et, Ph   41-93%

VI.A.19-6   G. H. Posner and M. J. Chapdelaine, *Tetrahedron Lett.*, 3227 (1977).

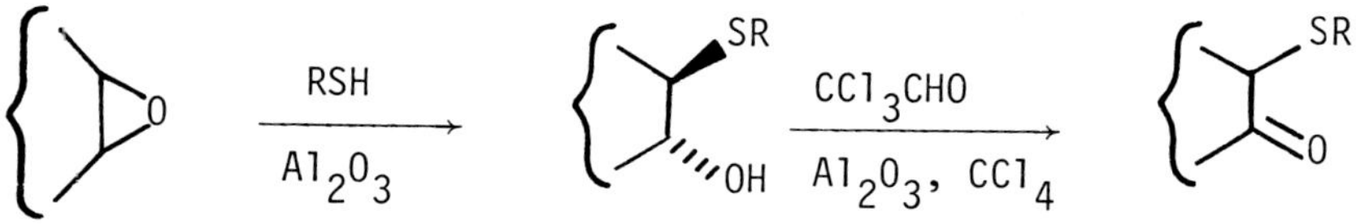

[epoxide] $\xrightarrow[\text{Al}_2\text{O}_3]{\text{RSH}}$ [β-hydroxy sulfide] $\xrightarrow[\text{Al}_2\text{O}_3, \text{CCl}_4]{\text{CCl}_3\text{CHO}}$ [α-SR ketone]

R = Et, Ph   29-75%

USEFUL SYNTHETIC PREPARATIONS

VI.A.19-7   C. T. Goralski and G. A. Burk, J. Org. Chem., 42, 3094 (1977).

$$ArSH + NaOH + BrCH_2Cl \xrightarrow{BzNEt_3^{\oplus} Br^{\ominus}} ArSCH_2Cl$$

Ar = substituted benzenes, heterocycles   ~60-90%

VI.A.19-8   T. Takamoto et al., Synthesis, 884 (1977).

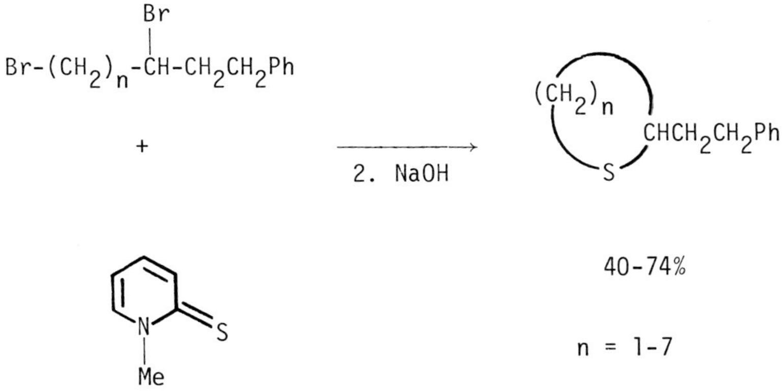

$$Br-(CH_2)_n-\overset{Br}{\underset{|}{CH}}-CH_2CH_2Ph$$

+

[N-methyl-2-thiopyridone]

$\xrightarrow{\text{2. NaOH}}$

[cyclic: $(CH_2)_n$ with S–CHCH$_2$CH$_2$Ph]

40-74%

n = 1-7

VI.A.19-9   K. C. Mattes and O. L. Chapman, J. Org. Chem., 42, 1814 (1977).

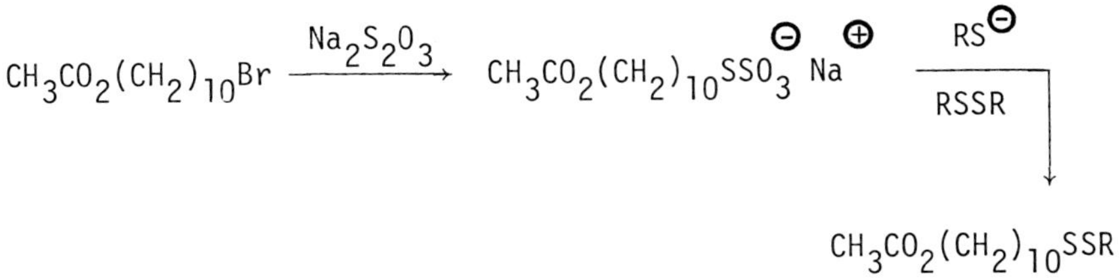

$$CH_3CO_2(CH_2)_{10}Br \xrightarrow{Na_2S_2O_3} CH_3CO_2(CH_2)_{10}SSO_3^{\ominus} Na^{\oplus} \xrightarrow[RSSR]{RS^{\ominus}}$$

$$CH_3CO_2(CH_2)_{10}SSR$$

89%, R = Me
93%, R = Et

VI.A.19-10 P. G. Gassman et al., J. Org. Chem., 42, 3233 and 3236 (1977).

$$\underset{(CH_2)_n}{\overset{O}{\|}}\!\!\!\bigg\rangle\!\!=\!\!\bigg\langle\!\!\underset{R'}{R} \quad \xrightarrow[\text{2. } CH_3SSCH_3]{\text{1. Li, NH}_3(l), \underline{t}\text{-BuOH}} \quad \underset{(CH_2)_n}{\overset{O}{\|}}\!\!\!\bigg\rangle\!\!-\!\!\underset{H}{\overset{SCH_3}{\underset{R'}{|}}}\!\!R$$

R,R' = H, Me, alkyl   ~40-60%
n = 3, 4, 5

$$RCH_2\overset{O}{\overset{\|}{C}}NR'R'' \quad \xrightarrow[\text{2. } CH_3SSCH_3]{\text{1. NaNH}_2,\ NH_3(l)} \quad R\overset{CH_3S}{\underset{|}{C}}H\overset{O}{\overset{\|}{C}}NR'R''$$

R = alkyl   40-60%
R',R''= Me, Ph

VI.A.19-11 I. Degani, R. Fochi, and M. Santi, Synthesis, 873 (1977).

$$R\text{-}X \ + \ K^{\oplus}\ \overset{S}{\underset{S}{\diagup}}\!\!\overset{\ominus}{C}\!\!-\!\!O\text{-}\underline{t}\text{-Bu} \quad \xrightarrow{\text{phase-transfer cat.}} \quad R\text{-}SH$$

R = 1°,2° alkyl, Bz   60-91%
X = Br, Cl

VI.A.19-12 E. Vedejs, D. A. Engler, and M. J. Mullins, J. Org. Chem., 42, 3109 (1977).

$$R\text{-}OH \quad \xrightarrow[\text{pyridine}]{TfOTf} \quad R\text{-}OTf$$

R = allylic, propargylic, $\underline{\alpha}$-carbonyl   ~40-80%

USEFUL SYNTHETIC PREPARATIONS

VI.A.19-13   W. Schroth et al., Z. Chem., **17**, 411 (1977).

$$\text{X-C}_6\text{H}_4\text{-H} \xrightarrow[\text{2. KOH, H}_2\text{O}]{\text{1. Cl-S-C(=O)-OMe}} \text{X-C}_6\text{H}_4\text{-SH}$$

X = alkyl, alkoxyl, Br,   80-98%
    fused aromatic

VI.A.19-14   K. Hojo, H. Yoshino, and T. Mukaiyama, Chem. Lett., 133 (1977).

$$R^*\text{-OH} \xrightarrow[\text{2. CH}_3\text{COSH, Et}_3\text{N}]{\text{1. [2-F-1-Me-pyridinium] OTs}^-,\ \text{Et}_3\text{N}} R^*\text{-S-C(=O)-CH}_3 \xrightarrow{\text{LiAlH}_4} R^*\text{-SH}$$

                                              79-93%   >95%

R = alkyl, benzyl   Proceeds with inversion of
                                  configuration.

VI.A.19-15   K. Hojo, H. Yoshino, and T. Mukaiyama, Chem. Lett., 437 (1977).

$$\text{ROH} + \text{[2-F-1-Me-pyridinium] OTs}^- \xrightarrow[\text{2. NaSC(=S)NMe}_2]{\text{1. Et}_3\text{N, CHCl}_3} \text{RSC(=S)NMe}_2 \xrightarrow{\text{LiAlH}_4} \text{RSH}$$

R = alkyl, benzyl, cholestanol,   ~60-97%
    protected sugar, etc.   (inversion
                                         of configuration)

VI.A.19-16   K. Horiki, Synth. Commun., 7, 251 (1977).

R-COOH $\xrightarrow{\begin{array}{c}1.\ \text{benzotriazole-OH, DCC} \\ 2.\ \text{R'SH, Et}_3\text{N}\end{array}}$ R-C(=O)-SR'

~80-90%

VI.A.19-17   Y. Yokoyama, T. Shioiri, and S. Yamada, Chem. Pharm. Bull., 25, 2423 (1977).

$$\text{RCOOH} + \text{R'SH} \xrightarrow[\text{Et}_3\text{N, DMF}]{\text{DEPC or DPPA}} \text{R-C(=O)-SR'}$$

~50-90%

R,R' = alkyl, aryl, benzyl, etc.

DEPC = Diethyl Phosphorocyanidate; DPPA = Diphenyl Phosphorazidate

---

VI.A.19-18   H.-J. Gais, Angew. Chem. Int. Ed., 16, 244 (1977).

$$\text{R-C(=O)-OH} \xrightarrow{\begin{array}{c}1.\ \text{carbonyl di-imidazole} \\ \text{or carbonyl-1,2,4-triazole} \\ 2.\ \text{R'SH}\end{array}} \text{R-C(=O)-SR'}$$

>80%

R = alkyl, Ph, thioester, etc.

R' = Et, Ph, i-Pr, t-Bu

VI.A.19-19 A. Pelter et al., J. Chem. Soc., Perkin I, 1672 (1977).

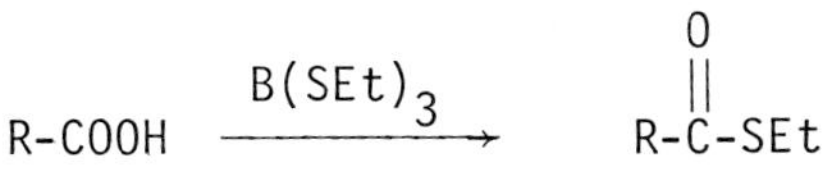

$$R\text{-COOH} \xrightarrow{B(SEt)_3} R\text{-C(=O)-SEt}$$

R = 1°, 2°, 3° alkyl, Ph   ~70-80%

VI.A.19-20 D. H. Lucast and J. Wemple, Tetrahedron Lett., 1103 (1977).

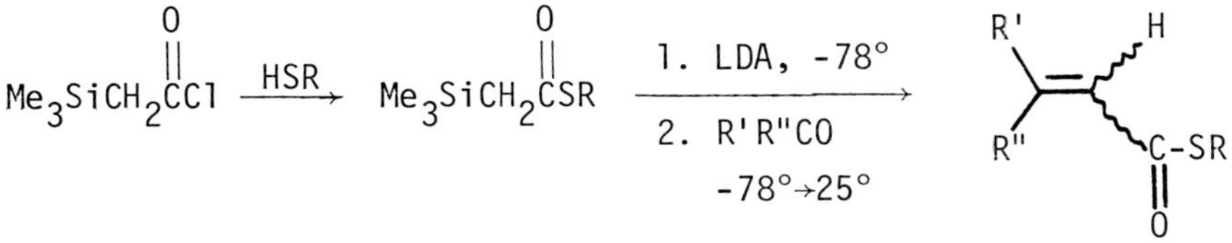

$$Me_3SiCH_2\overset{O}{\overset{\|}{C}}Cl \xrightarrow{HSR} Me_3SiCH_2\overset{O}{\overset{\|}{C}}SR \xrightarrow[\substack{2.\ R'R''CO \\ -78°\to 25°}]{1.\ LDA,\ -78°} \underset{R''}{\overset{R'}{>}}C=C(H)\text{-C(=O)SR}$$

R = 2°, 3° alkyl, Bz

R', R'' = H, Ph, i-Pr, cyclic

49-77%

VI.A.19-21 T. Takamoto et al., J. Org. Chem., 42, 2180 (1977).

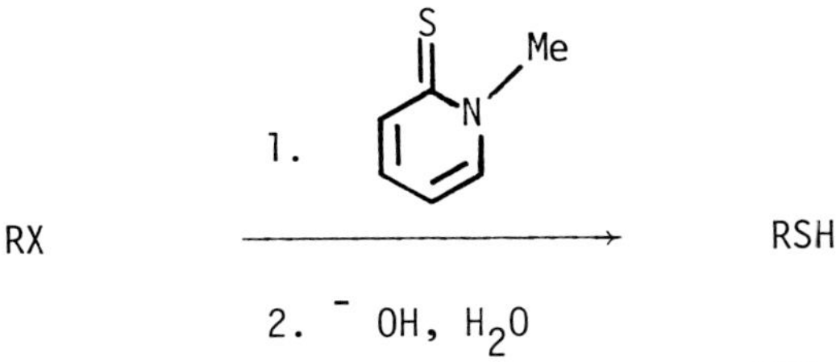

$$RX \xrightarrow[2.\ ^-OH,\ H_2O]{1.\ \text{(N-Me-pyridine-2-thione)}} RSH$$

~60% overall

R = Bz, 1°, 2° alkyl,
    containing ketones,
    ethers, esters

VI.A.19-22 R. D. Howells and J. D. McCown, Chem. Rev., 77, 69 (1977).

Review: "Trifluoromethanesulfonic Acid and Derivatives"

VI.A.19-23 J. B. Hendrickson, D. D. Sternbach, and K. W. Bair, Accounts Chem. Res., 10, 306 (1977).

Review: "Trifyl Activation in Organic Synthesis"

VI.A.19-24 C. S. Rao et al., Indian J. Chem., 14B, 999 (1976).

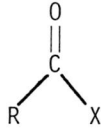 $\xrightarrow{\text{Et}_3\text{N-P}_4\text{S}_{10}}{\text{CH}_3\text{CN or CH}_2\text{Cl}_2}$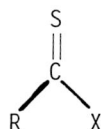

R = Me, subst. Ph   ~50-70%

X = NHR', NR'$_2$, Ph

VI.A.19-25 M. Hojo et al., Synthesis, 789 (1977).

$$\text{Ar-}\overset{\overset{O}{\|}}{\text{S}}\text{-CH}_2\text{Cl} \xrightarrow{\text{RMgBr}} \text{Ar-}\overset{\overset{O}{\|}}{\text{S}}\text{-R}$$

Ar = Ph, p-MePh   R = Et, i-Pr   55-99%

USEFUL SYNTHETIC PREPARATIONS

VI.A.19-26   P. Messinger and H. Greve, Synthesis, 259 (1977).

$$R-CH_2NMe_3^{\oplus} \xrightarrow{HOCH_2SO_2^{\ominus} Na^{\oplus}} R-CH_2-SO_2-CH_2-R$$

R = a wide range of functional groups   19-66%

VI.A.19-27   J. B. Hendrickson and K. W. Bair, J. Org. Chem., 42, 3875 (1977).

$$RLi \xrightarrow[Et_2O, -78°]{(CF_3SO_2)_2O \text{ or } PhN(SO_2CF_3)_2} RSO_2CF_3$$

R = 1°, 2° alkyl   ∿60-90%

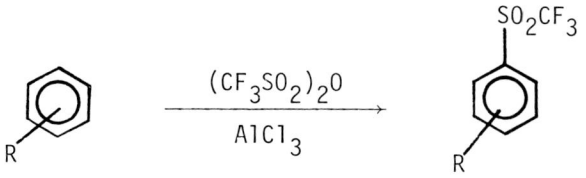

R = alkyl

(Incompatible with R = OCH_3, Cl, or NO_2)   ∿60-70%

VI.A.19-28   P. D. Magnus, Tetrahedron, 33, 2019 (1977).

Review: "Recent Developments in Sulfone Chemistry"

VI.A.19-29   A. van der Gen et al., Tetrahedron Lett., 885 (1977).

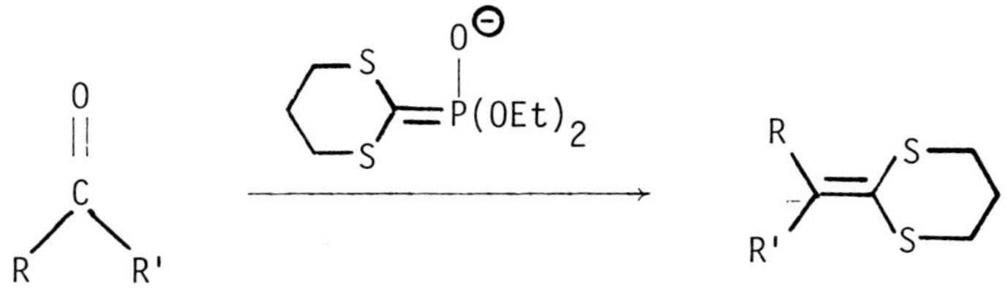

R,R' = H, alkyl, aryl, vinyl, cyclic

~70-90%

VI.A.19-30   D.P.N. Satchell, Chem. Soc. Rev., 6, 345 (1977).

Review: "Metal-ion-promoted Reactions of Organo-sulphur Compounds"

VI.A.19-31   T. L. Gilchrist and C. J. Moody, Chem. Rev., 77, 409 (1977).

Review: "The Chemistry of Sulfilimines"

VI.B.1-1   P. Boontanonda and R. Grigg, J. Chem. Soc., Chem. Commun., 583 (1977).

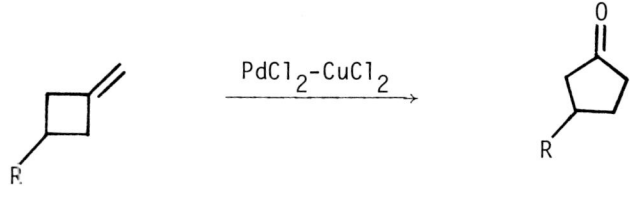

R = H, CN, CH_2NHCOMe

65-82%

VI.B.1-2 B. M. Trost et al., J. Am. Chem. Soc., 99, 3088 (1977).

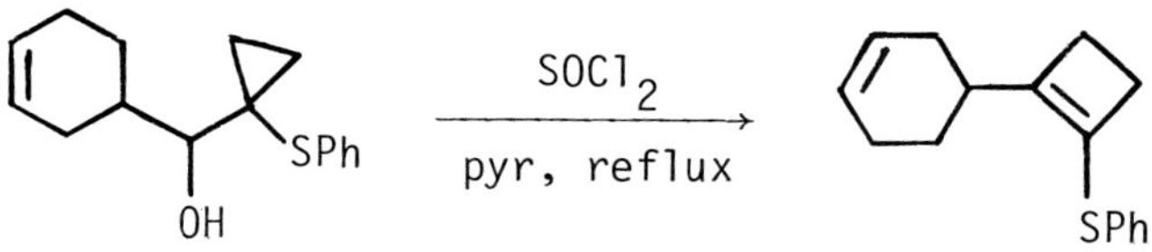

$$\text{cyclohexenyl-CH(OH)-C(SPh)(cyclopropyl)} \xrightarrow[\text{pyr, reflux}]{SOCl_2} \text{cyclohexenyl-cyclobutenyl-SPh}$$

94%

VI.B.1-3 B. M. Trost and P. H. Scudder, J. Am. Chem. Soc., 99, 7601 (1977).

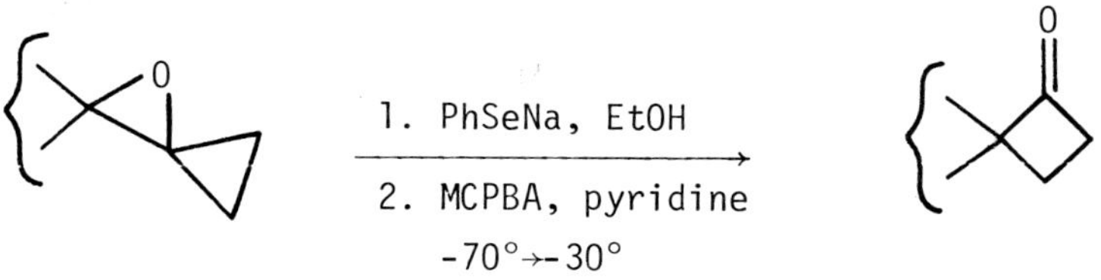

$$\text{spiro oxaspiropentane} \xrightarrow[\substack{2.\ MCPBA,\ pyridine \\ -70° \to -30°}]{1.\ PhSeNa,\ EtOH} \text{spiro cyclobutanone}$$

48-85%

(stereochemistry may be controlled)

VI.B.1-4 H.J.J. Loozen, J. W. de Haan, and H. M. Buck, J. Org. Chem., 42, 418 (1977).

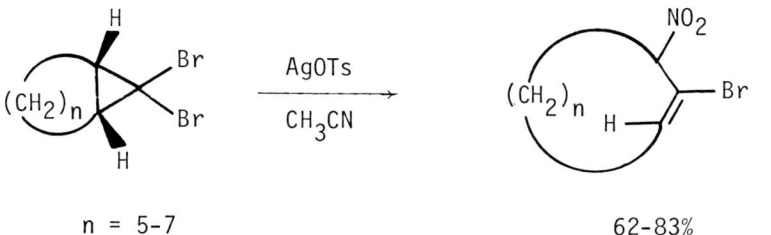

n = 5-7

62-83%

VI.A.1-5   A. E. Greene et al., Tetrahedron Lett., 2365 (1977).

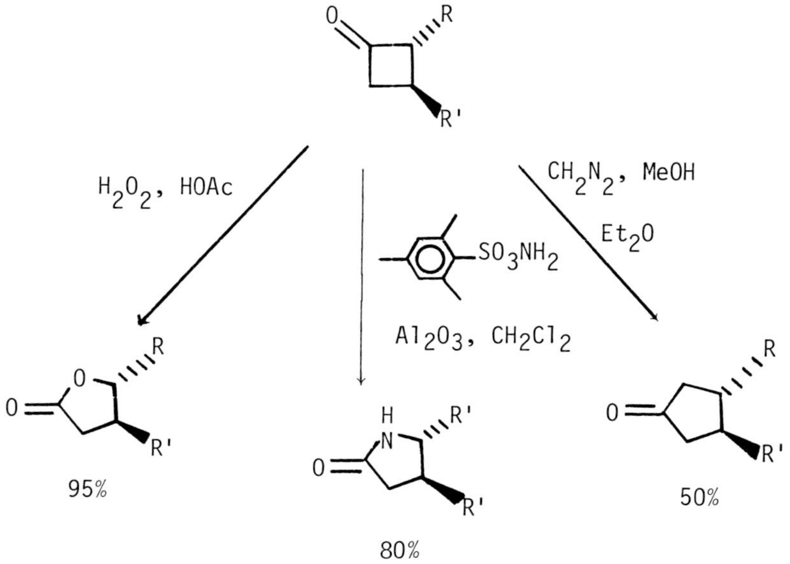

VI.B.1-6   E. C. Taylor and C.-S. Chiang, Tetrahedron Lett., 1827 (1977).

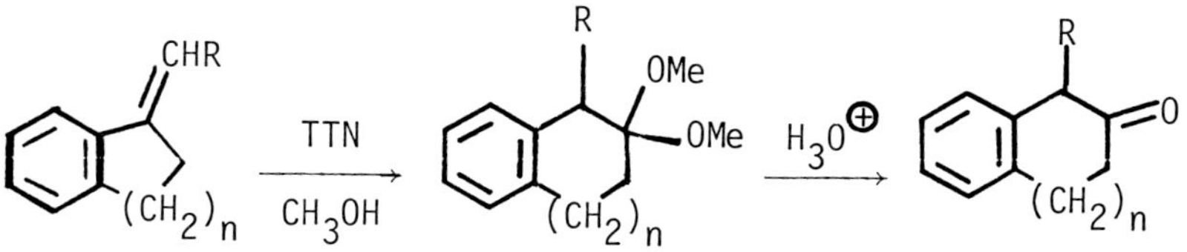

(from Wittig reactions)

81-96%

n = 1,2,3

R = H, Me, Et

VI.B.1-7 R. Pellicciari and B. Natalini, J. Chem. Soc., Perkin I, 1822 (1977).

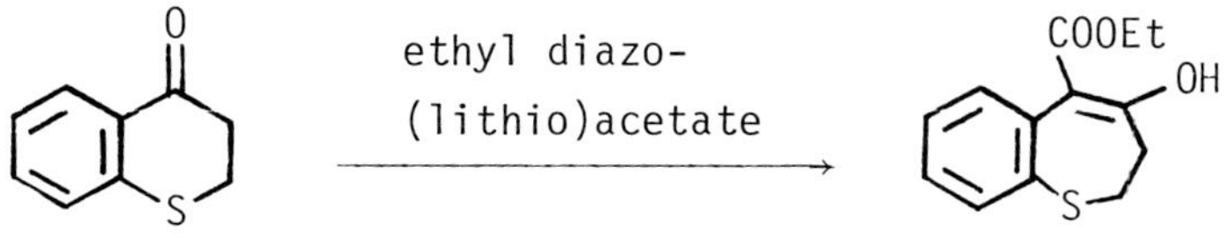

[thiochroman-4-one] + ethyl diazo-(lithio)acetate → [benzothiepine with COOEt, OH]

33%

VI.B.1-8 W. Schroth and W. Kaufmann, Z. Chem., 17, 331 (1977).

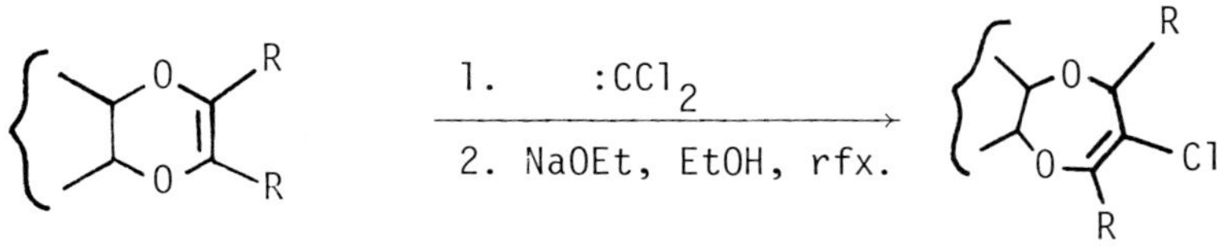

[bicyclic dioxene with R groups] 1. :CCl$_2$ 2. NaOEt, EtOH, rfx. → [product with R, Cl]

R = H, Ph

∼60% (second step)

VI.B.1-9 P.F.S. Filho and U. Schuchardt, Angew. Chem. Int. Ed., 16, 647 (1977).

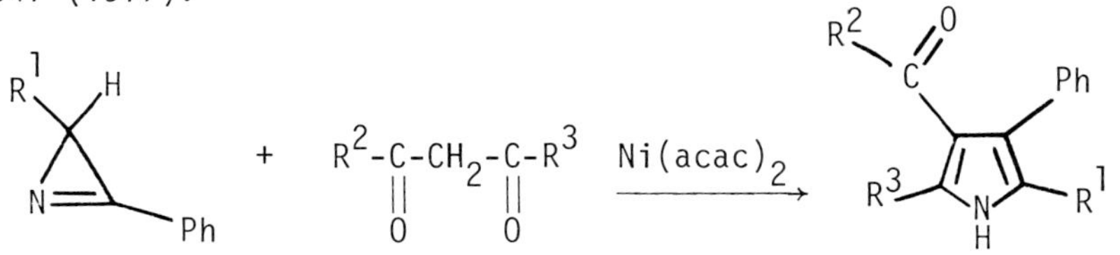

[azirine with R^1, H, Ph] + R^2-C(=O)-CH$_2$-C(=O)-R^3 Ni(acac)$_2$ → [pyrrole with R^{2}C(=O), Ph, R^3, R^1, NH]

R^1 = H, Me, Ph
R^2 = Me, Et, Ph, ⎫
R^3 = Me, Ph,   ⎬ cyclic
                    ⎭

generally 50-100%

VI.B.2-1   Y. Ohfune et al., Tetrahedron Lett., 279 (1977).

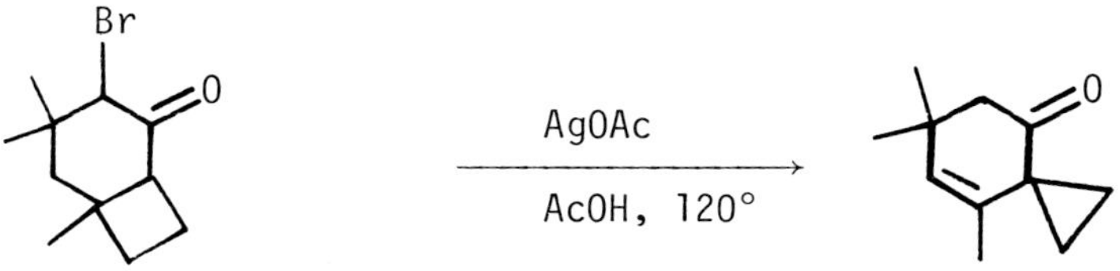

[Reaction: bromo bicyclic ketone → AgOAc, AcOH, 120° → spirocyclopropane cyclohexenone, 57%]

VI.B.2-2   D.H.R. Barton et al., J. Chem. Soc., Perkin I, 1107 (1977).

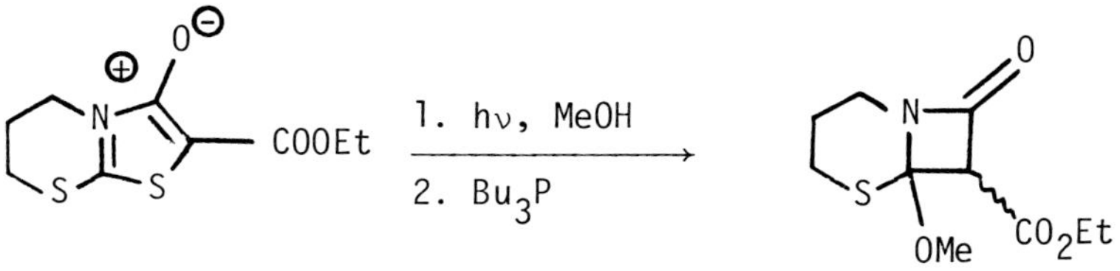

[Reaction: thiazinium-3-olate with COOEt → 1. hν, MeOH; 2. Bu$_3$P → β-lactam bicyclic product with OMe and CO$_2$Et, 9%]

Many additional examples

VI.B.2-3   N. Minami and I. Kuwajima, Tetrahedron Lett., 1423 (1977).

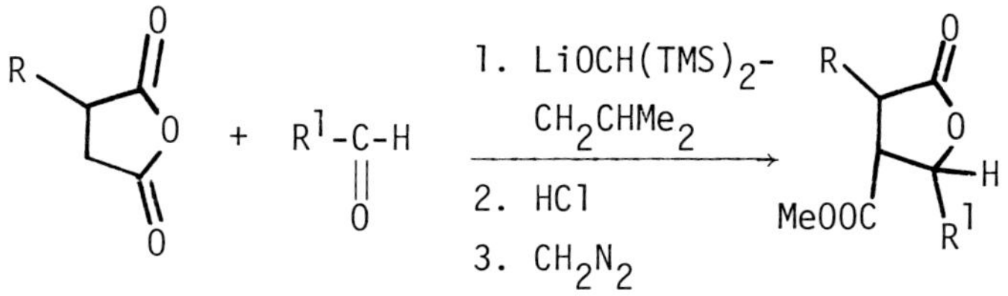

[Reaction: R-substituted succinic anhydride + R^1-CHO → 1. LiOCH(TMS)$_2$-CH$_2$CHMe$_2$; 2. HCl; 3. CH$_2$N$_2$ → lactone with MeOOC and R^1 substituents, 51-85%]

R = H, Me

R^1 = Ph, alkyl, vinyl

USEFUL SYNTHETIC PREPARATIONS

VI.B.2-4 J. J. Eisch and K. R. Im, J. Organomet. Chem., 139, C51 (1977).

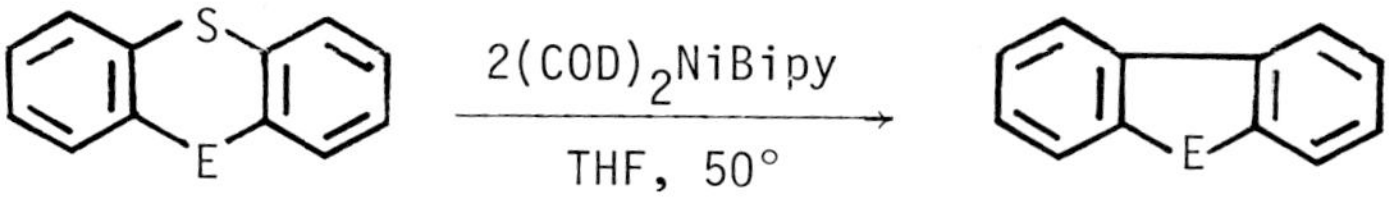

[dibenzo-fused heterocycle with E] → (2(COD)$_2$NiBipy, THF, 50°) → [dibenzofused 5-membered E heterocycle]

E = O, NH, S   50-70%

VI.B.2-5 W. Pfeiffer, E. Dilk, and E. Bulka, Synthesis, 196 (1977).

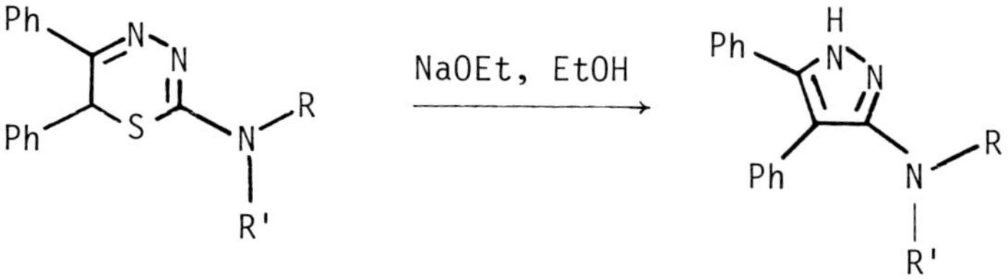

[Ph, Ph-substituted thiadiazine with N(R)R'] → (NaOEt, EtOH) → [Ph, Ph-substituted pyrazole with N(R)R']

R = H, Me, allyl, Ph, etc.   ~80-90%

R' = H, Me

VI.C.1-1 M. E. Jung et al., J. Org. Chem., 42, 3961 (1977).

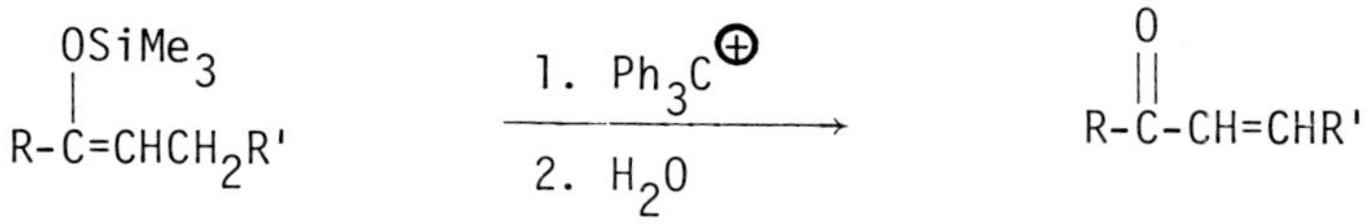

$$\underset{R-C=CHCH_2R'}{\overset{OSiMe_3}{|}} \quad \xrightarrow[2.\ H_2O]{1.\ Ph_3C^{\oplus}} \quad R-\overset{O}{\overset{\|}{C}}-CH=CHR'$$

R,R' = alkyl, cyclic   49-85% for cyclic cases

23-45% for acyclic cases

VI.C.1-2 B. M. Trost and G. Lunn, J. Am. Chem. Soc., 99, 7079 (1977).

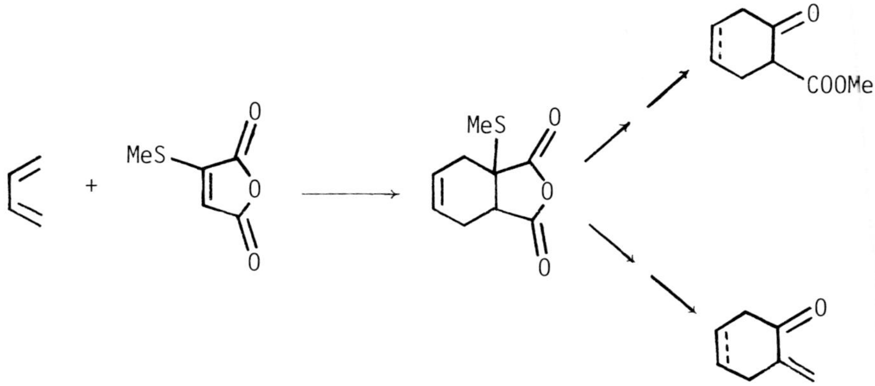

VI.C.1-3 M. E. Jung and J. P. Hudspeth, J. Am. Chem. Soc., 99, 5508 (1977).

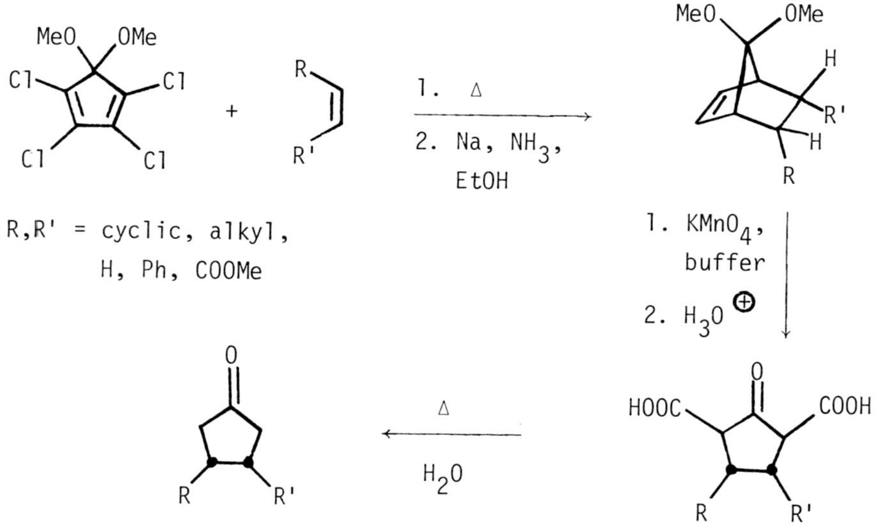

R,R' = cyclic, alkyl, H, Ph, COOMe

22-40% overall

VI.C.1-4 W. G. Dauben and D. M. Michno, J. Org. Chem., 42, 682 (1977).

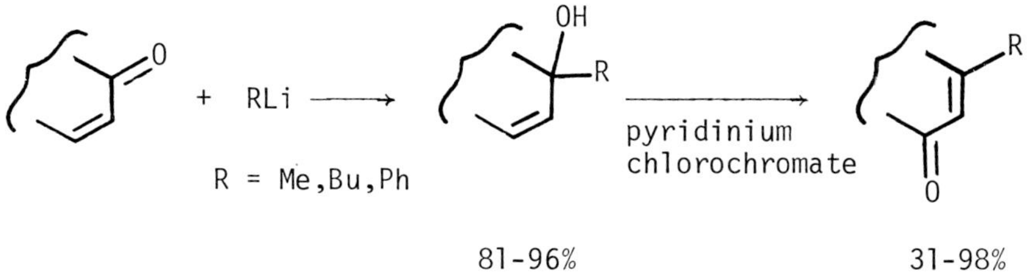

cyclohexenone + RLi → cyclohexenyl alcohol (OH, R) → (pyridinium chlorochromate) → cyclohexenone with R

R = Me, Bu, Ph

81-96%   31-98%

VI.C.1-5 T. Takeda, M. Ueda, and T. Mukaiyama, Chem. Lett., 245 (1977).

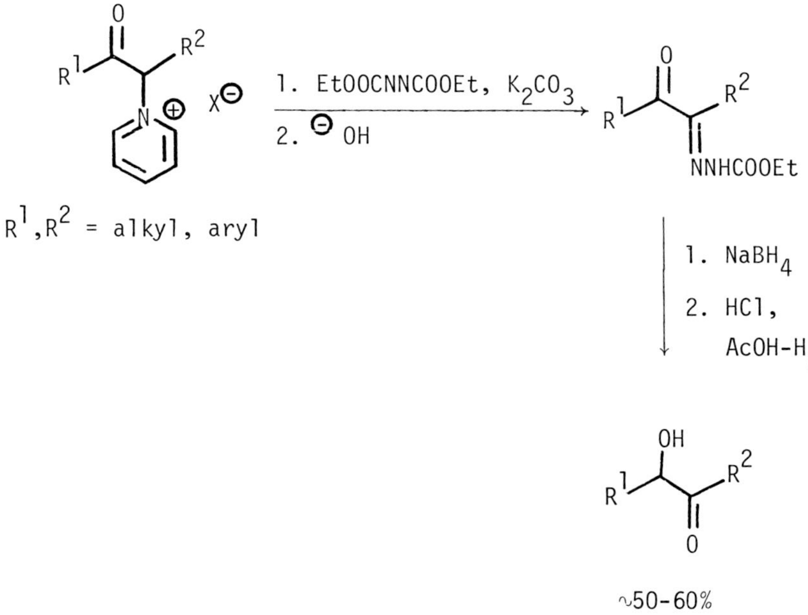

R^1, R^2 = alkyl, aryl

Starting pyridinium salt with R^1C(O)CH(R^2)-N(pyridinium) $X^\ominus$

1. EtOOCNNCOOEt, K_2CO_3
2. $^\ominus$OH

→ R^1C(O)C(R^2)=NNHCOOEt

1. $NaBH_4$
2. HCl, AcOH-H_2O

→ R^1CH(OH)C(O)R^2

∼50-60%

VI.C.1-6 E. McDonald and R. T. Martin, Tetrahedron Lett., 1317 (1977).

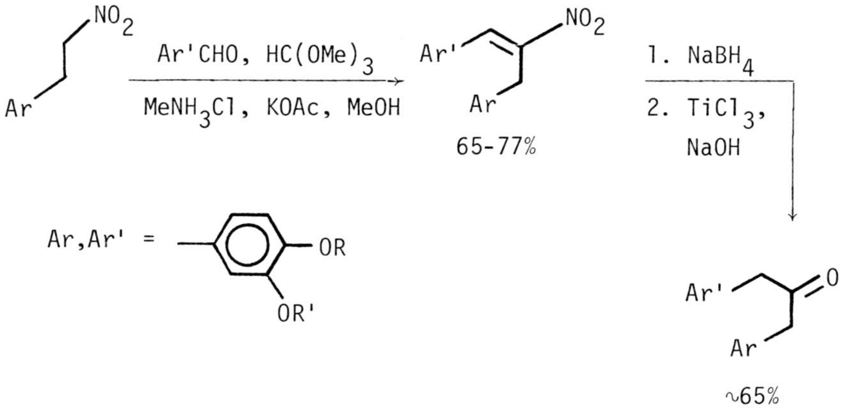

$$Ar\text{-}CH_2CH_2NO_2 \xrightarrow[\text{MeNH}_3\text{Cl, KOAc, MeOH}]{Ar'\text{CHO, HC(OMe)}_3} Ar'\text{-}CH=C(NO_2)\text{-}CH_2\text{-}Ar \xrightarrow[\text{2. TiCl}_3, \text{NaOH}]{\text{1. NaBH}_4} Ar'\text{-}CH_2\text{-}C(=O)\text{-}CH_2\text{-}Ar$$

65-77%   ~65%

Ar, Ar' = 3,4-(OR)(OR')-C$_6$H$_3$—

VI.C.1-7 R. G. Carlson and W. W. Cox, J. Org. Chem., 42, 2382 (1977).

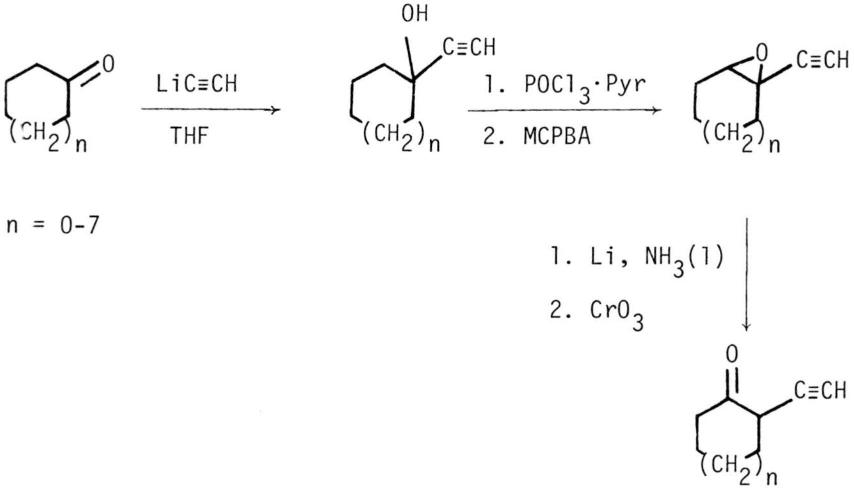

n = 0-7

Reagents: LiC≡CH / THF; 1. POCl$_3$·Pyr, 2. MCPBA; 1. Li, NH$_3$(l), 2. CrO$_3$

VI.C.1-8   D. Seebach, M. S. Hoekstra, and G. Protschuk, Angew. Chem. Int. Ed., 16, 321 (1977).

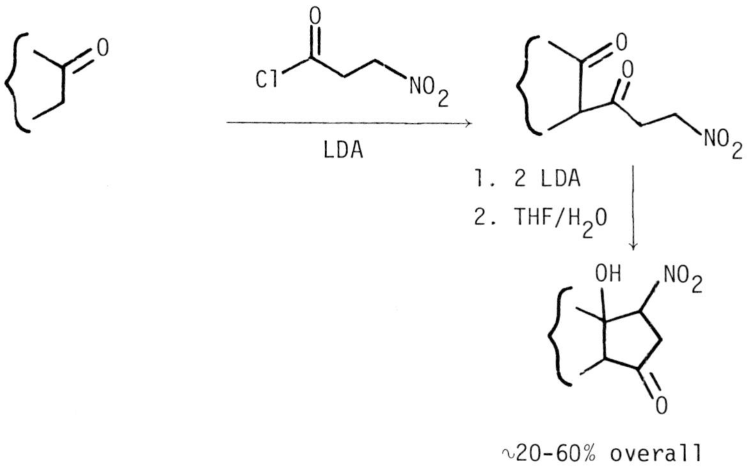

~20-60% overall

VI.C.1-9   M. Montury and J. Gore, Tetrahedron Lett., 219 (1977).

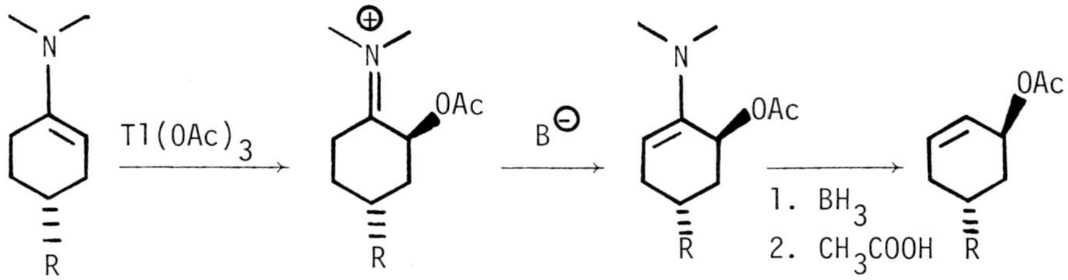

VI.C.1-10 J.L.C. Kachinski and R. G. Salomon, Tetrahedron Lett., 3235 (1977).

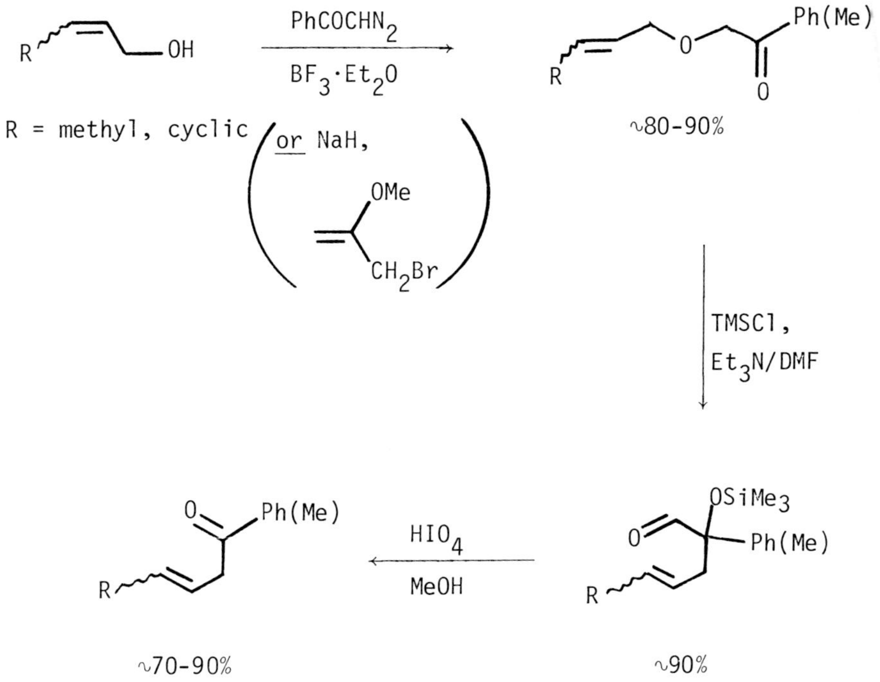

$R\text{-CH=CH-CH}_2\text{-OH} \xrightarrow[BF_3 \cdot Et_2O]{PhCOCHN_2} R\text{-CH=CH-CH}_2\text{-O-CH}_2\text{-CO-Ph(Me)}$

R = methyl, cyclic (or NaH, $CH_2=C(OMe)CH_2Br$)

~80-90%

↓ TMSCl, Et_3N/DMF

$R\text{-CH=CH-CH}_2\text{-C(=O)-Ph(Me)}$ (with O=) ←[MeOH]{HIO_4}— $R\text{-CH=CH-CH}_2\text{-C(OSiMe}_3\text{)=CH-Ph(Me)}$

~70-90%   ~90%

VI.C.2-1 C. A. Bunnell and P. L. Fuchs, J. Am. Chem. Soc., 99, 5185 (1977).

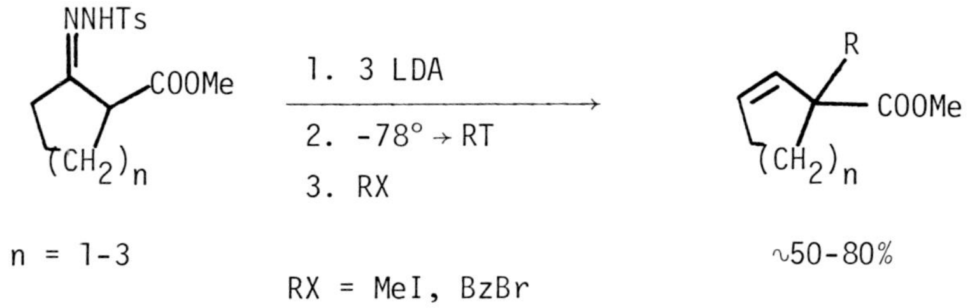

Cyclic ketone tosylhydrazone (NNHTs) with COOMe, $(CH_2)_n$

1. 3 LDA
2. -78° → RT
3. RX

→ Cyclopentene with R, COOMe, $(CH_2)_n$

n = 1-3
RX = MeI, BzBr

~50-80%

VI.C.2-2   S. Terashima and S. Jew, Tetrahedron Lett., 1005 (1977).

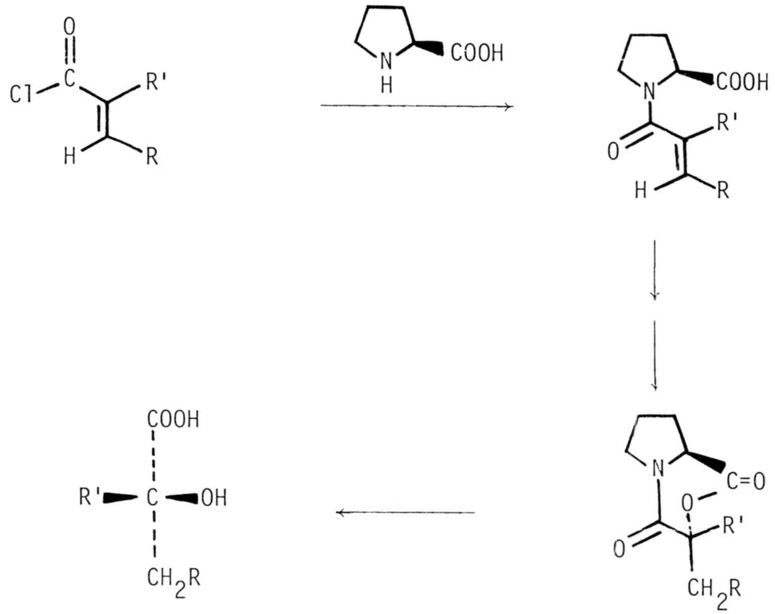

VI.C.2-3   R. M. Scarborough, Jr. and A. B. Smith III, Tetrahedron Lett., 4361 (1977); D. Liotta and H. Santiesteban, Tetrahedron Lett., 4369 (1977).

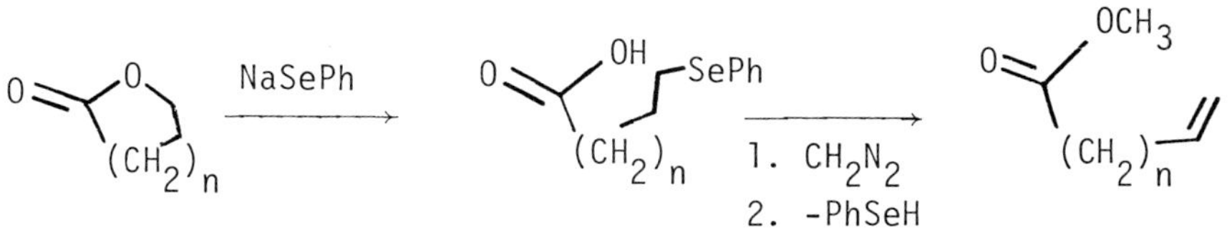

n = 1-8

60-70%

VI.C.2-4   D. S. Watt et al., J. Am. Chem. Soc., 99, 182 (1977).

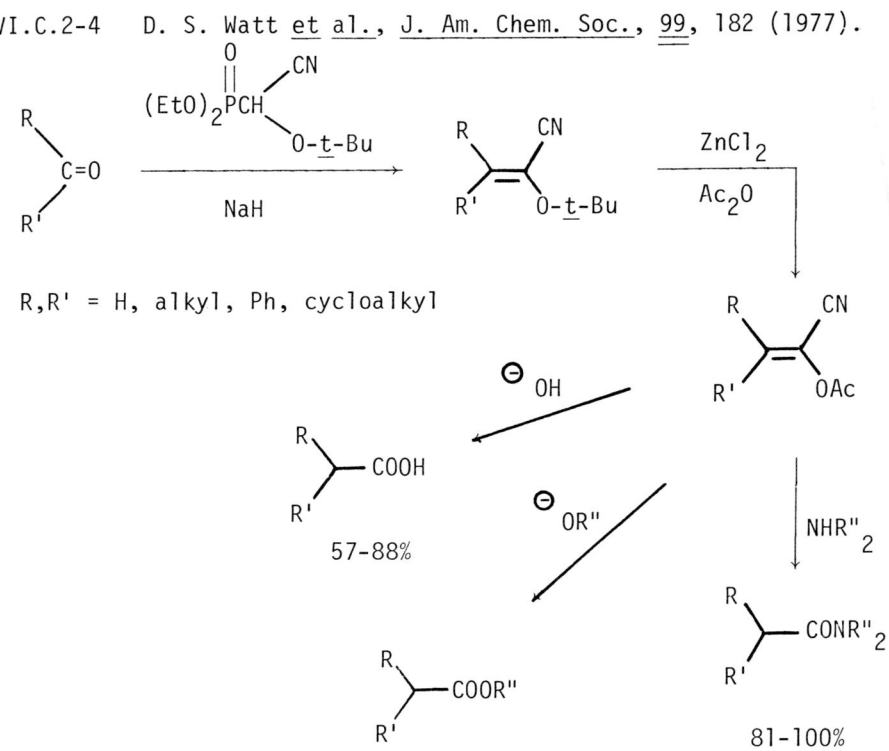

VII-1 T, Inoue, T. Uchimaru, and T. Mukaiyama, <u>Chem. Lett.</u>, 1215 (1977).

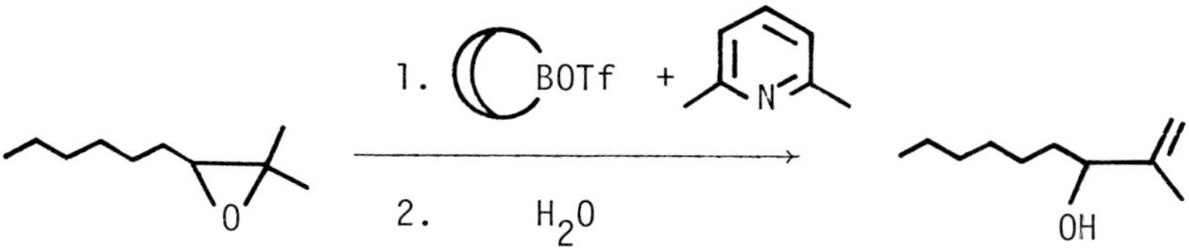

1. ⟨ ⟩BOTf + pyridine derivative
2. H_2O

96%

VII-2 J. S. Yadav, H.P.S. Chawla, and S. Dev, <u>Tetrahedron Lett.</u>, 201 (1977).

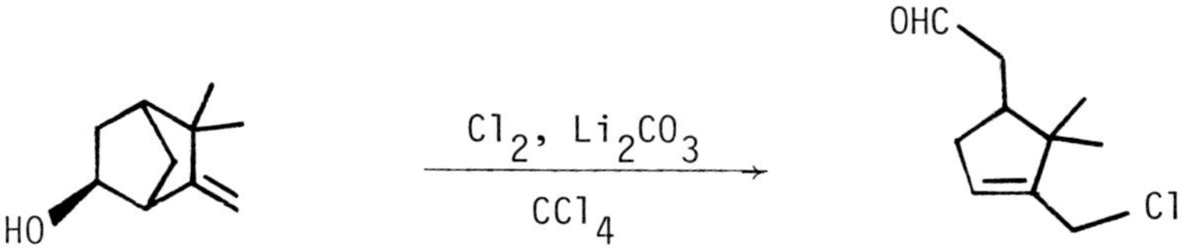

Cl_2, Li_2CO_3 / CCl_4

~90%

Several additional cleavages of homoallylic alcohols.

VII-3 G. Piancatelli and A. Scettri, <u>Synthesis</u>, 116 (1977).

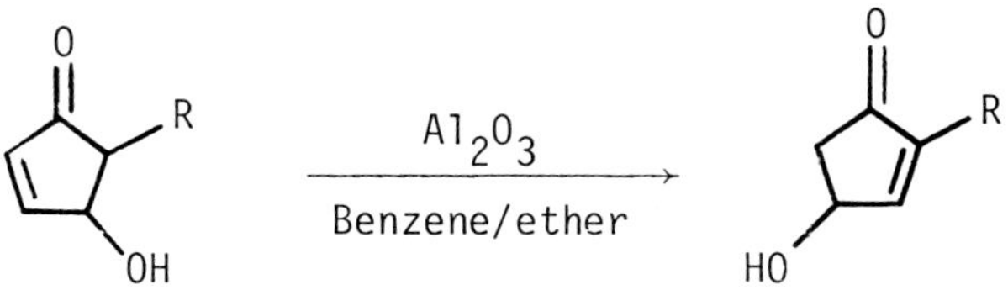

Al_2O_3 / Benzene/ether

R = Me, Ph, <u>n</u>-Hx

~90%

VII-4 R. G. Firth, G. Phillipou, and C. J. Seaborn, Tetrahedron Lett., 3403 (1977).

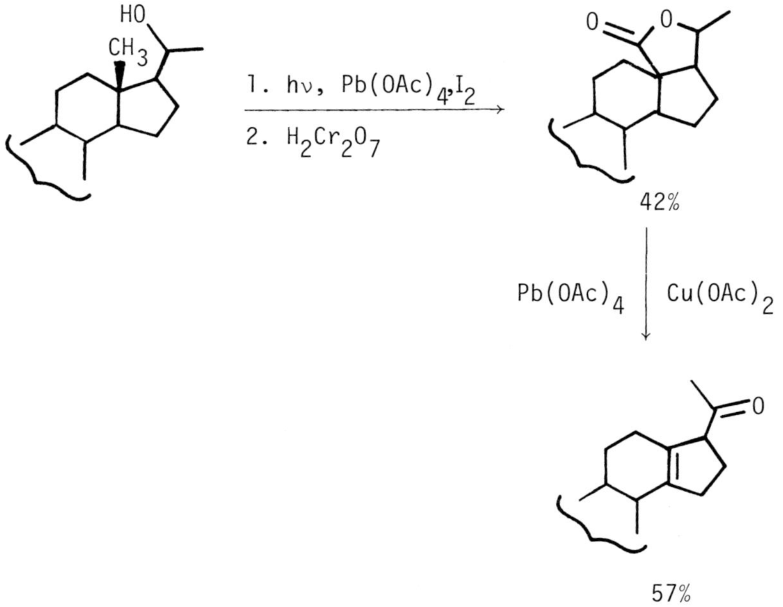

VIII-1   E. V. Dehmlow, Angew. Chem. Int. Ed., 16, 493 (1977).

Review: "Advances in Phase-Transfer Catalysis"

VIII-2   C. A. Bunton, Pure and Appl. Chem., 49, 969 (1977).

Review: "Micellar Catalysis and Inhibition"

VIII-3   N. G. Polyanskii and V. K. Sapozhnikov, Russ. Chem. Rev., 46, 226 (1977).

Review: "New Advances in Catalysis by Ion-Exchange Resins"

VIII-4   E. M. Cherkasova, S. V. Bogatkov, and Z. P. Golovina, Russ. Chem. Rev., 46, 246 (1977).

Review: "Tertiary Amines in Acyl Transfer Reactions"

VIII-5   F. E. Ziegler, Accounts Chem. Res., 10, 227 (1977).

Review: "Stereo- and Regiochemistry of the Claisen Rearrangement: Applications to Natural Products Synthesis"

VIII-6   G. B. Bennett, Synthesis, 589 (1977).

   Review: "The Claisen Rearrangement in Organic Synthesis: 1967 to January 1977"

VIII-7   B.-T. Gröbel and D. Seebach, Synthesis, 357 (1977).

   Review: "Umpolung of the Reactivity of Carbonyl Compounds Through Sulfur-Containing Reagents"

VIII-8   W. A. Szabo, Aldrichimica Acta, 10, 23 (1977).

   Review: "Chlorosulfonyl Isocyanate"

VIII-9   H. Hagemann, Angew. Chem. Int. Ed., 16, 743 (1977).

   Review: "Synthesis and Reactions of N-Chlorocarbonyl Isocyanate"

VIII-10   R. D. Rieke, Accounts Chem. Res., 10, 301 (1977).

   Review: "Preparation of Highly Reactive Metal Powders and Their Use in Organic and Organometallic Synthesis"

MISCELLANEOUS REVIEWS

VIII-11   K.P.C. Vollhardt, Accounts Chem. Res., 10, 1 (1977).

Review: "Transition-Metal-Catalyzed Acetylene Cyclizations in Organic Synthesis"

VIII-12   L. M. Jackman and B. C. Lange, Tetrahedron, 33, 2737 (1977).

Review: "Structure and Reactivity of Alkali Metal Enolates"

VIII-13   W. Kantlehner et al., Synthesis, 73 (1977).

Review: "The Preparative Chemistry of O- and N-functional Orthocarbonic Acid Derivatives"

VIII-14   E. Lukevics, Russ. Chem. Rev., 46, 264 (1977).

Review: "Latest Research on the Hydrosilylation Reaction"

VIII-15   J. K. Rasmussen, Synthesis, 91 (1977).

Review: "O-Silylated Enolates--Versatile Intermediates for Organic Synthesis"

VIII-16   V. T. Orekhov, Russ. Chem. Rev., 46, 420 (1977).

   Review: "Reactions of Organic Compounds with Higher
             Fluorides of f- and d-Elements in Groups
             V and VI.

VIII-17   R. Kh. Freidlina and F. K. Velichko, Synthesis, 145 (1977).

   Review: "Synthetic Applications of Homolytic Addition
             and Telomerisation Reactions of Bromine-
             Containing Addends with Unsaturated Compounds
             Containing Electron-Withdrawing Substituents"

VIII-18   S. M. Ali, T. V. Lee, and S. M. Roberts, Synthesis, 155 (1977).

   Review: "The Use of Bicyclo[3.2.0]heptanones as
             Versatile Synthons in Organic Chemistry"

VIII-19   U. Schöllkopf, Angew. Chem. Int. Ed., 16, 339 (1977).

   Review: "Recent Applications of α-Metallated Isocya-
             nides in Organic Synthesis"

MISCELLANEOUS REVIEWS

VIII-20   S. Ranganathan, Synthesis, 289 (1977).

Review: "Ketene Equivalents"

VIII-21   A. C. Pratt, Chem. Soc. Rev., 6, 63 (1977).

Review: "The Photochemistry of Imines"

VIII-22   A. Padwa, Chem. Rev., 77, 37 (1977).

Review: "Photochemistry of the Carbon-Nitrogen Double Bond"

VIII-23   A. V. El'tsov, O. P. Studzinskii, and V. M. Grebenkina, Russ. Chem. Rev., 46, 93 (1977).

Review: "Photoinitiation of the Reactions of Quinones"

VIII-24   G. W. Kirby, Chem. Soc. Rev., 6, 1 (1977).

Review: "Electrophilic C-Nitroso-compounds"

VIII-25 E. Kühle and E. Klauke, Angew. Chem. Int. Ed., 16, 735 (1977).

   Review: "Fluorinated Isocyanates and their Derivatives as Intermediates for Biologically Active Compounds"

VIII-26 H. O. Huisman, Pure and Appl. Chem., 49, 1307 (1977).

   Review: "Synthesis of (highly reactive) Functionalized Isoprene Building Blocks and their Application to di- and poly-isoprenoid Syntheses"

VIII-27 S. Hanessian and G. Rancourt, Pure and Appl. Chem., 49, 1201 (1977).

   Review: "Approaches to the Total Synthesis of Natural Products from Carbohydrates"

VIII-28 R. V. Stevens, Accounts Chem. Res., 10, 193 (1977).

   Review: "General Methods of Alkaloid Synthesis"

VIII-29 C. A. Henrick, Tetrahedron, 33, 1845 (1977).

   Review: "The Synthesis of Insect Sex Pheromones"

MISCELLANEOUS REVIEWS

VIII-30   R. Rossi, Synthesis, 817 (1977).

　　Review: "Insect Pheromones: I. Synthesis of Achiral Components of Insect Pheromones"

VIII-31   T. M. Harris and C. M. Harris, Tetrahedron, 33, 2159 (1977).

　　Review: "Synthesis of Polyketide-type Aromatic Natural Products by Biogenetically Modeled Routes"

VIII-32   S. H. Wilen, A. Collet, and J. Jacques, Tetrahedron, 33, 2725 (1977).

　　Review: "Strategies in Optical Resolutions"

AUTHOR INDEX

Abdipranoto, A. - 227
Abdulla, R. F. - 185
Abe, K. - 322
Abramovitch, R. A. - 370
Achiwa, K. - 222, 369
Achmatowicz, O. - 376
Acker, R.-D. - 181
Adam, W. - 196
Adams, E. - 368
Ahluwalia, V. K. - 278
Akiba, K. - 319
Akiyama, F. - 411
Albarella, J. P. - 42
Albonico, S. M. - 292
Alper, H. - 161, 170, 172, 231, 232, 248, 250
Anastassiou, A. G. - 284
Ando, W. - 91
Anselme, J.-P. - 299, 388
Ashby, E. C. 68, 179, 240, 246
Aumann, R. - 276
Auteri, S. C. - 377
Bäckvall, J.-E. - 255
Baggett, N. - 221
Baiker, A. - 359
Balanson, R. D. - 22, 50
Banks, A. R. - 375
Barany, G. - 324
Barluenga, J. - 78
Bartlett, P. A. - 201, 207, 382

Barton, D.H.R. - 96, 268, 336, 337, 338, 343, 350, 424
Basha, A. - 226
Baumstark, A. L. - 118
Beak, P. - 16, 27, 49, 61
Beam, C. F. - 47
Beck, T. G. - 280
Becker, K. B. - 84
Belsky, I. - 64
Belzecki, C. - 265
Ben-Ishai, D. - 137, 368
Bennett, G. B. - 130, 436
Bensoam, J. - 201
Bergbreiter, D. E. - 11
Berlin, K. D. - 320
Bertrand, M. - 317
Bestmann, H.-J. - 74
Bethell, D. - 351
Bey, P. - 11
Bird, C. W. - 277
Birkhofer, L. - 263
Bischoff, C. - 308
Blomberg, C. - 191
Blum, J. - 168
Boeckman, Jr., R. K. - 9, 29, 58
Bondon, D. - 216
Bonjouklian, R. - 125
Bonnett, R. - 195
Boontanonda, P. - 420

Bose, A. K. - 264
Böshagen, H. - 313
Bosnich, B. - 237, 366
Botteghi, C. - 281
Boudjouk, P. - 146, 402
Boykin, D. W. - 7, 36
Bradshaw, T. K. - 398
Braterman, P. S. - 162
Breslow, R. - 198, 207
Breugelmans, M. - 117
Brinkmeyer, R. S. - 219
Brown, A. D. - 398
Brown, D. J. - 305
Brown, H. C. - 150, 151, 152, 159, 221, 223, 226, 253, 344
Bryson, T. A. - 19, 69
Büchi, G. - 254, 262
Bulka, E. - 304, 425
Bunton, C. A. - 435
Burton, D. J. - 114, 369
Butula, I. - 354
Buzas, A. - 161
Cacchi, S. - 184
Calderazzo, F. - 165
Calo, V. - 200
Cardillo, G. - 210
Carlier, P. - 255
Carlson, R. G. - 428
Casey, C. P. - 116
Casini, G. - 389
Casiraghi, G. - 145

Cassar, L. - 161
Castro, B. - 331
Caubere, P. - 218, 233, 239, 240, 246
Cavazza, M. - 380
Cella, J. A. - 206
Chan, T. H. - 3, 104, 336, 400
Chapman, D. D. - 287
Charles, G. - 361
Chastrette, M. - 53, 363
Chatterjee, A. - 277
Chatterjee, B. G. - 266
Chebyshev, A. V. - 249
Cheng, K.-F. - 45
Cherkasova, E. M. - 435
Chottard, J. C. - 376
Cinquini, M. - 363
Clive, D.L.J. - 97, 229, 258, 274
Coates, R. M. - 131
Cohen, T. - 345
Coke, J. L. - 112, 401
Cole, T. E. - 247, 385
Coleman, R. A. - 176
Collignon, N. - 55
Colvin, E. W. - 113, 401
Compere, E. L. - 365
Consiglio, G.- 236
Cooke, Jr., M. P. - 69, 160
Corey, E. J. - 279, 331
Corriu, R.J.P. - 162
Couffignal, R. - 39

AUTHOR INDEX

Cousseau, J. - 407
Cowlagi, B. S. - 361
Crabbe', P. - 58
Crabtree, R. H. - 234
Cragg, G.M.L. - 159
Cresson, P. - 111
Crivello, J. V. - 196
Dalla Croce, P. - 89
Danishevsky, S. - 124
Dardoize, F. - 322, 329, 330
Dauben, W. G. - 15, 123, 427
Dauphin, G. - 282
Davis, F. A. - 61
Deady, L. W. - 202
Debal, A. - 6
Dehmlow, E. V. - 435
DeKimpe, D. - 119, 255
de Meijere, A. - 81, 348
Deno, N. C. - 195, 215
Derguini-Boumechal, F. - 180
de Waard, E. R. - 83
Di Bello, C. - 404
Dickson, R. S. - 189
Dimworth, K. - 216
Divakar, K. J. - 217
Dolbier, W. R. - 129
Doleschall, G. - 386
Dötz, K. H. - 189
Douglass, J. E. - 269
Dowle, M. G. - 389
Doyle, M. P. - 143, 249, 353, 399

Driquez, H. - 211, 358
Dubois, J.-E. - 9
Duhamel, L. - 30, 371
Ege, G. - 318
Eisch, J. J. - 294, 425
El'tsov, A. V. - 439
El-Zaru, R. A. - 194
Emert, J. - 380
Epsztein, R. - 46
Eschenmoser, A. - 105
Evans, D. A. - 138, 337
Everhardus, R. H. - 295
Faragher, R. - 282
Fauvarque, J. F. - 142
Fedoryński, F. - 114
Fedoryński, M. - 370
Felkin, I. E. - 244
Ficini, J. - 57
Fisher, C. - 238
Fitjer, L. - 118
Fitt, J. J. - 367
Flood, T. C. - 169
Fochi, R. - 414
Fourrey, J.-L. - 397
Francis, G. W. - 399
Franck-Neumann, M. - 122
Fraser-Reid, B. - 73, 398
Freidlina, R. K. - 76, 117, 438
Friedrich, K. - 136, 386
Frimer, A. A. - 208
Fry, J. L. - 379

Fuchs, P. L. - 10, 243, 430
Fuji, K. - 21
Fujita, E. - 203
Fujiwara, Y. - 144
Gais, H.-J. - 416
Gaoni, Y. - 251
Garanti, L. - 310
Gareev, G. A. - 410
Gassman, P. G. - 148, 331, 414
Gelbard, G. - 379, 395
Geneste, P. - 204
Gewald, K. - 304
Giam, C.-S. - 281
Gibson, H. W. - 224
Giese, B. - 71
Gilbert, E. E. - 202
Gilchrist, T. L. - 420
Gill, G. B. - 133
Giordano, C. - 320
Glover, E. E. - 268
Gokel, G. W. - 42, 140, 353
Gold, E. H. - 412
Goralski, C. T. - 413
Gossauer, A. - 85, 270
Goto, G. - 96
Goyal, S. C. - 1
Graham, R. - 295
Granik, V. G. - 340
Greene, A. E. - 270, 422
Greenhill, J. V. - 372
Greuter, H. - 282

Gribble, G. W. - 244, 260, 365
Grieco, P. A. - 65, 206, 277, 337
Grunwell, J. R. - 297
Gschwend, H. W. - 12
Guignard, A. - 312
Gusarova, N. K. - 203
Hagemann, H. - 436
Hagihara, N. - 167
Haimova, M. - 263, 269
Hall, S. S. - 62, 210
Hambrecht, J. - 256, 290
Hamilton, G. A. - 208
Hanessian, S. - 440
Hansen, H. J. - 131
Harada, K. - 366
Harada, T. - 222
Hardy, P. M. - 326
Harris, R.L.N. - 306
Harris, T. M. - 441
Hart, H. - 7, 81
Harvey, R. G. - 242
Hashem, A. I. - 258
Hauser, F. M. - 43, 342
Hayakawa, Y. - 258
Hayashi, T. - 102, 238
Heathcock, C. H. - 37, 128
Heck, R. F. - 144, 231, 246
Hegedus, L. S. - 183, 190, 276, 283
Helmchen, G. - 356
Helquist, P. - 67

```
   756.01
   748.68
   _____
     7.33
```

```
  8.7⁹    61.3⁹
  7.33    56.5⁰
  ____    _____
 16.12    4.8⁹
   8.1
```

```
    764.⁸⁰
    756.79
    _____
      8.01
```

1500 |

```
742.33
736.47
_____
  5.86
```

```
733.05
726.70
_____
  6.35
```

CHOCK 1

2-6 ppm

6-10 ppm

no integration

CHOCK 2

2 - 5.5 ppm

5.5 - 9.0 ppm

no integration

CHOCK

AUTHOR INDEX

Hendrickson, J. B. - 384, 418, 419
Henrick, C. A. - 440
Hergrueter, C. A. - 289
Herz, J. E. - 377
Herz, W. - 374
Hinze, R. P. - 85
Hirai, K. - 12
Hiyama, T. - 30, 31, 54, 110
Ho, T.-L. - 204, 228, 353, 357
Hodge, P. - 389
Hoffmann, H.M.R. - 76
Hojo, M. - 35, 418
Holand, S. - 373
Holder, R. W. - 243
Holton, R. A. - 172, 173, 322
Holy, A. - 396
Holy, N. L. - 241, 360
Hommes, H. - 401
Hong, P. - 267
Hope, A. P. - 327
Hoppe, D. - 98
Horiki, K. - 416
Howells, R. D. - 418
Hruby, V. J. - 405
Hudrlik, P. F. - 97, 406
Huet, F. - 334
Huffman, J. W. - 1
Huisman, H. O. - 440
Hullot, P. - 36
Hutchins, R. O. - 192, 245
Hyatt, J. A. - 106

Ikeda, M. - 123
Imaizumi, H. - 140
Inamoto, Y. - 357
Ioffe, B. V. - 133
Isagawa, K. - 343, 346
Ishikawa, K. - 207
Ishikawa, N. - 53, 140, 341
Ishimaru, T. - 329
Isobe, M. - 67
Ito, H. - 403
Ito, Y. - 302
Itoh, K. - 34
Itoh, M. - 327
Ivanov, C. - 273
Ivanshchenko, A. V. - 306
Iwai, K. - 271
Iwata, C. - 147
Jackman, L. M. - 4, 437
Jackson, W. R. - 187
Jacobson, R. M. - 2
Jäger, V. - 82, 211, 275, 396
Jallabert, J. - 192
James, B. R. - 237
Johncock, P. - 408
Jolly, W. L. - 227
Jones, Jr., D. A. - 406
Jones, J. H. - 356
Joullie', M. M. - 318
Julia, S. - 20, 111, 134
Jung, G. - 404

Jung, M. E. - 3, 321, 330, 333, 346, 425, 426
Jutz, C. - 283, 364
Kabalka, G. W. - 254
Kada, R. - 257
Kadin, S. B. - 284
Kagan, H. B. - 333
Kagan, J. - 351
Kalk, W. - 138
Kalkote, U. R. - 311
Kambe, S. - 313
Kametani, T. - 391
Kane, V. V. - 70
Kantlehner, W. - 437
Karnojitzky, V. - 217
Kasahara, A. - 145, 272
Katritzky, A. R. - 348, 374
Kauffmann, T. - 25, 28, 47, 61, 90, 93 347
Kay, I. T. - 108
Keinam, E. - 201, 381
Kellog, R. M. - 218
Kemp, D. S. - 231
Keumi, T. - 136
Khan, N. H. - 252, 362, 366
Kikugawa, Y. - 362
Kimura, M. - 206
Kirby, G. W. - 439
Kitatani, K. - 311
Klausner, Y. S. - 328
Knowles, W. S. - 238, 367

Knunyants, I. L. - 109
Kobayashi, T. - 266
Kobayashi, Y. - 178
Koch, T. H. - 121
Kochloefl, K. - 233
Kolbina, N. M. - 412
Kornblum, N. - 104, 197, 250 326
Köster, R. - 159, 254
Kosugi, M. - 143, 167
Kotake, H. - 256
Kozikowski, A. P. - 127
Krapcho, A. P. - 36, 215
Kraus, G. A. - 64, 70, 373
Kreutzberger, A. - 305
Krief, A. - 51, 103
Kriz, O. - 219
Kröhuke, F. - 293
Kropf, H. - 205
Krysin, A. P. - 137
Kuehne, M. E. - 190, 225
Kühle, E. - 440
Kunieda, T. - 112
Kunuda, T. - 247
Kurono, M. - 352
Kurozumi, S. - 248
Kuwajima, I. - 39, 77, 78, 94, 171, 274, 424
Kyba, E. P. - 362
Landor, S. R. - 300
Lane, C. F. - 254
Lang, Jr., S. A. - 310

AUTHOR INDEX

Lange, B. C. - 4
Lange, M. - 303
Lantzsch, R. - 308
Lapkin, I. I. - 306
Larock, R. C. - 177
Lavielle, G. - 90
Le Borgne, J.-F. - 2
Le Corre, M. - 84
Lee-Ruff, E. - 217
Leir, C. M. - 286
Lenoir, D. - 170
Levkoeva, E. I. - 320
Levy, A. B. - 408
Lewis, M. J. - 345
Ley, S. V. - 271
Leznoff, C. C. - 323, 338
Liotta, C. L. - 392
Liotta, D. - 330, 431
Liu, S.-H. - 66
Loozen, H.J.J. - 421
Loupy, A. - 87, 392
Louw, R. - 198
Lucast, D. H. - 417
Lukevics, E. - 437
Lund, H. - 63
Lutz, W. - 106, 334
Magdesieva, N. N. - 120
Magid, R. M. - 347
Magnus, P. - 59, 60, 256, 386, 419
Maione, A. M. - 225
Maitlis, P. M. - 242

Makosza, M. - 5, 188
Málek, J. - 14
Mali, R. S. - 278
Malpass, D. B. - 182
Mandell, L. - 245
Manning, M. J. - 213
Manning, W. B. - 126
Marquet, A. - 41
Marshall, J. A. - 79, 95, 250, 251, 274
Martin, S. F. - 74, 88
Maruyama, K. - 32
Masamune, T. - 82
Matsuda, I. - 312
Matsumoto, K. - 302
Matsuura, T. - 216
Mattes, K. C. - 413
Matteson, D. S. - 158
Mauze', B. - 176
Mayer, R. - 394
Mazur, Y. - 202
Mc Cabe, P. H. - 100
Mc Donald, E. - 428
Mc Intosh, J. M. - 63, 296
Mc Killop, A. - 139, 212, 288
Mc Murry, J. E. - 38, 169
Meienhofer, J. - 324, 402
Merrifield, R. B. - 405
Merrill, R. E. - 154
Merz, A. - 56
Messinger, P. - 419
Mestroni, G. - 218

Meyer, H. - 286
Meyers, A. I. - 364
Michl, J. - 129
Midland, M. M. - 156, 252, 402
Miginiac, L. - 175, 176
Miller, J. M. - 13, 378
Miller, R. B. - 101, 408
Minato, H. - 243
Mincione, E. - 216
Mitin, Yu.V. - 406
Mitra, R. B. - 45
Miyano, S. - 116
Mizoroki, T. - 226
Mock, W. L. - 79, 377
Modarai, B. - 199
Montury, M. - 429
Mori, M. - 261, 288
Morin, C. - 320
Morley, J. O. - 135
Mornet, R. - 347
Morozova, L. V. - 163
Mosterd, A. - 272
Muchowski, M. M. - 65
Mukaiyama, T. - 33, 40, 97, 191, 221, 245, 279, 348, 349, 359, 383, 395, 415, 427, 433
Mukherjee, R. - 355
Mulzer, J. - 93
Murahashi, S.-I. - 179

Murai, S. - 117, 162, 163, 164, 387
Nagai, Y. - 341
Nakai, T. - 35
Nanjo, K. - 237
Narayanan, C. R. - 335
Nasipuri, D. - 297
Nazir, M. - 339
Negishi, E. - 141, 145, 154, 159, 182
Newaz, S. S. - 280
Newcomb, M. - 11
Newkome, G. R. - 320
Nicholas, K. M. - 184
Nicolaou, K. C. - 273, 381
Nicoletti, R. - 259, 360
Niedenzu, K. - 365
Nishimura, S. - 220
Nishinaga, A. - 213
Normant, J. F. - 25, 26, 32, 34, 54, 56, 175, 181, 182
Norris, R. K. - 18
Northrup, Jr., R. C. - 224
Noyori, R. - 39
Nozaki, S. - 403
Oae, S. - 204, 205, 228, 229
Odinokov, V. N. - 214, 342
Ogilvie, K. K. - 378
Ohfune, Y. - 424
Ohta, A. - 231, 385

AUTHOR INDEX

Ojima, I. - 170, 222, 265, 376
Oka, K. - 196, 200
Olah, G. A. - 101, 165, 178, 193, 211, 230, 252, 323, 332, 334, 349, 381
Olofson, R. A. - 251, 321, 328
Olsson, L.-I. - 111
Ono, N. - 5, 15
Oppolzer, W. - 130
Orekhov, V. T. - 438
Overman, L. E. - 267
Owsley, D. C. - 232, 356
Padwa, A. - 130, 439
Palecek, J. - 355
Palmer, D. C. - 298
Pandit, U. K. - 249, 293
Papadopoulos, E. P. - 299, 300, 312, 313
Paquette, L. A. - 100, 104, 385
Parker, K. A. - 110, 359
Paust, J. - 108
Pearson, A. J. - 185, 186
Pedersen, E. B. - 285, 309
Pellicciari, R. - 297, 423
Pelter, A. - 152, 388, 417
Pete, J.-P. - 244
Phillipou, G. - 434
Piancatelli, G. - 209, 294, 433
Piers, E. - 26

Piper, J. R. - 309
Pirkle, W. H. - 323
Pitacco, G. - 125
Pittman, Jr., C. U. - 168
Pochat, F. - 393
Pollini, G. P. - 310
Polyanskii, N. G. - 435
Pommer, H. - 93
Pommier, J. C. - 372
Posner, G. H. - 194, 224, 364, 379, 399, 412
Potts, K. T. - 290, 314
Pracejus, H. - 239
Pratt, A. C. - 439
Prokof'ev, M. A. - 339
Prota, G. - 316
Pulst, M. - 372
Rajappa, S. - 296
Ramirez, F. - 397, 398
Rananathan, S. - 439
Rao, C. S. - 418
Rasmussen, J. K. - 4, 391, 437
Rasteikiene, L. - 211
Rathke, M. W. - 8, 37, 158, 183
Raucher, S. - 409
Ravid, U. - 273
Reese, C. B. - 80, 324, 390
Reetz, M. T. - 85, 107, 134
Regel, E. - 299
Reutrakul, V. - 98, 352, 388

Rieke, R. D. - 436
Riobe, O. - 29
Roberts, B. W. - 186
Roberts, S. M. - 438
Rondestvedt, C. S. - 232
Romeo, A. - 220
Rosenblum, M. - 187, 264
Rosini, G. - 315
Rossi, R. - 241, 441
Roustan, J. L. - 160
Rubottom, G. M. - 105, 209, 410
Russell, T. W. - 235
Saegusa, T. - 178, 220, 259
Sahni, M. K. - 354, 375, 403
Saîhi, M. L. - 166
Sainsbury, M. - 289
Saito, K. - 267
Sakakibara, T. - 179
Sakan, T. - 34, 52
Sakurai, H. - 72, 121, 410
Salmond, W. G. - 86, 408
Salomon, R. G. - 132, 134, 387, 430
Sammes, P. G. - 284
San Filippo, Jr., J. - 230
Sarel, S. - 129, 161
Sarkar, S. - 242
Sarkhel, B. K. - 278
Satchell, D.P.N. - 420
Sato, F. - 234
Savel'yanov, V. P. - 346

Schaumann, E. - 133
Schegolev, A. S. - 76
Schiavelli, M. D. - 110
Schick, H. - 341
Schlatter, J. M. - 406
Schlosser, M. - 99
Schmidt, R. R. - 6, 398
Schmitz, E. - 269
Schöllkopf, U. - 18, 19, 47, 66, 360, 438
Schroth, W. - 200, 319, 415, 423
Schuchardt, U. - 291, 423
Schulman, E. M. - 199, 351, 371
Schwartz, J. - 166, 173, 174, 193
Schwartz, M. A. - 147
Schweizer, E. E. - 307
Scott, L. T. - 370
Screttas, C. G. - 411
Seebach, D. - 16, 17, 20, 22, 23, 24, 27, 40, 44, 46, 48, 49, 51, 59, 71, 91, 92, 316, 429, 436
Sekiya, M. - 115
Semmelhack, M. F. - 102, 142, 146, 148, 236
Sendrik, V. P. - 275

AUTHOR INDEX

Senga, K. - 302
Severin, T. - 108, 291
Seybold, G. - 109
Seyferth, D. - 86, 91
Shackelford, S. A. - 350
Shamma, M. - 149
Shanmugan, P. - 287
Sharf, V. Z. - 219, 236
Sharpless, K. B. - 197, 210
Shatzmiller, S. - 280
Sheradsky, T. - 141
Shone, R. L. - 397
Shono, T. - 63, 72, 382
Shvarts, I. Sh. - 396
Siegl, W. O. - 262
Singh, B. - 305
Sivanandaiah, K. M. - 332
Slanovnik, B. - 319
Sliwa, H. - 94, 313
Slobbe, J. - 35
Smith, III, A. B. - 431
Smith, K. - 48, 157, 343
Snider, B. B. - 75
Sonoda, N. - 233
Sorokin, V. I. - 114
Speckamp, W. N. - 270
Spencer, A. - 164
Spitzner, D. - 289
Sraga, J. - 44
Srinivasan, R. - 81
Stadlbauer, W. - 318
Stammer, C. H. - 308

Stang, P. J. - 115, 136
Steckhan, E. - 195
Steglich, W. - 83, 292
Stenberg, V. I. - 192, 215
Stetter, H. - 75
Stevens, R. V. - 440
Still, I.W.J. - 229
Still, W. C. - 132
Stoodley, R. J. - 8
Stork, G. - 80
Stradi, R. - 298, 301
Straub, H. - 190
Strauss, M. J. - 262
Strohmeier, W. - 223, 233
Sucrow, W. - 24, 51
Suhr, H. - 112
Suzuki, A. - 153, 154, 155, 156, 165
Suzuki, H. - 396
Swierczewski, G. - 180
Szabo, W. A. - 436
Tabushi, I. - 214
Taguchi, T. - 21
Takagi, K. - 177, 304
Takahashi, K. - 43
Takahashi, T. T. - 336
Takaki, K. - 120
Takamoto, T. - 413, 417
Takei, H. - 257, 384
Tamura, Y. - 261, 361, 373
Tanaka, K. - 38, 60, 113
Taniguchi, H. - 260

Tanikaga, R. - 102, 228, 400
Tanimoto, S. - 124
Tashiro, M. - 247
Taylor, E. C. - 1, 198, 333, 422
Teichmann, H. - 109
Teramoto, T. - 403
Terashima, S. - 431
Thompson, D. W. - 176
Thuillier, A. - 52
Tilak, M. I. - 404
Tishchenko, I. G. - 374
Tomasik, P. - 265
Trass, P. C. - 387
Traxler, J. T. - 227
Trost, B. M. - 27, 50, 105, 188, 191, 197, 214, 215, 342, 383, 421, 426
Tsuda, Y. - 225
Tsuji, J. - 223, 234
Tsujikawa, T. - 14
Turner, R. W. - 358
Ucciani, E. - 235
Ueda, S. - 295
Ueki, M. - 405
Uemura, S. - 205, 407
Ueno, Y. - 5, 203
Ugi, I. - 394
Umani-Ronchi, A. - 7, 24
Utimoto, K. - 29, 151, 153, 174

Vakatkar, V. V. - 335
van der Gen, A. - 14, 86, 420
van Koten, G. - 138
van Leusen, A. M. - 298, 315, 382, 383, 390
van Tamelen, E. E. - 339
Vasil'ev, A. E. - 406
Veber, D. F. - 325, 327
Vedejs, E. - 62, 103, 414
Vialle, J. - 118
Viehe, H. G. - 307, 317, 371
Volkov, A. N. - 394
Vollhardt, K.P.C. - 188, 191, 283, 437
Wadia, S. - 358
Wakatsuki, Y. - 126
Walker, K.A.M. - 411
Walter, W. - 106
Wang, S.-S. - 329, 375
Warren, S. - 87, 88, 95, 271, 409
Warrener, R. N. - 129
Washburne, S. S. - 191
Wasserman, H. H. - 264
Watanabe, T. - 294
Watt, D. S. - 432
Weber, W. P. 127, 128
Weiler, L. - 13
Weinreb, S. M. - 354, 355, 357

AUTHOR INDEX

Welch, W. M. - 260
Wemple, J. - 92
Wender, P. A. - 68
Wermuth, C.-G. - 390
Westmijze, H. - 33, 174, 181, 407
White, D. A. - 121
White, J. D. - 99, 400
Wilen, S. H. - 441
Williams, F. J. - 380
Wolfbeis, O. S. - 303
Wollenberg, R. H. - 57
Woodgate, P. D. - 209
Wuest, J. D. - 354
Wynberg, H. - 139
Yadav, J. S. - 433
Yagupol'skii, L. M. - 349
Yajima, H. - 326
Yajima, T. - 198
Yale, H. L. - 372
Yamada, S. - 416
Yamada, Y. - 212
Yamamoto, A. - 168
Yamamoto, H. - 4, 41, 55
Yamamoto, K. - 172
Yamamoto, T. - 230
Yamamoto, Y. - 155
Yamamura, K. - 391
Yamashiro, D. - 405
Yamashita, M. - 160, 161, 375, 378
Yamazaki, H. - 268
Yanovskaya, L. A. - 77
Yokoyama, K. - 129
Yoneda, F. - 193, 301
Yoshida, H. - 89
Yoshida, K. - 392
Yoshida, Z. - 194, 296, 303
Yoshii, E. - 235
Yoshikoshi, A. - 275, 321
Youssefyeh, R. - 397
Zamojski, A. - 257
Zbiral, E. - 119
Zehavi, U. - 325
Zheltukhina, G. A. - 404
Ziegler, F. E. - 130, 393, 435
Zimmerman, H. E. - 130
Zimmerman, M. P. - 10
Zupan, M. - 350
Zweifel, G. - 157, 241, 345
Zwierzak, A. - 361, 363